Advanced Machining Processes

Advanced Machining Processes

V.K. Jain

Department of Mechanical Engineering
Indian Institute of Technology
Kanpur

ALLIED PUBLISHERS PVT. LTD.

New Delhi • Mumbai • Kolkata • Chennai • Bangalore • Hyderabad

ALLIED PUBLISHERS PRIVATE LIMITED

Regd. Off.: 1/13-14 Asaf Ali Road, **New Delhi**–110002
Ph.: 011-23239001 • E-mail: delhi.books@alliedpublishers.com

17 Chittaranjan Avenue, **Kolkata**–700072
Ph.: 033-22129618 • E-mail: cal.books@alliedpublishers.com

15 J.N. Heredia Marg, Ballard Estate, **Mumbai**–400001
Ph.: 022-42126969 • E-mail: mumbai.books@alliedpublishers.com

No. 25/10 Commander-in-Chief Road, Ethiraj Lane (Next to Post Office) Egmore, **Chennai**–600008
Ph.: 044-28223938 • E-mail: chennai.books@alliedpublishers.com

P.B. 9932, 1st Floor, Sunkada Chickkana Complex, Loop Lane Off Race Course Road, Next to Hotel Maurya, Gandhi Nagar, **Bangalore**–560009
Ph.: 080-22262081 • E-mail: bngl.books@alliedpublishers.com

Sri Jayalakshmi Nilayam, No. 3-4-510, 3rd Floor (Above More Super Market) Barkatpura, **Hyderabad**–500027
Ph.: 040-27551811, 040-27551812 • E-mail: hyd.books@alliedpublishers.com

Website: www.alliedpublishers.com

First Published: 2002
First Reprint: 2004
Twenty Fifth Reprint: 2015
Second Revised Edition, 2021
Reprint: 2024

ISBN: 81-7764-294-4

Published by Sunil Sachdev and printed by Ravi Sachdev at Allied Publishers Pvt. Ltd., 1/13-14 Asaf Ali Road, New Delhi–110002

Dedicated
to

My late parents and brother

Sri Sriyans Kumar Jain, Smt. Kamla Bai Jain

&

Sri Saroj Kumar Jain

To achieve ambitious goals of your life:

"Be honest, sincere, hardworking and
have *junoon* (full dedication) towards your work".

FOREWORD

The international academic community has always respected Professor V.K. Jain for his many contributions to the subject of unconventional machining. The key to his success lies in his personal commitment to the advancement of knowledge, and his enthusiasm for research a field that continues to give rise to new industrial applications and be driven by fresh needs of industry. These personal qualities of Professor Jain are evident in his new book. He and all of us in higher education are aware of our responsibilities towards our students. From us they have to learn how to use their "grey matter," for their own development, in order to advance the social, economic and industrial well-being of local, national and international society. Through "Advanced Machining Processes," Professor Jain has helped us all in this mission.

That industry needs unconventional machining processes is understood by colleges and universities everywhere: the subject has its own place in under- and post-graduate engineering curriculum that deal with mechanical and manufacturing engineering and in research laboratories. An effective transfer of technology has already taken place with the adoption by industry of many of the processes that hitherto were a matter of academic curiosity. Professor Jain has recognised this transition. He has focussed his book on those methods that are still undergoing investigation, or are not well understood, or lack appreciation. In order to clarify the different types of processes available, the author has divided the text into: mechanical, thermo-electric and electrochemical and chemical techniques, all useful subdivisions of a highly cross-disciplinary subject. We are then provided with a treatment of the principles that govern each process, a presentation of the effects of the main process variables on engineering performance, a discussion of the capabilities and applications, and a bibliography for further reading. Every chapter carries an innovative "At-A-Glance" summary of the method discussed. A textbook on advanced machining has long been needed that properly provides for learning this subject. The acquisition of knowledge has to be tested and Professor Jain takes heed by providing three types of questions for each process: multiple-choice, 'self-test' for understanding and descriptive and numerical calculations based on working principles. Industrialists and scholars are indeed well-served.

I began this foreword by drawing attention to our regard for Professor Jain. This new book is a testimony to his standing in Kanpur, India and the world.

Joseph McGeough
Regius Professor of Engineering,
University of Edinburgh, Scotland

PREFACE

Books on this subject available in the market are entitled as non-*conventional, non-traditional, or modern machining processes.* In my opinion, majority of these processes have already crossed the doors of the research labs. They are higher level machining processes than conventional ones. They are being commonly and frequently used in medium and large scale industries. This book therefore has been named as *"Advanced Machining Processes (AMPs)."*

This book on *"Advanced Machining Processes"* is intended primarily for the undergraduate and postgraduate students who plan to take up this course as one of their majors. The objective of writing this book is to provide a thorough knowledge of the principles and applications of these processes. This book aims at bringing the readers up-to-date with the latest technological developments and research trends in the field of *AMPs.* As a result, some of the processes yet to get popularity amongst the common industrial users have been included and discussed.

The contents of the book have been broadly divided into three major parts. Part-I deals with the *mechanical type AMPs,* viz.; ultrasonic machining (USM), abrasive jet machining (AJM), water jet machining (WJM), abrasive water jet machining (AWJM) and abrasive flow machining (AFM). Part-II describes thermoelectric type AMPs viz., electric discharge machining (EDM), laser beam machining (LBM), plasma arc machining (PAM), and electron beam machining (EBM). Part-III of the book contains details about the electrochemical and chemical type AMPs viz.; electrochemical machining (ECM) and chemical machining (ChM). Relvant enough recent developments have been included at appropriate places in different chapters to keep the interest of the researchers alive.

Keeping in view the trends in many universities and technical institutions at home and abroad specially in large classes, three kinds of questions given at the end of each chapter. The first category includes *multiple choice questions* to test the thorough understanding of the subject. The second category of questions are *descriptive long answer type*. The third category includes the *questions based on calculations*. An attempt has been made to provide enough number of numerical problems for practice to be done by the students and a few solved problems to understand how to attack such problems.

The technology developed in research organizations can't be brought to the shop floor unless its applications are realized by the user industries. With this in view, diversified industrial applications of different AMPs cited in available literature have been included. This would help the readers in evolving more and more new areas of applications to make the fullest possible exploitation of capabilities of AMPs.

The review section given at the end of each chapter is unusually large. It is prepared to the students for quick revision of a chapter, to the teachers for preparing transparencies for teaching in a class, and 'at a glance' look for the practicing engineers to decide about the specific process to be used for machining a particular component.

I hope the readers of this book will enjoy learning AMPs to a great extent.

Dr. V.K. Jain

CONTENTS

PART-II: THERMOELECTRIC ADVANCED MACHINING PROCESSES

PART-III: ELECTROCHEMICAL AND CHEMICAL ADVANCED MACHINING PROCESSES

PART–I
Mechanical Advanced Machining Processes

- Ultrasonic Machining (USM)
- Abrasive Jet Machining (AJM)
- Abrasive Finishing Processes
- Water Jet Cutting (WJC)
- Abrasive Water Jet Machining (AWJM)

Chapter One

INTRODUCTION

WHY DO WE NEED ADVANCED MACHINING PROCESSES (AMPs)?

Technologically advanced industries like aeronautics, nuclear reactors, automobiles, etc. have been demanding materials like high strength temperature resistant alloys having high "*strength to weight*" ratio. Researchers in the area of materials science are developing materials having higher level of strength, hardness, toughness and other diverse properties. This also needs similar kind of developments in the area of cutting tool materials so that the productivity is not hampered. However, there is always an upper limit to these properties that can be created in the materials.

It is a well established fact that during conventional machining processes (turning, milling, drilling, etc.) higher hardness of work material results in a decrease in economic cutting speed. It is no longer possible to find tool materials which are sufficiently hard and strong to cut (at economic cutting speeds) materials like titanium and its alloys, stainless steel, nimonic and similar other high strength temperature resistant alloys, fiber-reinforced plastic composites, stellites (cobalt based alloys), ceramics and difficult-to-machine materials (*DeBarr and Oliver*, 1975). Production of complex shapes in such materials by traditional machining methods is still more difficult. Other higher level requirements are surface finish in nanometer range, low values of tolerances, higher production rates, machining of complex shapes, automated data transmission, miniaturezation, etc. (*Snoeys et al.*, 1986). Making of holes (shallow entry angles, non-circular, micro sized, high aspect ratio, a large number of very small holes in one workpiece (say, filter), contoured cooling holes in a gas turbine blade, hole without burrs, etc.) in difficult-to-machine materials is another area where appropriate machining processes are very much in demand. Aforesaid characteristics of materials are commonly required in the products used in industries like aerospace, nuclear reactors, missiles, turbines, automobiles, etc. To meet such demands, a different class of machining processes (*i.e.,* Non-traditional machining processes or more correctly named as ***Advanced Machining Processes*** (AMPs)) has been developed.

There is a need for machine tools and processes which can accurately and easily machine (*Merchant*, 1962; *Krabacher*, 1962) the most difficult-to-machine materials to intricate and accurate shapes. These machine tools should be easily adaptable for automation as well. In order to meet these challenges, a number of newer material removal processes have been developed to the level of commercial utilization. These newer methods are also called **unconventional** in the sense that conventional cutting tools (wedge shaped tools) are not employed for machining of materials. Instead, the energy in its direct form is used to remove material from the workpiece. The range of applications and performance of the newly developed machining processes are determined by the constraints imposed by the process on the work material properties, namely, electrical and thermal conductivity, melting temperature, electrochemical equivalent, magnetic properties, etc. Some of these newly developed processes can also machine workpieces in the areas which are inaccessible for conventional machining methods. The use of these processes is becoming increasingly unavoidable and popular at the shop floor of industries. These machining processes become still more important when one considers **precision machining and ultra-precision machining**. *Taniguchi* (1983) has concluded that such high accuracy cannot be achieved by conventional machining methods in which material is removed in the form of chips. However, such accuracy can be achieved by some of the Advanced Machining Techniques (AMTs) whereby the material is removed in the form of atoms or molecules, individually or in groups.

ADVANCED MACHINING AND FINISHING PROCESSES

Advanced machining processes can be classified into three basic categories, *i.e.* mechanical, thermoelectric, and electrochemical & chemical machining processes (Figure 1.1). None of these processes is the best under all machining situations. Some of them can be used only for electrically conductive materials while others can be used for both electrically conductive and electrically non-conductive materials. Performance of some of these processes is not very good while machining materials like aluminum having very high thermal conductivity and reflectivity. Also, these machining processes have their distinct characteristic features. Hence, selection of an appropriate machining process for a given situation (or product requirements) becomes very important.

During machining, one tries to create a feature on the workpiece while during finishing, the surface finish, surface texture and other surface characteristics are changed. With this in view, mechanical type processes are divided as machining and finishing processes (Figure 1.1).

Mechanical Advanced Machining Methods like '*Abrasive Jet Machining*' (AJM), '*Ultrasonic Machining*' (USM), '*Water Jet Machining*' (WJM) have been developed. Here, Kinetic Energy (KE) of either abrasive particles or *water jet* is utilized to remove material from the workpiece. In '*Abrasive Water Jet Machining*' (AWJM), KE of mainly abrasive particles flowing along with water jet is utilized.

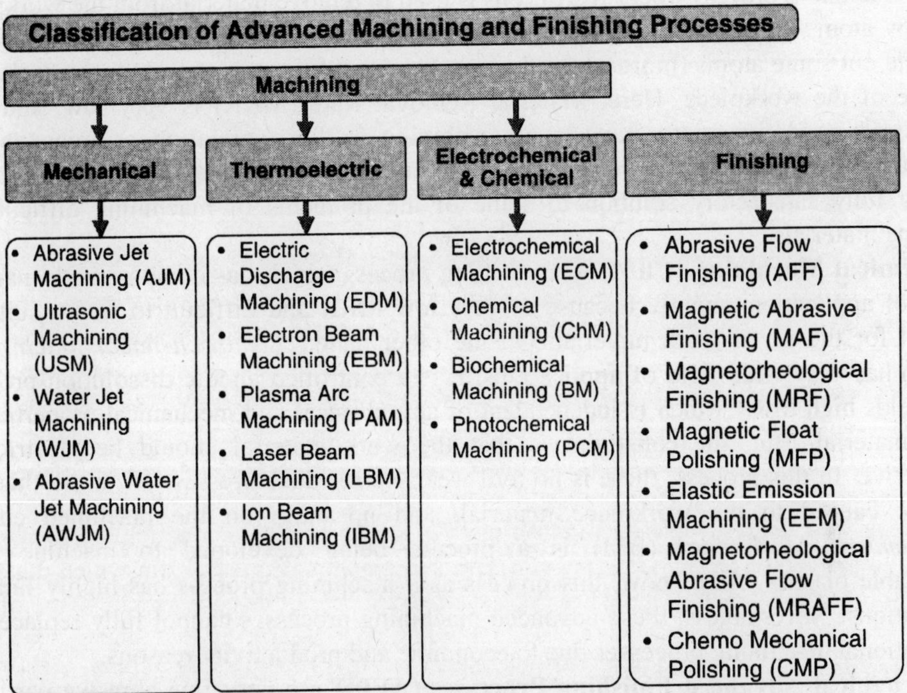

Fig. 1.1: Classification of advanced machining and finishing processes

However, performance of the above discussed processes depends upon hardness, strength, and other physical and mechanical properties of work material. What is really needed is the development of machining processes whose performance is unaffected by physical, metallurgical and mechanical properties of work material. Thermoelectric, electrochemical and chemical methods are able to overcome some of these barriers. Therefore, these processes are more and more deployed in metal working industries.

In **Thermoelectric Methods,** the energy is supplied in the form of heat (*Plasma Arc Machining*–PAM), light (*Laser Beam Machining*–LBM), or electron bombardment (*Electron Beam Machining*–EBM). This energy is concentrated onto a small area of

'*Workpiece*' (WP) resulting in melting, or melting and vaporization both. PAM, in general, has been identified as a rough machining process. LBM and EBM are good enough for making very fine cuts or holes and commonly used for micromachining also. However, '*Electric Discharge Machining*' (EDM) is a process which is more commonly used on the shop floor as compared to other AMPs, and it is capable to machine materials economically and accurately. This process is widely used for machining hard and tough but electrically conductive materials only. It is unsuitable for many applications where very good surface finish, low thermal damage to the machined surface, and high Material Removal Rate (MRR) are the requirements. Ion Beam Machining (IBM) is a nano-/micro-machining process where KE of ions is used to remove material from the workpiece atom by atom. In ion beam machining, ions hit the workpiece at an energy sufficient to separate out some atoms (more correctly, one ion should remove one atom) from the top surface of the workpiece. Here, Material Removal Rate (MRR) is very low, and this process is good for micro-/nano-level machining or for fabrication of micro-/nano-structures or features. Thus, mechanical and thermo-electric methods of AMPs also do not offer a fully satisfactory solution to some of the problems of machining difficult-to-machine materials.

Chemical Machining (ChM) is an etching process which has comparatively narrow range of applications mainly because of very low MRR and difficult to find a suitable etchant for the given work material. On the other hand, '*Electrochemical Machining*' (ECM) has very wide field of applications. It is a controlled anodic dissolution process that yields high MRR which is independent of any physical and mechanical properties of work material. But, the constraint is that the work material should be electrically conductive. In this process, there is no tool wear, no residual stresses left in, no thermal damage caused to the workpiece material, and no burrs on the machined edges. '*Biochemical Machining*' (BM) is a process being developed to machine bio-degradable plastics. As of now, this process as a machining process has highly limited applications. Nevertheless, these advanced machining processes cannot fully replace the conventional machining processes due to economic and productivity reasons.

Mechanical Advanced Finishing Processes (AFPs) use very fine abrasive particles to improve surface finish and other characteristics of the surface. There are various abrasive based nano-finishing processes in use. Only some of them have been discussed in this book. '*Magnetic Abrasive Finishing*' (MAF) is a process in which flexible ferromagnetic abrasive brush is employed to reduce surface irregularities from the premachined surfaces. Another finishing process called '*Abrasive Flow Finishing*' (AFF) has also been developed and comparatively it is more commonly used by the industries. Processes like '*Magnetic Float Polishing*' (MFP), '*Magnetorheological Finishing*' (MRF), '*Chemomechanical Polishing*' (CMP), and '*Elastic Emission Machining*' (EEM) are the finishing processes that have very specific applications in industries. The best surface finish that can be achieved by these processes is in the range of few tens of nanometer and in some cases sub-nanometer as well.

While selecting a machining or finishing process for a given work, following factors should be taken care of: process capabilities, physical parameters, shape to be machined, properties of workpiece material to be cut, and economics of the process.

MICROMACHINING

Miniaturization of the parts and/or features is getting importance in industries due to their inherent advantages, such as, low volume of material used, low power requirement, less manpower requirement, better performance and low cost. Now-a-days, a pocket size mobile has all the features of a big land line phone, video and still camera, Laptop/Computer, internet, games and what not. All this is possible due to miniaturization of the products and evolution of micromanufacturing processes (*Jain et al.*, 2012). Micromachining (*Jain*, 2010) is one of the micromanufacturing processes, and it is useful for small and medium size batch production. Micromanufacturing processes can be classified as shown in Figure 1.2(a).

Micromachining can be classified mainly in two categories: Traditional micromachining processes (say, Microturning, Micromilling, Microdrilling, etc.) and advanced micromachining processes (say, ultrasonic micromachining, laser beam micromachining, electrochemical micromachining, etc.). Advanced micromachining processes can be classified mainly in three classes as shown in Figure 1.2(b).

Nanofinishing processes also come in this category of micromachining because the material is removed at micro/nano level. However, both are different from the objective point of view. In the first case (micromachining), some features are created by removing comparatively more amount of material while in the second case (nanofinishing) surface

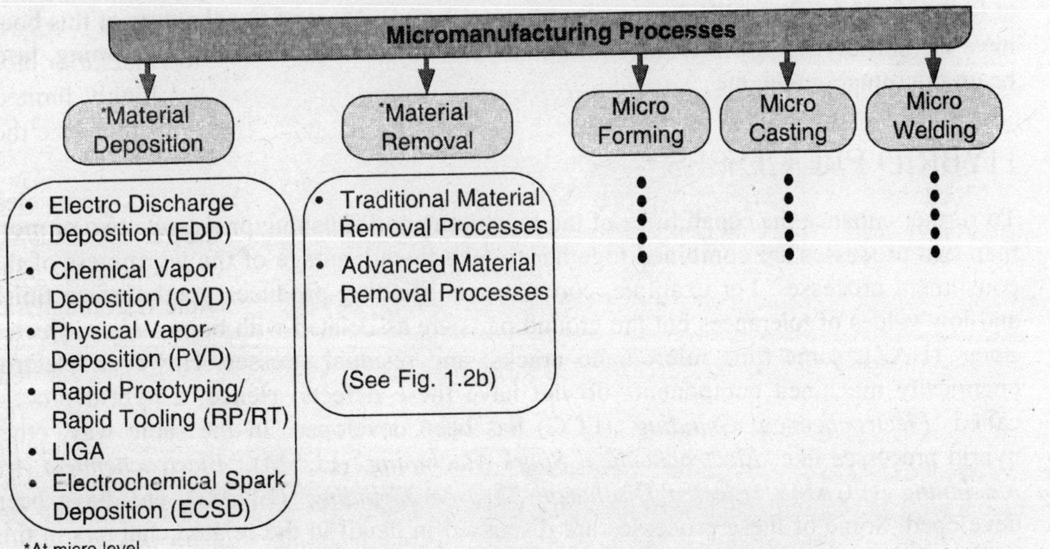

Fig. 1.2(a): Classification of micromanufacturing processes

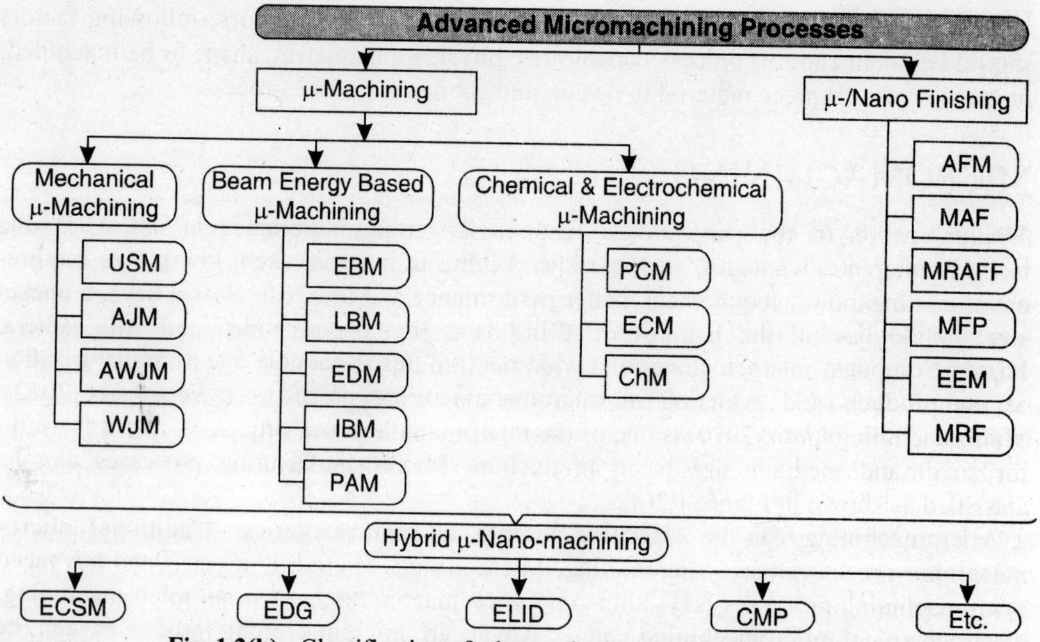

Fig. 1.2(b): Classification of advanced micromachining processes

characteristics (surface finish, surface texture, etc.) are changed by removing comparatively smaller amount of material from larger area of workpiece. The important point to note about micromachining (micro + machining) is that there is no constraint on the size of the product which can be of macro, meso or even micro size but the feature created will have at least one dimension in the range of micron (< 1 mm). Most of the chapters of this book have a small section on micromachining, for example, ultrasonic micromachining, laser beam micromachining, etc.

HYBRID PROCESSES

To further enhance the capabilities of the machining and finishing processes, two or more than two processes are combined together to take the advantage of the worthiness of the constituent processes. For example, conventional grinding produces good surface finish and low values of tolerances but the ground parts are associated with burrs, '*Heat Affected Zone*' (HAZ), some time micro/nano cracks, and residual stresses. However, electrochemically machined components do not have these defects. Hence, a hybrid process called '*Electrochemical Grinding*' (ECG) has been developed. In the same way, other hybrid processes like '*Electrochemical Spark Machining*' (ECSM), '*Electrochemical Arc Machining*' (ECAM), '*Electro Discharge Abrasive Grinding*' (EDAG), etc. have been developed. Some of these processes are discussed in detail in the related chapters in this book.

EVOLUTION OF SOME ADVANCED MACHINING PROCESSES

It is interesting to understand the working principles of some of the AMPs from the events that we come across in everyday life. It also tells possibly how some of these processes might have evolved. To understand how the mechanical type of advanced machining processes work, let us study Figure 1.3 which is based on simple examples. In Figure 1.3(a), water from the tap at a certain height, is falling on a wooden block and a soil block. After a certain period of time, it forms a crater of different size on each block whose dimensions would depend upon mechanical properties of the soil and wooden block, and definitely the velocity at which water stream is hitting these blocks. Larger sized crater in the soil block is observed as compared to the wooden block (assumption, wooden block hardness > soil block hardness). This water stream falling from the water tap cannot so easily form a crater on a stone or aluminum block. If the water stream contains sufficiently fine sand particles in it and falls from a reasonable height at comparatively very high velocity, it would be possible to have a crater formed in the stone and aluminum block as well (Figure 1.3(b)). In both these examples, '*Kinetic Energy*' (KE) of the stream/jet (water in Figure 1.3(a), or water and sand particles in Figure 1.3(b)) is utilized to remove material from the block/workpiece. Figure 1.3(a) explains in a very simple way, how '*Water Jet Machining*' (WJM) works while Figure 1.3(b) describes '*Abrasive Water Jet Machining*' (AWJM) principle. In '*Abrasive Jet Machining*' (AJM), again kinetic energy of abrasive particles is used to cut brittle and/or hard materials. When a jet of air mixed with abrasive particles attains very high velocity and strikes the workpiece (say, glass or brittle metallic piece), it leads to a crater formation (Figure 1.3c). In '*Ultrasonic Machining*' (USM), a tool vibrating at ultrasonic frequency ($\geq$ 16 kHz) hits abrasive particles suspended in the carrier liquid (say, water) (Figure 1.3(d)). These high velocity abrasive particles (at high KE) hit the workpiece surface and result in a crater formation even in the material like ceramic.

Mechanical type '*Advanced Finishing Processes*' (AFPs) include magnetic abrasive finishing, abrasive flow finishing, '*Magnetic Float Polishing*' (MFP) and others as shown in Figure 1.1. In magnetic abrasive finishing, magnetic force is utilized to form a flexible magnetic abrasive brush consisting of ferromagnetic particles and abrasive particles (or sintered magnetic abrasive particles), Figure 1.3(e). This flexible magnetic abrasive brush when comes in contact with sharp peaks on the work surface it flattens the sharp surface irregularities on the workpiece by shearing them, and reduces their heights after finishing (Figure 1.3(f)), on the axially vibrating workpiece. In abrasive flow finishing, the medium is made up of long chains polymer (viscoelastic), additives and abrasive particles. It is passed to and fro over the workpiece surface to be finished, Figure 1.3(g), using hydraulic, electric or other drive. This process can be used for deburring, radiusing, and polishing at the same time.

USSR scientists (Lazerenko brothers) observed that during putting an electric switch to off/on position several times, there were small dark pits formed on both parts (male and

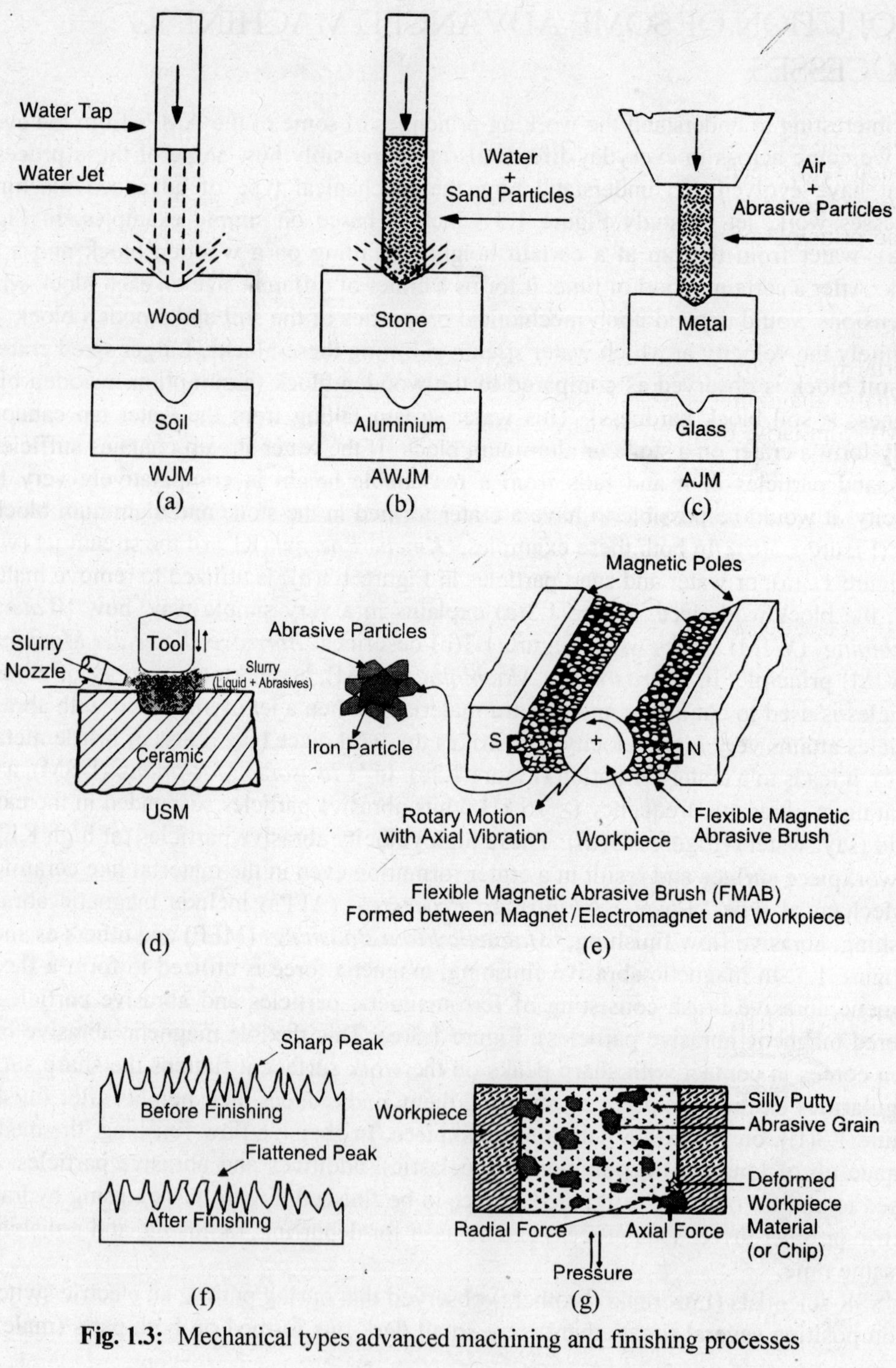

Fig. 1.3: Mechanical types advanced machining and finishing processes

female) of the switch. 'Switching on' or 'switching off' results in sparking, *i.e.* very high localized temperature leading to thermal erosion of both parts of the switch (Figure 1.4(a)). After making this observation, both brothers started thinking how to make constructive use of this destructive phenomenon. As a results of this, Lazerenko brothers developed and patented first 'electro spark machining' machine which was later on named as '*Electric Discharge Machining*' (EDM) m/c.

The principle of using laser for material removal by thermal erosion could be easily understood if we recall the 'fun game' played in the childhood by focusing sun rays by means of a convex lens on a sheet of paper (Figure 1.4(b)), to burn it. In the same way, highly intense heat of laser beam is concentrated on a workpiece surface to remove material from it by melting, or by melting and vaporization both. Figure 1.4(c) shows a schematic diagram to understand the working principle of '*Laser Beam Machining*' (LBM). Similarly, one can understand the working principle of '*Chemical Machining*' (ChM) being practiced from centuries by the Jewelers for making gold and silver

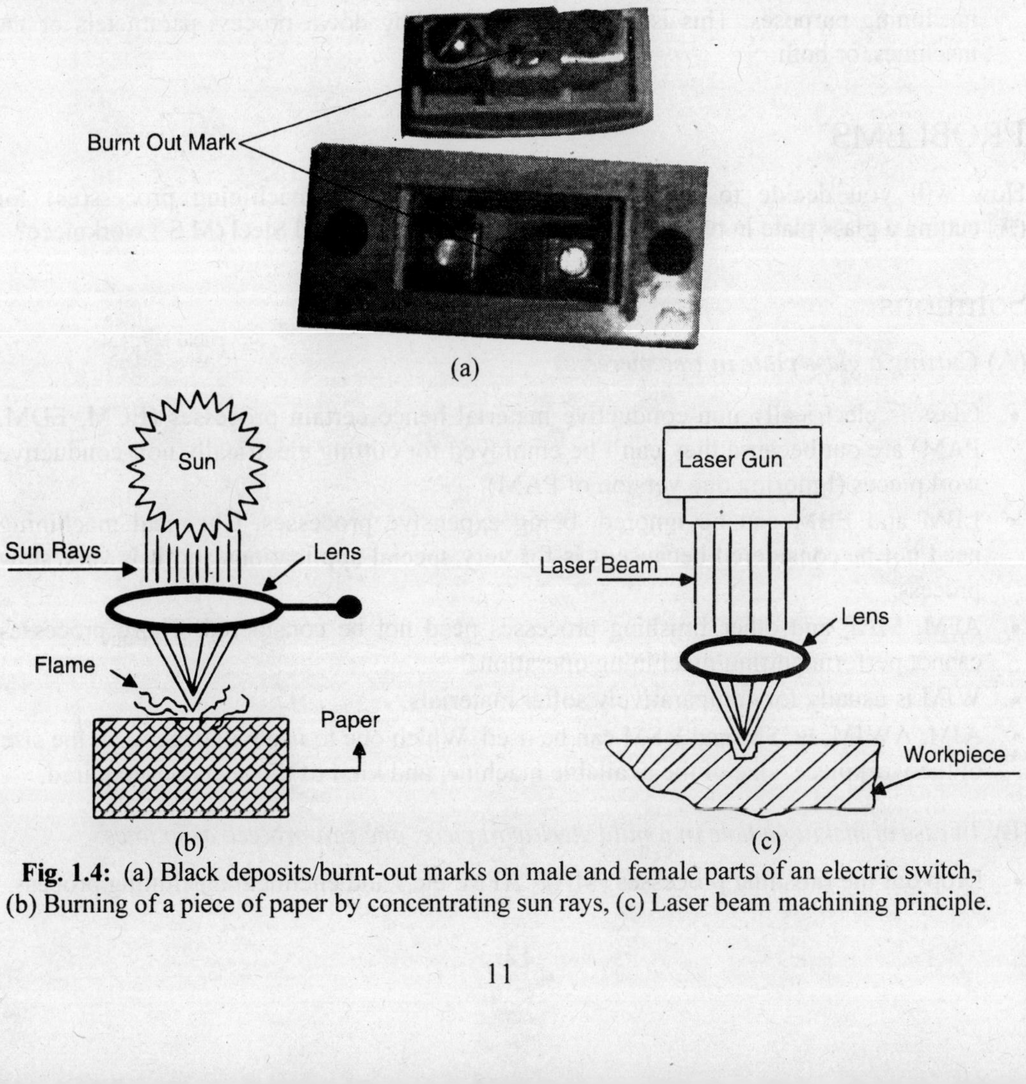

Fig. 1.4: (a) Black deposits/burnt-out marks on male and female parts of an electric switch, (b) Burning of a piece of paper by concentrating sun rays, (c) Laser beam machining principle.

REMARKS

Most of these AMPs have experienced a steady growth since their inception. In some cases, productivity as compared to conventional methods, can be increased either by reducing total number of manufacturing operations required or by performing operations faster. A review of recent literature has revealed the following facts:

- The trend shows that the capabilities of different advanced machining and finishing processes are being enhanced through research efforts.
- Machine tools for most of these processes are equipped with a computer control which means higher rate of acceptance by users, higher reliability, higher productivity, better repeatability, and higher accuracy.
- Application of Adaptive Control (AC) to these processes and in-process inspection techniques being employed are helping in widening their area of applications and leading towards the unmanned machining modules and automated factories.
- Most of the AMPs are being scaled down so that they can be employed for micro-machining purposes. This is being done by scaling down process parameters or the machines, or both.

PROBLEMS

How will you decide to recommend specific advanced machining process(es) for (A) cutting a glass plate in two parts, (B) making a hole in a Mild Steel (M.S.) workpiece?

Solutions

(A) *Cutting a glass plate in two pieces:*

- Glass is electrically non-conductive material hence certain processes (ECM, EDM, PAM) are out because they can't be employed for cutting electrically non-conductive workpieces (Ignoring one version of PAM).
- LBM and EBM can be ignored, being expensive processes. Chemical machining need not be considered because it is for very special applications and it is very slow process.
- AFM, MAF and other finishing processes need not be considered. These processes cannot perform cutting/machining operation.
- WJM is usually for comparatively softer materials.
- AJM, AWJM, ECSM and USM can be used. Which one to use will depend on the size of the workpiece, size of the available machine, and kind of the accuracy required.

(B) *In case of making a hole in a mild steel workpiece, one can proceed as follows:*

- Drop out the finishing processes (MAF, AFM, etc.) and chemical machining process.

- More suitable for comparatively brittle materials, one can drop out AJM and USM also. WJM is good for softer materials hence it can be excluded. AWJM system is quite expensive however it is quite fast and can be used if the AWJM machine is available.
- M.S. being electrically conductive, ECM, EDM and PAM can be employed. LBM and EBM can also be employed but they will be much more expensive than others. At this point, one should know the requirements of the hole in terms of dimensions, tolerances and surface integrity. If it is not a micro hole, one can easily adopt ECM or EDM. If surface integrity required is high, ECM should be used, and so on.

Thus, by 'elimination process' one can select appropriate machining and finishing processes to be used for a given work.

BIBLIOGRAPHY

[1] Bellows, Guy and Kohls, John B. (1982), "Drilling without Drills", *Am. Machinist*, pp. 173–188.

[2] Benedict, G.F. (1987). *Nontraditional Manufacturing Processes*, Marcel Dekker Inc., New York.

[3] Bhattacharyya, A. (1973). *New Technology*, The Institution of Engineer (I), Calcutta.

[4] DeBarr, A.E. and Oliver, D.A. (1975). (Ed.), *Electrochemical Machining*, McDonald and Co. Ltd.

[5] Jain, V.K. (2010). *Micromachining processes: An Introduction*; in Introduction to Micromachining (Ed. V.K. Jain), Narosa publisher, New Delhi.

[6] Jain, V.K., Sidpara, Ajay; Ravisankar, M. and Das, M. (2012). *Micro-manufacturing: An Introduction*; in Micro-manufacturing processes (Ed. V.K. Jain), CRC press, New York.

[7] Kaneeda, T., Yokomizo, S., Miwa, A., Mitsuishi, K., Uno, Y. and Marjoka, H. (1997). "Biochemical Machining-Biochemical Removal Process of Plastic", *Precision Engineering*, Vol. 21, No. 1, pp. 57–63.

[8] Krabacher, E.J. *et al.* (1962). "Electrical Methods of Machining", *Proc. Int. Con. on Production Res.* (ASME), pp. 232–241.

[9] McGeough, J.A. (1988). *Advanced Machining Methods*, Chapman and Hall.

[10] McGeough, J.A., McCarthy, W.J. and Wilson, C.B. (1987). "Electrical Methods of Machining", in Machine Tools *Encycl. Brit.*, Vol. 28, pp. 712–736.

[11] Merchant, M.E. (1962). "Newer Methods for the Precision Working of Metals – Research and Present Status", *Proc. Int. Con. on Production Res.* (ASME), pp. 93–107.

[12] Pandey, P.C. and Shan, H.S. (1980). *Modern Machining Processes,* Tata McGraw Hill Publishing Co. Ltd., New Delhi.

[13] Snoeys, R., Stallens, F. and Dakeyser, W. (1986). "Current Trends Non-conventional Material Removal processes", *Annals of the CIRP*, Vol. 35, No. 2, pp. 467–480.

[14] Taniguchi, N. (1983). "Current Status in and Future Trends of Ultraprecision Machining and Ultrafine Materials Processing", *Annals of the CIRP*, Vol. 32, No. 2, pp. 1–8.

SELF TEST QUESTIONS

Q1. Multiple Choice Type Questions.

1. While selecting thermoelectric advanced machining process to be used, which of the following is/are not considered?
   (a) Process capability (b) Workpiece hardness
   (c) Shape to be machined (d) None of these
2. Which of the following is/are not hybrid process(es)?
   (a) Electric discharge grinding (b) Wire-cut EDM
   (c) Electrochemical drilling (d) AWJM (e) ECSM
3. Which of the following process(es) can be used for drilling a hole in a glass plate:
   (a) AFM (b) AJM (c) EBM
   (d) AWJM (e) Ultrasonic Assisted EDM
4. Why do we need advanced machining processes?
   (a) To minimize time for machining.
   (b) To achieve higher MRR.
   (c) To machine complex shapes with low tolerance.
   (d) To achieve high surface finish and better surface integrity.
   (e) To reduce tool wear.
5. Which of the following is/are not mechanical advanced machining process(es)?
   (a) WJM (b) USM (c) MAF
   (d) LBM (e) EBM
6. Performance of electrochemical machining process is independent of:
   (a) Thermal properties of work material.
   (b) Mechanical (strength) properties of work material.
   (c) Physical and metallurgical properties of work material. (d) All of the above.
7. Which of the following is/are not a thermal advanced machining process(es)?
   (a) MAF (b) EDM (c) ECM (d) EBM
8. Which of the following come under the category of micromachining?
   (a) Drilling a hole of 0.5 mm in steel workpiece of size 50 mm × 50 mm × 5 mm.
   (b) Creating a channel of 0.3 mm × 10 mm × 2 mm on a glass piece.
   (c) Making a metallic filter having hundreds of holes of 1.2 mm diam.
   (d) A turbine blade having 40 cooling holes of diameter 900 μm and depth of 60 mm.

Answers

1. (d)	2. (b), (c) & (d)	3. (b), (c) & (d)	4. (c) & (d)
5. None	6. ?	7. (a) & (c)	8. ?

(? – Find answer yourself)

Q2. Review Questions.

1. How the developments in the area of '*materials*' are partly responsible for evolution of advanced machining processes?
2. Enlist the requirements with examples from different fields that demand the use of AMPs.
3. Write the constraints that limit the performance of different kinds of AMPs. Also, write the conditions under which individual process will have advantages over others.
4. What do you understand by the word "*unconventional*" in unconventional machining processes? Is it justified to use this word in context to the present day utilization of these processes on the shop floor?
5. Name the important factors that should be considered during the selection of an advanced machining process for a given job.
6. Classify advanced machining processes on the basis of the type of energy employed. Also, state the mechanism of material removal, transfer media, and energy source used in each case. Answer in table form in brief.
7. Write the names of different types of finishing processes. Can you finish electrically non-conducting workpieces using these processes?
8. Which machining process cannot be efficiently used for machining Aluminium? Also state why?
9. Define the term "micromachining". Differentiate between 'machining' and 'finishing' operations.
10. Sketch a magnified view of the surface it looks like before finishing and after finishing.

ACRONYMS

AC	:	Adaptive Control
AFF	:	Abrasive Flow Finishing
AFPs	:	Advanced Finishing Processes
AJM	:	Abrasive Jet Machining
AMPs	:	Advanced Machining Processes
AMTs	:	Advanced Machining Techniques
AWJM	:	Abrasive Water Jet Machining
BM	:	Biochemical Machining
ChM	:	Chemical Machining
CMP	:	Chemomechanical Polishing
EBM	:	Electron Beam Machining
ECM	:	Electrochemical Machining
ECG	:	Electrochemical Grinding
ECSM	:	Electrochemical Spark Machining
ECAM	:	Electrochemical Arc Machining
EDG	:	Electric Discharge Grinding
EDM	:	Electric Discharge Machining
EDAG	:	Electro Discharge Abrasive Grinding
EDDG	:	Electric Discharge Diamond Grinding
EEM	:	Elastic Emission Machining
HAZ	:	Heat Affected Zone
IBM	:	Ion Beam Machining
KE	:	Kinetic Energy
LBM	:	Laser Beam Machining
MAF	:	Magnetic Abrasive Finishing
MFP	:	Magnetic Float Polishing
MRR	:	Material Removal Rate
PAM	:	Plasma Arc Machining
WJM	:	Water Jet Machining
WP	:	Workpiece
USM	:	Ultrasonic Machining
USSR	:	Union of Soviet Socialist Republics

CHAPTER 1
AT-A-GLANCE INTRODUCTION

Need of Advanced Machining Processes

- ➢ Limitations of conventional machining methods
- ➢ Rapid improvements in the properties of materials (Both metals & non-metals)
  - Tool hardness >> WP hardness
- ➢ Product requirements
  - Machining of WPs made of advanced engg. materials
  - Complex shapes
  - Machining in inaccessible areas
  - Low tolerances
  - Better surface integrity
  - Nanometer level of surface finish
- ➢ High production rate
- ➢ Low cost of production
- ➢ Precision & ultraprecision machining
- ➢ Nanometer finishing/machining

↓

Requires material removal in the form of atoms or molecules.

Some Important Characteristics of AMPs

- Performance is independent of strength of WP
- Performance depends on thermal, electrical, magnetic, or/and chemical properties of WP
- Use of energy in its direct form to remove material from the WP
- In general, low MRR but better quality products
- Comparatively high initial investment cost.

Classification of Advanced Machining and Finishing Processes

Mechanical	Thermoelectric	Electrochemical & Chemical	Nanofinishing Processes
AJM, USM, WJM, AWJM	EDM, LBM, EBM, PAM, IBM	ECM, ChM, BM, PCM	AFM, CMP, MRF, MAF, MRAFF, EEM, MFP

Micro-machining

- Applications
- Traditional and advanced micromachining processes
- No constraint on the size of WP
- Feature: At least one dimension < 1 mm
- Material removal at micro level
- Macro components but material removal at micro/nano level
- Micro / nano components and material removal at micro/nano level (ex. MEMS, NEMS).

Mechanical Micromachining processes

- Abrasive jet micromachining (AJMM)
- Ultrasonic micromachining (USMM)
- Water jet micromachining
- Abrasive water jet micromachining (AWJMM)

Thermal Micromachining Processes

- Electrodiscarge Micromachining (EDMM)
- Electron Beam Micromachining (EBMM)
- Laser Beam Micromachining (LBMM).

Chemical Micromachining Processes

- Electrochemical Micromachining (ECMM)
- Photo Chemical Micromachining (PCMM)
- Electrochemical Spark Micromachining (ECSMM)
- Electrochemical Microdeburring (ECMDe).

Applications

- Some of the products include devices such as computer hard disc drive heads, inkjet printer heads, sensors, infrared images
- Reduced nozzle diameter
- Lower hole pitch and longer head help in better quality printing.

Some Hybrid Processes

- Electrical Discharge Grinding (EDG)
- Electrical Discharge Abrasive Grinding (EDAG)
- Electrochemical Grinding (ECG)
- Electrochemical Spark Machining (ECSM)
- Ultrasonic Assisted EDM (UA-EDM).

Evolution of APMs

- Day to day life examples
- Tap water creating crater → KE of water in WJM, water + abrasive in AWJM, Air + Abrasive particles in AJM
- Female and male parts with carbon deposit → EDM
- Burning of a piece of paper by concentrating sun rays → LBM.

Remarks

- Enhanced volumetric material removal rate
- Computer control of the processes resulting in better performance
- Application of adaptive control → unmanned machining/factory of future.

Chapter Two

ULTRASONIC MACHINING (USM)

INTRODUCTION

Ultrasonic Machining (*USM*) is a *mechanical type advanced machining process*. It is employed to machine hard and/or brittle materials (both electrically conductive and non-conductive) having hardness usually greater than 40 RC. It uses abrasive slurry and a shaped tool vibrating with high frequency and low amplitude. In *USM*, material is removed by the abrasive particles which are driven into the work surface by the shaped tool oscillating normal to the work surface. To understand the working principle of USM, consider that a small ball 'B' is thrown on the soft wax wall (Figure 2.1) with a certain force F_1. It will penetrate into the wall to a depth equal to ℓ_1. It will create a crater on the wall, having the crater diameter as D_1. If the same ball is thrown with force F_2 ($F_2 > F_1$) then it will penetrate deeper in the wall ($\ell_2 > \ell_1$, and $D_2 > D_1$). USM also works almost in the similar way. In USM, the throwing force is contributed by the tool oscillating at ultrasonic frequency. The abrasive particles are of different sizes (size varying within a small range) and they are thrown many thousands times per second. In some cases, abrasive particles in the slurry are hammered also. Figure 2.2 shows a schematic diagram of USM system.

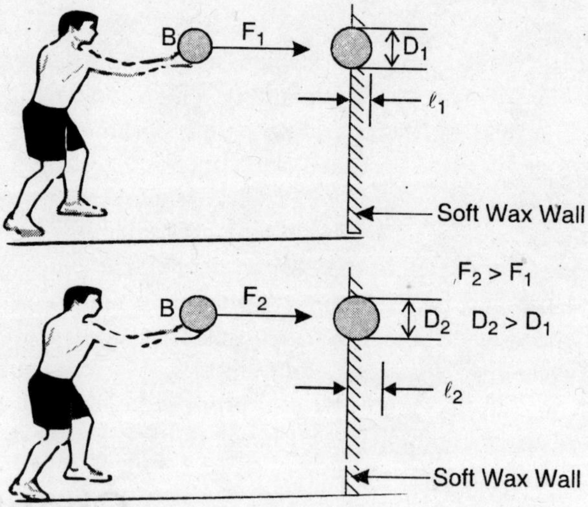

Fig. 2.1: A person throwing a ball on a soft wax wall

The word *ultrasonic* describes vibratory waves having frequency larger than upper frequency limit of human ear (usually greater than 16 kc/s). Waves are usually classified as shear waves and longitudinal waves. High velocity longitudinal waves can easily propagate in solids, liquids and gases. They are normally used in ultrasonic applications.

In USM, the principle of *longitudinal magnetostriction* is used. The transducer based on this principle is known as *magnetostriction type transducer*. When an object made of ferromagnetic material is placed in the continuously changing magnetic field, a change in its length takes place. The coefficient of magnetostrictive elongation (E_m) is defined as,

$$E_m = \Delta L/L.$$

Where, ΔL is change in length and L is the length of magnetostrictor coil.

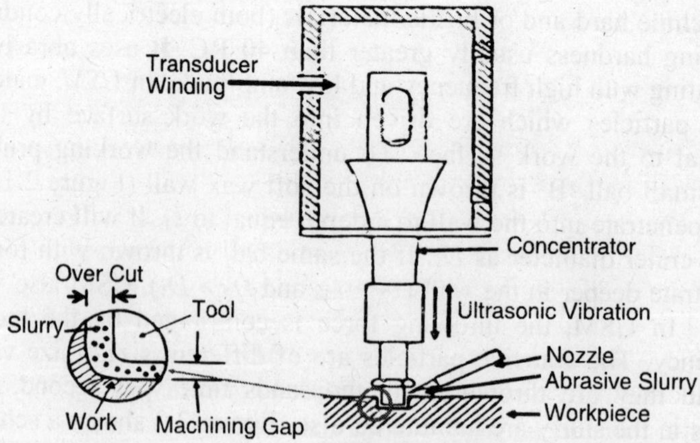

Fig. 2.2: A schematic diagram of ultrasonic machining [*Kaczmarck*, 1976]

A device that converts any form of energy into ultrasonic waves is called *ultrasonic transducer*. In USM, a transducer converts high frequency electrical signals into high frequency linear mechanical motion (or vibration). These high frequency vibrations are transmitted to the tool *via* tool holder. For achieving optimum Material Removal Rate (MRR), tool and tool holder are designed such that the resonance can be achieved. *Resonance* (or maximum amplitude of vibration) is achieved when the frequency of vibration matches with the natural frequency of tool and tool holder assembly.

The *tool shape* is made converse to the desired cavity. The tool is placed very near to the work surface and the gap between the vibrating tool and the workpiece (called as machining gap) is flooded with abrasive *slurry* made up of fine abrasive particles and suspension medium (usually water). As the tool vibrates, it strikes the abrasive particles in its downward stroke. This impact from the tool propels the abrasive particles across the machining gap. These particles attain Kinetic Energy (KE) and strike the work surface with a force much higher than their own weight. This force is sufficient to remove material from the brittle workpiece surface and results in a crater formation on it. Each down stroke of the tool accelerates numerous abrasive particles resulting in the formation of thousands of tiny chips per second. A very small percentage (about 5%) of material is also believed to be removed by a phenomenon known as *cavitation erosion*. To maintain a very low constant machining gap, a constant downward feed is usually given to the tool. The downward feed rate depends upon the linear material removal rate.

USM gives low *MRR* but it is capable to machine intricate cavities in single pass in fragile or/and hard materials. In USM, there is no direct contact between the tool and workpiece hence it is a good process for machining very thin and fragile components. By this process, a brittle material can be machined more easily than a ductile one. It is considered as a safe process because it does not involve high voltage, chemicals, high mechanical forces and large amount of heat.

ULTRASONIC MACHINING SYSTEM

USM machines are available in the range of 40 W to 2.4 kW. USM system has power supply, transducer, tool holder, tool, and abrasives slurry.

High power sine wave generator converts low frequency (60 Hz) electrical power to high frequency ($\approx$20 kHz) electrical power. This high frequency electrical signal is transmitted to the transducer which converts it into high frequency low amplitude vibrations. In USM, either of the two types of transducers are used, *i.e.* piezoelectric transducer or magnetostrictive transducer. *Piezoelectric crystals* (say, quartz) generate a small electric current when they are compressed. Also, when an electric current is passed through the crystal, it expands; when the current is removed the crystal attains its original size. This effect that is expansion and contraction of the crystal is known as *piezoelectric effect*. Such transducers are available up to the capacity of 900 W or so.

Magneto-strictive transducer also changes its length when subjected to a strong magnetic field. These transducers are made of nickel, or nickel alloy sheets. Their conversion efficiency (20–25%) is much lower than the piezoelectric transducer efficiency (up to 95%) hence their cooling is essential to remove the waste heat. These magnetostrictive type transducers are available with power capacity as high as 2.4 kW. The maximum change in length (or amplitude of vibration) that can be usually achieved is 25 μm.

Tool holder holds and connects the tool to the transducer. It virtually transmits the energy and in some cases, amplifies the amplitude of vibration. Material of the tool therefore, should have good acoustic properties, and high resistance to fatigue cracking. Due measures should be taken to avoid ultrasonic welding between the transducer and tool holder. For example, attaching tool holder with transducer using loose fitting screws and/or half hard copper washers between the transducer and tool holder would result in ultrasonic welding [*Benedict*, 1987].

Commonly used *materials for the tool holder* are monel, titanium and stainless steel. Because of good brazing as well as acoustic properties, monel is commonly used material for low amplitude applications. High amplitude application requires good fatigue strength of tool holder material. Further, tool-holders may be *amplifying or non-amplifying*. Non-amplifying tool holders have circular cross section and give the same amplitude at both the ends, *i.e.* input end and output end. Amplifying tool holders give as much as 6 times increased tool motion at the output end. It is achieved by stretching and relaxing the tool holder material. Such a tool holder yields MRR upto 10 times higher than non-amplifying tool holder [*Pandey and Shan*, 1980]. However, amplifying tool holders are more expensive, demand higher operating cost and yield poorer surface quality.

Tools are usually made of relatively ductile materials (brass, stainless steel, mild steel, etc.) so that the Tool Wear Rate (TWR) can be minimized. The value of the ratio of *TWR* and *MRR* also depends upon the kind of abrasive, workpiece material, and tool material used during USM. Surface finish of the tool is important because it affects the surface finish obtained on the workpiece. To *safeguard tool* and *tool holder* against their early fatigue failure, they should not have scratches or machining marks on the surface. Tools should be properly designed to account for overcut (Figure 2.2). Silver brazing of the tool with tool holder minimizes fatigue problems which are associated with screw attachment method.

Tool feed system is an important element of any USM system. It is supposed to facilitate a constant low feed (≈ linear material removal rate, mm/s) to the tool, and to provide adequate cutting force on the abrasive particles, during USM process. The feed system should also over run the tool to a small distance to ensure the required hole size at the exit (if it is a through hole), and to withdraw the tool from the cavity/hole at the end of the operation.

Hardness, particle size, usable life time and cost should be considered as criteria for selecting *abrasive particles* to be used in USM. Commonly used abrasives in the order of their increasing hardness, are Al_2O_3, SiC and B_4C (Boron carbide). In order to have high usable life time of abrasives, their hardness should be more than workpiece material.

MRR and surface finish obtained during USM are also the function of abrasive particles size. Coarser *particles* result in higher MRR and poorer surface finish while reverse is true with finer *particles*. Mesh sizes of grits generally used, range from 240 to 800. Abrasive slurry consists of water and abrasive particles usually in 1:1 (by weight). However, it can vary depending upon the type of operation to be performed, *viz.*, thinner mixtures (*i.e.*, low abrasive particles concentration) are used while drilling deep holes, or machining complex cavities so that the slurry flow is more efficient. The slurry stored in a reservoir, is pumped to the gap formed between the tool and work (Figure 2.2). In case of heavy duty machines, a cooling system may be required to remove heat from the slurry. Figure 2.3 shows different USM elements and working principle, in brief.

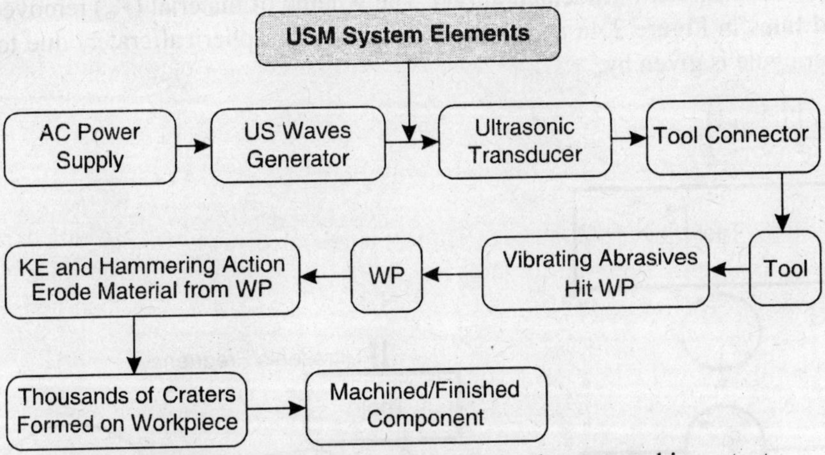

Fig. 2.3: Block diagram of an USM system elements and important events taking place during USM (WP-Workpiece, US-Ultrasonic)

MECHANICS OF CUTTING

Various mechanisms of material removal have been proposed by different researchers [*Miller*, 1957; *Shaw*, 1956; *Kazantsev and Rosenberg*, 1965; *Kainth, et al.*, 1979]. A comprehensive review of different models for USM process proposed by researchers has been presented by *Jain and Jain* [2001]. Theory proposed by *Shaw* [1956] for explaining mechanism of material removal in USM has been briefly explained as follows.

Model Proposed by Shaw

Material removal during *USM* due to cavitation under the tool and chemical corrosion due to interaction between the workpiece and slurry are considered insignificant. Hence, material removal due to these two factors has been ignored. Contributions to the material removal only by abrasive particles due to *'throwing'* and *'hammering'* actions have been analyzed.

Abrasive particles are assumed to be spherical in shape having diameter as '*d*' units. Abrasive particles (suspended in carrier liquid) move under the high frequency vibrating tool. There are two possibilities when the tool hits an abrasive particle. If the size of the particle is small and the gap between the bottom of the tool and top of the work surface is large enough, then the particle will be thrown by the tool, to hit the work surface (throwing model). Under the reverse conditions (that is when the machining gap becomes smaller than the diameter of the abrasive particle), the abrasive particle will be hammered by the tool, over the work surface. In both cases, let us assume that a particle after hitting the work surface generates a crater of depth '*h*' and radius '*r*' (Figure 2.4a, b). It is also assumed that the volume of the material removed from the workpiece surface is approximately proportional to the diameter of indentation (2r). The volume of material (V_g) removed (shown by dashed lines in Figure 2.4a and 2.4b, assuming hemi-spherical crater) due to fracture per grit per cycle is given by,

$$V_g = \frac{1}{2}\left(\frac{4}{3}\pi r^3\right). \qquad \text{... (2.1)}$$

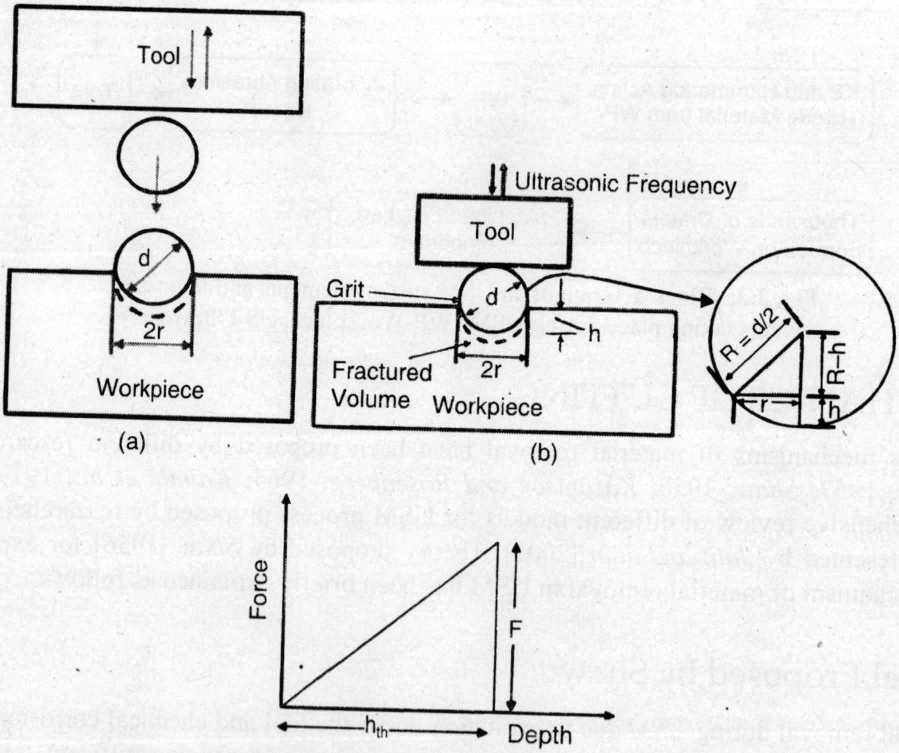

Fig. 2.4: Development of fracture in the workpiece by an abrasive particle (a) by Throwing, (b) by Hammering, (c) Assumed variation of force (F) with a change in depth of penetration during throwing action.

From the geometry of Figure 2.4b, it can be shown that,

$$r^2 = \left(\frac{d}{2}\right)^2 - \left(\frac{d}{2} - h\right)^2 \approx dh \text{ (neglecting } h^2 \text{ terms. } h << d) \quad \text{... (2.2)}$$

From Eqs. (2.1) and (2.2),

$$V_g = K_1\,(hd)^{3/2} \quad \text{... (2.3)}$$

where, K_1 is a constant.

Number of impacts (N) on the workpiece surface by the grits in each cycle will depend upon the number of grits beneath the tool at that time. This number depends upon the diameter of the abrasive particles that can be accommodated under the given tool diameter (Figure 2.5). This is inversely proportional to the diameter of the grit (assumed spherical) as given below.

$$N = K_2 \frac{1}{d^2} \quad \text{... (2.4)}$$

where, K_2 is a constant of proportionality.

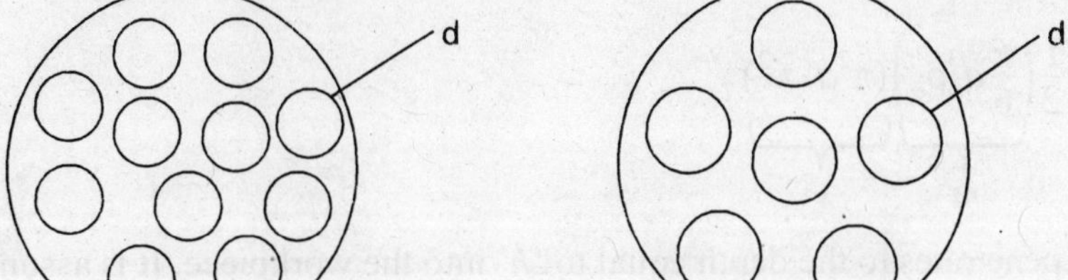

Fig. 2.5: Different number of particles of two different diameters accommodated in the same area of tool

All abrasive particles under the tool need not be necessarily effective. Let K_3 be the probability of an abrasive particle under the tool being effective. Then, the volume of material removed per second (V_s) is given by Eq. (2.5), and it will be equal to the frequency (f) times the amount of material removed per cycle (V_c),

$$V_s = V_c \times f = (V_g \times N) \times f$$

$$\boxed{V_s = K_1 K_2 K_3 \sqrt{\frac{h^3}{d}} \cdot f} \quad \text{... (2.5)}$$

To evaluate the depth of penetration 'h' of an abrasive particle, *Shaw* [1956] proposed two models. Model 1 considers that when a particle is hit by the tool, it is thrown (Figure 2.4a) on the workpiece surface. Model 2 assumes that an abrasive particle is hammered (Figure 2.4b) by the tool, and it penetrates into the workpiece and tool both. Both these models are discussed below.

Model 1 (Particle Throwing Model)

It is assumed that a particle is hit and thrown by the tool onto the workpiece surface. Assuming sinusoidal vibration, the displacement (Y) of the tool is given by Eq. (2.6) in which 't' is time period and $a/2$ is amplitude of oscillation.

$$Y = \frac{a}{2}\sin(2\pi ft). \qquad \text{... (2.6)}$$

By differentiating both sides of Eq. (2.6), velocity of the tool is obtained as follows,

$$\dot{Y} = \pi af\cos(2\pi ft). \qquad \text{... (2.7)}$$

The maximum velocity of the tool $\dot{Y}_{max}$ is given by:

$$\dot{Y}_{max} = \pi af \text{ (for } \cos(2\pi ft) = 1 \text{ in Eq. (2.7)).} \qquad \text{... (2.8)}$$

Let us assume that the abrasive particles also leave the tool with the same maximum velocity, *i.e.* $\dot{Y}_{max}$. Then, Kinetic Energy (KE) of an abrasive particle is given by

$$KE = \frac{1}{2}m(\pi^2 a^2 f^2),$$

$$= \frac{1}{2}\underbrace{\left(\frac{\pi}{6}d^3\rho_a\right)}_{m}\underbrace{(\pi^2 a^2 f^2)}_{v^2} \qquad \text{... (2.9)}$$

A particle penetrates to the depth equal to 'h' into the workpiece. It is assumed that full KE of the abrasive particle is absorbed by the workpiece before the particle comes to rest. Then, the work done by an abrasive particle (assuming triangular variation of force (F) with the depth of penetration) (Figure 2.4c) is given by,

$$\text{Work done, } W_g = \frac{1}{2}Fh_{th} \text{ (from Figure 2.4c).} \qquad \text{... (2.10)}$$

Work done by the particle (W_g) should be equal to the KE of the particle.

$$\frac{1}{2}F\ h_{th} = \frac{1}{2}\left(\frac{\pi}{6}d^3\rho_a\right)(\pi^2 a^2 f^2),$$

$$\text{Or, } h_{th} = \frac{\pi^3 a^2 f^2 d^3 \rho_a}{6F}. \qquad \text{... (2.11)}$$

The force 'F' required for a particle to penetrate h_{th} in the workpiece can be written in terms of workpiece property (workpiece hardness, H_w) that is known beforehand. Mean stress acting on the workpiece surface (σ_W) is given by,

$$\sigma_W = \frac{F}{A} = \frac{F}{\pi r^2} = \frac{F}{\pi h_{th} d} \text{ (Using Eq. (2.2))} \qquad \text{... (2.12)}$$

$$F = \pi\sigma_w h_{th} d.$$

From Eqs. (2.11) and (2.12),

$$h_{th} = \frac{\pi^3 a^2 f^2 d^3 \rho_a}{6\pi \ \sigma_w \ h_{th} \ d},$$

$$h_{th}^2 = \pi^2 a^2 f^2 d^2 . \frac{\rho_a}{6\sigma_w}.$$

$$\boxed{\text{or, } h_{th} = \pi afd\sqrt{\frac{\rho_a}{6\sigma_w}}.} \quad \text{... (2.13)}$$

Volumetric material removal rate due to throwing mechanism (V_{th}) can be obtained using Eqs. (2.5) and (2.13),

$$\boxed{V_{th} = K_1 K_2 K_3 \left[\frac{\pi^2 a^2 \rho_a}{6\sigma_w}\right]^{3/4} df^{5/2}.} \quad \text{... (2.14)}$$

Model 2 (Particle Hammering Model)

When the gap between the bottom position of the vibrating tool and the top surface of the workpiece is smaller than the diameter of the grit, then the grit will result into partial penetration in the tool (h_{tt}) as well as in the workpiece (h_w) (Figure 2.6a). In this situation, the gap between the mean position of the vibrating tool and the workpiece should always be larger than the (maximum) grit diameter so that the slurry can easily enter into the working gap. The values of h_w and h_{tt} will depend on the hardness of the tool as well as hardness of the workpiece material. Force '*F*' acts on the abrasive particles only for a short time ΔT during the cycle time '*T*' (Figure 2.6b). During this time period, the abrasive particle is in contact with the tool and workpiece both (Figure 2.6c). The mean force (F_{avg}) on the abrasive particle can be expressed by Eq. (2.15),

$$F_{avg} = \frac{1}{T}\int_O^T F(t)\, dt. \quad \text{... (2.15)}$$

Here, $F(t)$ is the force at any instant of time '*t*'. Force on the grit by the tool starts increasing as soon as the abrasive particle gets in contact with both the tool and the workpiece at the same time. It attains maximum value and then starts decreasing until attains zero value (Figure 2.6b). Hence, the momentum equation can be written as,

$$\int_O^T F(t)\, dt \approx \left(\frac{F}{2}\right)\Delta T. \quad \text{... (2.16)}$$

Total penetration due to hammering (h_h) action (Figure 2.6a) is given as follows,

$$h_h = h_w + h_{tl}. \quad \text{... (2.17)}$$

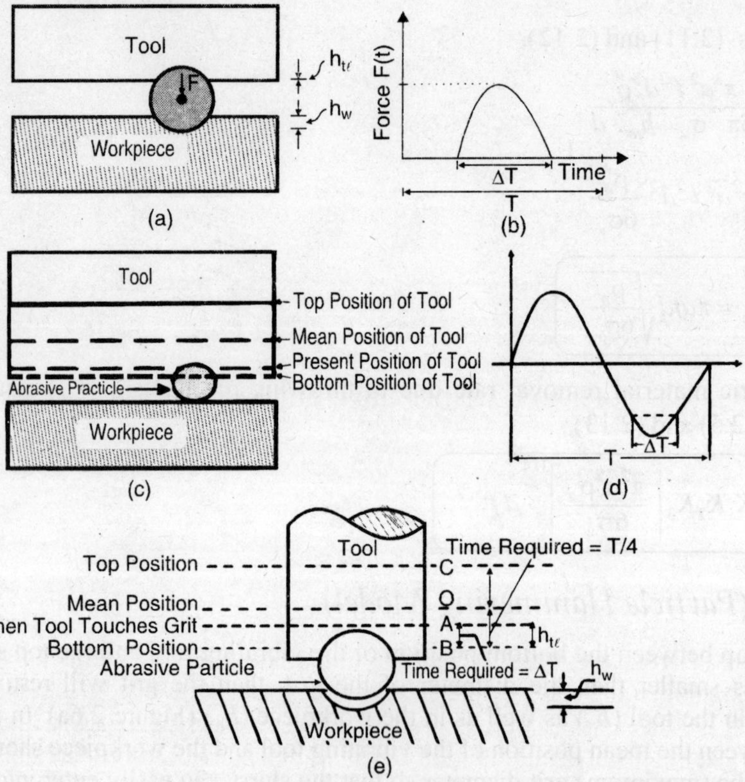

Fig. 2.6: (a) Partial penetration of an abrasive particle in the tool and workpiece, (b) Variation of Force (F) with Time (T), (c) Schematic diagram of an abrasive particle just starting hammering the workpiece, (d) Assumed mode of tool vibration, (e) Various positions of tool while hitting workpiece *via* an abrasive particle.

$a/2$ is amplitude of oscillation of the tool. The mean velocity of the tool during the quarter cycle (from O to B in Figure 2.6e) is given by $(a/2)/(T/4)$. Therefore, time (ΔT) required to travel from A to B is given by the following equation:

$$\Delta T \cong \frac{h_h}{(a/2)}.(T/4) = \frac{h_h}{a}\left(\frac{T}{2}\right). \qquad \text{... (2.18)}$$

From Eqs. (2.15), (2.16), and (2.18),

$$F = F_{avg}\frac{4a}{h_h}. \qquad \text{... (2.19)}$$

Let 'N' be the number of abrasive particles under the tool. Stress acting on the tool (σ_{tl}) and the workpiece (σ_w) can be found as follows:

$$\sigma_w = \frac{F}{N(\pi h_w d)} \qquad \text{... (2.20)}$$

$$\sigma_{t\ell} = \frac{F}{N(\pi h_{t\ell} d)}$$

$$= \sigma_w \frac{h_w}{h_{t\ell}} \text{ (from Eq. 2.20)} \qquad \text{... (2.21)}$$

From (Eqs. (2.4), (2.19) and (2. 20)),

$$\sigma_w = F_{avg} \frac{4\, ad^2}{h_h K_2 (\pi\, h_w d)}$$

$$= \frac{4\, F_{avg} ad}{\pi\, K_2\, h_w (h_w + h_{t\ell})}$$

$$= \frac{4\, F_{avg} ad}{\pi\, K_2\, h_w^2 \left(\frac{h_{t\ell}}{h_w} + 1\right)} \qquad \text{... (2.22)}$$

From Eq. (2.21),

$$h_{t\ell}/h_w = \sigma_w/\sigma_{t\ell} = j. \qquad \text{... (2.23)}$$

j can be taken as the ratio of hardness of workpiece material to the hardness of tool material. From Eqs. (2.22) and (2.23),

$$\boxed{h_w = \sqrt{\frac{4F_{avg} ad}{\sigma_w \pi K_2 (j+1)}}}. \qquad \text{... (2.24)}$$

Material removal rate from the workpiece due to hammering mechanism (V_h), can be evaluated using Eqs. (2.5) and (2.24) as follows:

$$\boxed{V_h = K_1 K_2 K_3 \left[\frac{4\, aF_{avg}}{\sigma_w\, \pi K_2 (j+1)}\right]^{3/4} d^{1/4} f} \qquad \text{... (2.25)}$$

From computational results obtained using Eqs. (2.14) and (2.25), it is observed that $V_h >> V_{th}$. Therefore, material removed by the throwing mechanism can be neglected as compared to the material removal by the hammering mechanism. Further, σ_w can be replaced by the known quantity that is workpiece hardness, H_w.

PARAMETRIC ANALYSIS

Performance (MRR, accuracy and surface finish) of the USM process depends upon different parameters as shown in Figure 2.7. Details of some of these parameters are as follows.

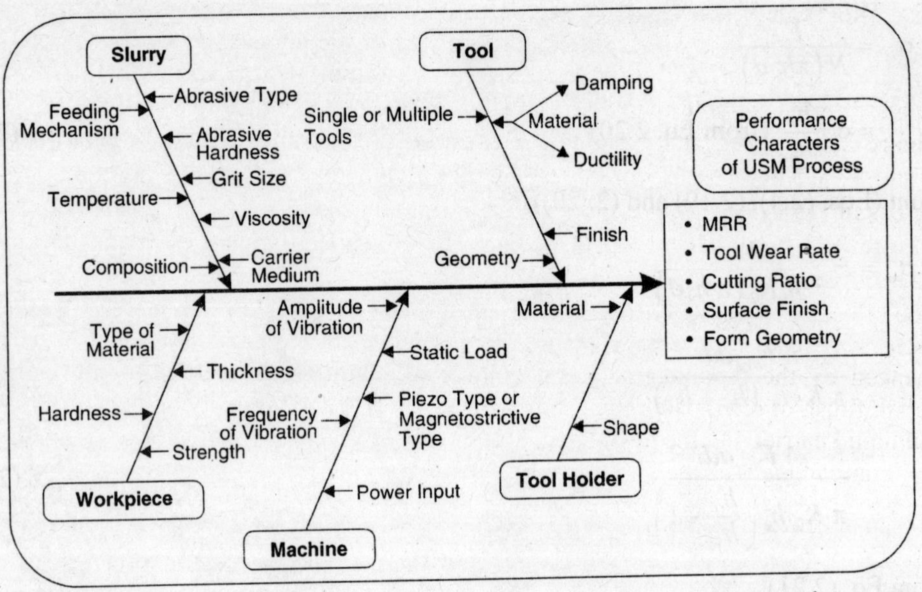

Fig. 2.7: Ultrasonic machining process performance affecting parameters and responses

The elongation obtained from a magnetostrictor operating at resonance, ranges from 0.001 to 0.1 μm. It is achieved by using a "concentrator" at the output end of the tool. To maximize the amplitude of vibration, the concentrator should operate at resonance condition. Figure 2.8(a) shows the effect of amplitude of vibration on linear MRR (on MRR_l), also called as penetration rate. Increase in the *amplitude of vibration* increases

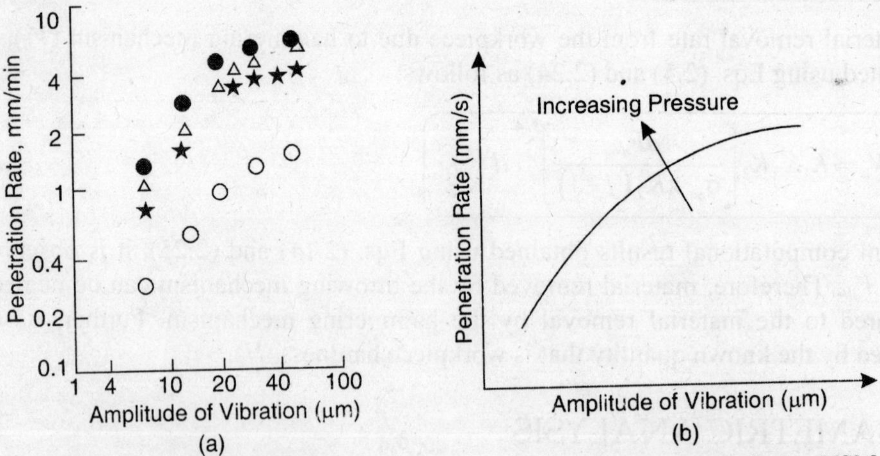

Fig. 2.8: (a) Effect of amplitude of vibration on material removal rate during USM. workpiece: glass; tool: steel; abrasive: B_2C (120 mesh size); Pressure: ● 0.20 MPa; Δ 0.16 MPa; ★ 0.10 MPa; O 0.04 MPa. [*Kremer et al.*, 1981], (b) Schematic representation of the effect of amplitude of vibration on penetration rate.

MRR_ℓ. This is the most significant process parameter that affects MRR_ℓ. Under certain circumstances, this also limits the maximum size of the abrasive particles to be used. It can also be seen that as the pressure increases (Figures 2.8(a) and 2.8(b)), MRR_ℓ also increases. Beyond a certain maximum value of MRR_ℓ, it attempts to stabilize.

Figure 2.8(b) shows a schematic relationship for the same. As can be seen from Eqs. 2.14 and 2.25, MRR_v varies non-linearly with a change in amplitude of vibration. *Neppiras and Foskett* [1956–57] also found non-linear relationship. *Miller* [1957] reported that this relationship is linear. However, a generalized conclusion cannot be made because the relationship depends on other variables also.

The effect of *abrasive particle size* is illustrated in Figure 2.9. An increase in abrasive *particle* size results in higher MRR_ℓ but poorer surface finish. Surface finish is also influenced by the parameters like amplitude of vibration, properties of the workpiece material, finish of the tool surface, abrasive particle size and viscosity of the slurry (mixture of the liquid carrier and the abrasives). Maximum MRR_ℓ is achieved when abrasive *particle* size is comparable with the amplitude of vibration of the tool. Machining rate is a function of KE (= (1/2) (mv^2)) of the active abrasive particle. Once KE increases, the machining rate also increases. Abrasive particle size affects the accuracy of the machined feature. The feature/hole size is always bigger than the tool size because of the flow of the slurry along the sides of the feature created. In case of highly accurate components, more than one tool and more than one size of abrasive particles can be used in different stages of machining (roughing, and finishing stages). Hardness of the abrasives and method of introducing slurry in the machining zone also affect the machining rates.

Frequency of Vibration also has a significant effect on MRR_ℓ (Figure 2.10). Efficient cutting is obtained at resonance frequency. Tool and tool holder combination having higher resonance frequency will yield higher MRR_ℓ provided machining is done at the resonance

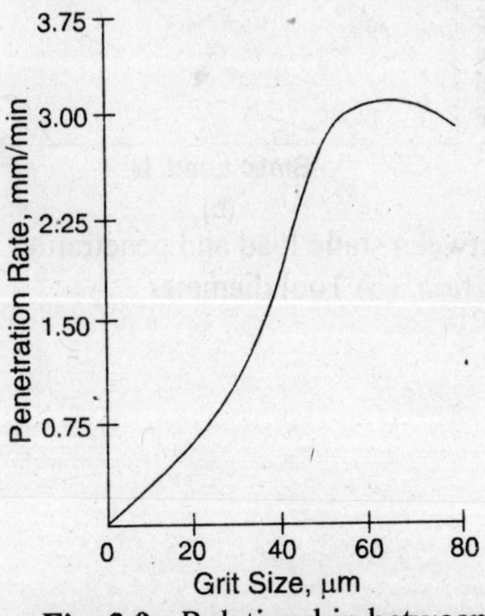

Fig. 2.9: Relationship between penetration rate and abrasive grit size [*Kazantsev*, 1956]

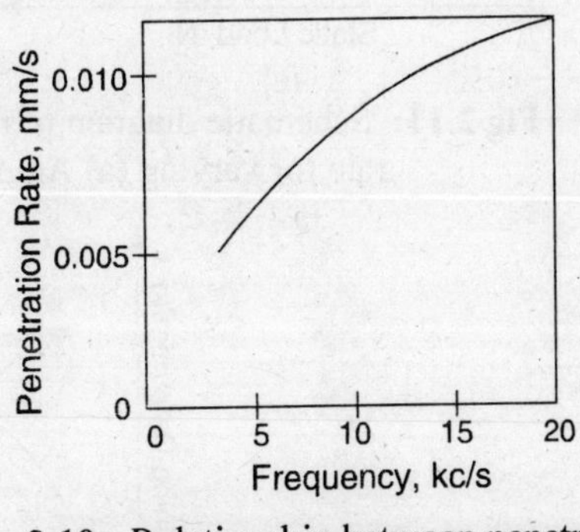

Fig. 2.10: Relationship between penetration rate and frequency of vibration of tool [*Neppiras and Foskett,* 1956]

frequency (Eq. 2.14). Contrary to *Nappiras and Foskett* [1956], some other researchers [*Shaw*, 1956] reported this relation as linear (see Eq. 2.25 if you ignore MRR by throwing mechanism, Eq. 2.14). Again, this relationship will be affected by machining conditions and workpiece material. MRR_l goes down as the depth of the hole increases. It happens so because of inefficient flow of slurry through the cutting zone at large depth.

The relationship between *static load* and *penetration rate* (Figure 2.11a) shows a maxima. This optimum penetration rate increases with an increase in amplitude of vibration. An increase in tool diameter decreases penetration rate (Figure 2.11b) because pressure decreases (assumption: force acting on the tool remains constant in two cases). Increase in static load increases the total force acting on the slurry or, the impact force acting on each abrasive particle. Static load can be changed by counter weight, pivoted counter weight, spring, pneumatic or hydraulic means [*Pandey and Shan,* 1980]. Beyond the maxima, the penetration rate starts decreasing.

As the *abrasive concentration increases*, the machining rate attains maxima (at about 30% abrasive concentration by volume) (Figure 2.12). If the concentration is too high then due to lower impact velocity (total force remains the same) and collision of abrasive particles with each other, result in reduction in machining rate (Figure 2.12). Viscosity of the slurry also affects machining rate. The pressure at which slurry is supplied in the machining zone has remarkable effect on machining rate in USM.

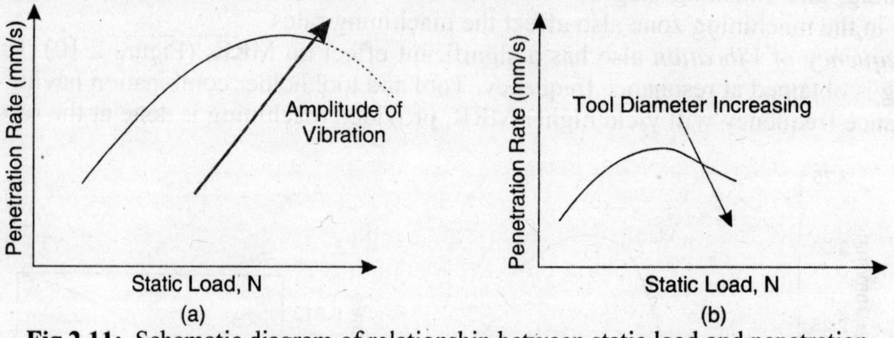

Fig 2.11: Schematic diagram of relationship between static load and penetration rate for varying (a) Amplitude of vibration, (b) Tool diameter

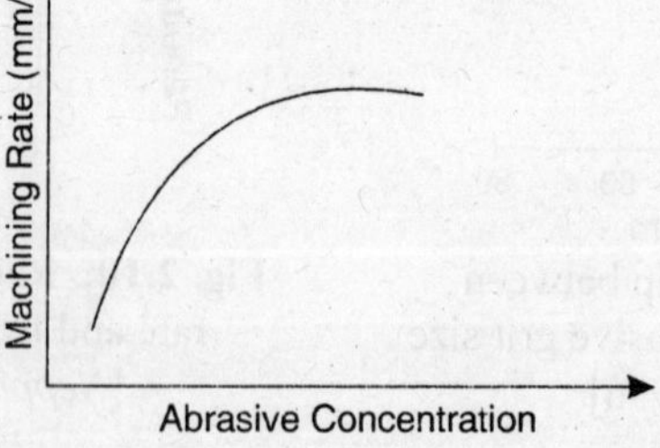

Fig. 2.12: Effect of abrasive concentration on machining rate

To *enhance the machining rate* in USM, raising the slurry temperature by 50°C optimizes the cavitation in water and as high as 70% increase in MRR can be obtained. On the other side, it is also reported that if the slurry is refrigerated to 0°C at room temperature then it causes embrittlement in the workpiece material which can enhance MRR by 20% to 30%.

PROCESS CAPABILITIES

USM works satisfactorily only when workpiece hardness is greater than HRC 40 (hardness on Rockwell scale '*C*'). It works very well if workpiece hardness is greater than HRC 60. Materials (carbides, ceramics, tungsten, glass, etc.) that can't be easily machined by conventional methods, can be easily machined by this technique.

Tolerances that can be achieved by this process range between 7 μm–25 μm. Holes as small as 76 μm have been drilled. Hole depths up to 51 mm have been easily achieved while 152 mm deep holes also have been drilled by using special flushing technique. The aspect ratio as high as 40:1 has been achieved in USM.

Linear material removal rate, MRR_l achieved during USM ranges from 0.025–25.0 mm/min, and it depends upon various parameters. Surface finish achieved during the process varies from 0.25 μm to 0.75 μm, and it is mainly governed by abrasive particles size. USM results in a *non-directional surface texture* compared to conventional grinding process.

Accuracy of the machined surface is governed by the size of the abrasive particles, tool wear, transverse vibration and machined depth. Usually, overcut (clearance between the tool and the machined surface wall on the workpiece) is used as a measure of accuracy. Radial overcut may be as low as 1.5 to 4.0 times the mean abrasive particle size. Overcut also depends on other parameters like workpiece material and method of tool feed. The overcut is not uniform along the machined depth and it results in the *conicity* in the machined cavity. Various means are suggested to reduce the conicity, *viz*., use of higher static load, direct injection of slurry into the machining zone, and use of a tool having negative taper.

Out-of-roundness is another criterion used to measure accuracy of the machined cylindrical holes. Inaccurate setting of the tool during USM is the main source of lateral vibration which results in out-of-roundness in the cavity. Out-of-roundness also depends on the type of workpiece material.

Depending upon the mass of tool holder, the power rating is determined. Trepanning type of tool can be used to reduce power requirements in the operations like drilling of large diameter and low depth holes.

APPLICATIONS

Most successful *USM application* is machining of cavities (or holes) in electrically non-conductive ceramics. It is quite successful in case of fragile components in which otherwise scrap rate is quite high. To increase productivity, it is used to drill multiple number of holes at a time, *viz*. 920 holes each of radius equal to 0.22 mm [*Benedict*, 1987]. For this purpose, hypodermic needles have been used as tools. USM has also been employed for multisteps processing for fabricating silicon nitride (Si_2N_4) turbine blades. This process can be used to cut a large disc by trepanning operation, complex shaped features, turning, etc.

ULTRASONIC MICROMACHINING (USMM)

Ultrasonic micromachining (USMM) is an unique process for the precision machining of difficult-to-machine brittle and hard materials [*Bellows and Kahl*, 1982], such as, glass, sapphire, quartz, silicon, etc. The working principle of USM and USMM remains the same however the size of the tool and machining parameters are quite different in both cases. In some cases, researchers have proposed to fabricate a microtool on a machine, and then perform USMM operation on the same machine without disturbing the orientation and mounting of the tool [*Sun et al*., 1996; *Jadoun*, 2013]. Here, wire-electric discharge grinding (Wire-EDG), EDM, electrochemical machining or chemical machining is used for preparing microtool. Then, this tool is used to perform USMM operation on the same machine by giving rotation (if required) to the tool or workpiece.

As discussed earlier in USM, the mechanism of material removal in USMM also is due to one or more of the following four modes: micro chipping (throwing model) by impact of the freely moving abrasive particles, direct hammering of the abrasive particles against the workpiece surface (hammering model), cavitation erosion by abrasive slurry and chemical action due to the fluid in the slurry. Figure 2.7 shows the input and output process parameters for USM process which are also applicable to the USMM process.

TWR (tool wear rate) can be defined as weight loss of tool during USMM per unit time. The cutting ratio can be defined as MRR per unit TWR (= MRR/TWR). Researchers have reported [*Egashira and Mizutani,* 2002] that USMM can be used to drill the holes as small as 5–15 μm diameter. *Sun et al.* (1996) drilled microholes using USMM, having roughness = 200 nm, out of roundness = 1.0 μm and conicity of 5%. Ultrasonic micromachining is also used for micro-deburring purposes [*Yeo*, 1997]. This has been achieved by ultrasonic cavitation with specially compounded acid slurry. This is suitable for *micro deburring* of inaccessible sections in small parts.

Ultrasonic vibration assisted cutting has been performed to drill a microhole (≈ 10 μm) in glass by preparing a drill type of tool using wire electro discharge grinding. The cutting was performed in the ductile regime by taking 50 nm as depth of cut. Due to ductile regime cutting, no fracture or cracks were observed around the edge of the hole.

PROBLEM

Find out the approximate time required to machine a hole of diameter equal to 6.0 mm in a tungsten carbide plate (fracture hardness = 6900 N/mm^2 = 6.9×10^9 N/m^2) of thickness equal to one and half times of hole diameter. The mean abrasive particle size is 0.015 mm diameter. The feed force is equal to 3.5 N. The amplitude of tool oscillation is 25 μm and the frequency is equal to 25 kHz. The tool material used is copper having fracture hardness equal to 1.5×10^3 N/mm^2. The Slurry contains one part abrasive to one part of water. Take the values of different constants as $K_1 = 0.3$, $K_2 = 1.8$ mm^2, $K_3 = 0.6$ and abrasive density = 3.8 g/cm^2.

Also calculate the ratio of volume removed by throwing mechanism to the volume removed by hammering mechanism.

Solution

Following data are given:

Hole diam. = 6×10^{-3} m, plate thickness = 1.5 × hole diam = 9×10^{-3} m, mean abrasive grain size (d) = 1.5×10^{-5} m, feed force (F) = 3.5 N, amplitude of tool oscillation ($a/2$) = 25×10^{-6} m, frequency of oscillation (f) = 25000 cps, fracture hardness of workpiece material, σ_w (= H_w) = 6.9×10^9 N/m^2, fracture hardness of tool material (H_{tl}) = 1.5×10^9 N/m^2, abrasive grain density (ρ_a) = 3.8×10^2 kg/m^3, $j = H_w/H_t = 4.6$, $K_1 = 0.3$, $K_2 = 1.8$ mm^2 = 1.8×10^{-6} m^2, $K_3 = 0.6$, $C = 1$.

Following procedure should be followed to solve the given problem:

What is the time required to machine the hole?

We know that the volume (V) of material removed during USM can be calculated using the following relationship:

$$V = K_1 K_2 K_3 \sqrt{\frac{h^3}{d}} \times f \qquad \text{... (2.5)}$$

Step 1: Calculate the value of "h" which is different for throwing model (h_{th}) and for hammering model (h_w).

Step 2: After knowing the values of h_{th} and h_w, calculate V_{th} and V_h by substituting these values in above Eq. 2.5. Find total volume of material removed per unit time (V_s) by adding V_{th} and V_h.

Step 3: Calculate the total amount of material to be removed to make the required hole. Divide it by V_s to find the total time required to make the hole.

Step 4: Find the ratio of V_{th}/V_h

Following above steps, all calculations are made as given below:

Step 1: In Eq. (2.5), except "h" all other parameters are known.

Let us calculate h_{th} as follow.

$$h_{th} = \pi a f d \sqrt{\frac{\rho_a}{6\sigma_w}}$$

$$= \pi \times \left(50 \times 10^{-6}\right) \times \left(2.5 \times 10^4\right) \times \left(1.5 \times 10^{-5}\right) \sqrt{\frac{3.8 \times 10^3}{6 \times \left(6.9 \times 10^9\right)}}$$

$$\boxed{h_{th} = 1.78 \times 10^{-5} \text{ mm}}$$

Penetration (h_w) in the workpiece due to hammering is given as,

$$h_w = \sqrt{\frac{4 F_{avg} a d}{\pi k_2 \sigma_w (j+1)}}$$

$$= \sqrt{\frac{4 \times 3.5 \times \left(2 \times 25 \times 10^{-6}\right) \times \left(1.5 \times 10^{-5}\right)}{\pi \times \left(1.8 \times 10^{-6}\right) \times \left(6.9 \times 10^{9}\right)(1+4.6)}}$$

$$\boxed{h_w = 2.192 \times 10^{-4} \text{ mm}}$$

Step 2: Now, volume of workpiece material removed per second by throwing mechanism (V_{th}) is given as,

$$V_{th} = K_1 K_2 K_3 \sqrt{\frac{h_{th}^3}{d}}.f$$

$$= 0.3 \times 1.8 \times 0.6 \sqrt{\frac{\left(1.78 \times 10^{-5}\right)^3}{1.5 \times 10^{-2}}}.\ 2.5 \times 10^4$$

$$\boxed{V_{th} = 4.97 \times 10^{-3} \text{ mm}^3/\text{s}}$$

Volume of workpiece material removed per second by hammering action is given by,

$$V_h = K_1 K_2 K_3 \sqrt{\frac{h_w^3}{d}}.f$$

$$= 0.3 \times 1.8 \times 0.6 \sqrt{\frac{\left(2.192 \times 10^{-4}\right)^3}{1.5 \times 10^{-2}}} \times 2.5 \times 10^4$$

$$\boxed{V_h = 0.2146 \text{ mm}^3/\text{s}}$$

Step 3: Time required to drill the desired hole $= \dfrac{\text{Volume of the hole to be drilled}}{\text{Volumetric } MRR\ (= V_h + V_{th})}$

$$= \frac{(\pi/4) \times 6^2 \times 9}{0.219}$$

$$\boxed{= 19.266 \text{ min}}$$

Step 4: Ratio, $\dfrac{V_{th}}{V_h} = \dfrac{0.00497}{0.2146} = 0.022$

Thus, it is evident that the material removed by hammering is much more than throwing (approximately 42 times), Hence, for approximate calculations, V_{th} can be ignored as compared to V_h.

BIBLIOGRAPHY

[1] Bellows, G. and Kohls, J.B. (1982). "Drilling Without Drills", *American Special Report*, 742, p. 187.
[2] Benedict G.F. (1987). "Nontraditional Manufacturing Processes", Marcel Dekker Inc., New York.
[3] Bhattacharyya, A. (1972). "New Technology", The Institution of Engineers (I), Calcutta.
[4] Egashira, K. and Mizutani, K. (2002). Ultrasonic vibration drilling of microholes in glass, *Annals of CIRP*, Vol. 51(1), pp. 339–342.
[5] Jadoun, R.S. (2014). Ultrasonic Micromachining, in, Introduction to Micromachining (Ed. V.K. Jain), Narosa Publishing House, pp. 7.1–7.36.
[6] Jadoun, R.S., Kumar, P. Mishra, B.K. and Mehta, R.C.S. (2006). Optimization of process parameters for ultrasonic drilling (USD) of advanced engineering ceramics using the Taguchi approach, *Engg. Optim.*, Vol. 38(7), pp. 771–787.
[7] Jain, N.K. and Jain, V.K. (2001). Modeling of material removal in mechanical type advanced machining processes: A State-of-the Art Review, *Int. J. Mach. Tool and Manuf.* Vol. 41(11), pp. 1573–1635.
[8] Kainth, G.S. Nandy, A. and Singh, K. (1979). "On the Mechanics of Material Removal in Ultrasonic Machining", *Int. J. Mach. Tool Des. and Res.*, Vol. 19, pp. 22–41.
[9] Kaczmarek, J. (1976). Principles of Machining by cutting Abrasion and Erosion, Peter Peregrinus Ltd., Stevenage, pp. 448–462.
[10] Kazantsev, V.F. and Rosenberg, L.D. (1965). The Mechanism of Ultrasonic Cutting, Ultrasonic, Vol. 2, pp. 166.
[11] Kennedy, D.C. and Grieve, R.J. (Sept. 1975). "Ultrasonic Machining: A Review", *Prod. Eng.*, Vol. 54, pp. 481–486.
[12] Kremer D., Saleh S., Ghabrial S. and Moison A. (1981). "The State of the Art of Ultrasonic Machining", *Annals CIRP*, Vol. 20(1), pp. 107–110.
[13] Markov, A.I. (1977). "Ultrasonic Drilling and Milling of Hard Non-Metallic Materials with Diamond Tools", *Mach. Tooling*, Vol. 48(9), pp. 33–35.
[14] Masuzava, R.T. (2000). Sate of the art of micromachining, *Annals of CIRP*, Vol. 49(2), pp. 473–488.
[15] McGeough, J.A. (1988). "Advanced Methods of Machining", Chapman & Hall, London.
[16] Miller, G.E. (1957). "Special Theory of Ultrasonic Machining", *J. Appl. Phy.*, Vol. 28(2), p. 149.
[17] Neppiras, E.A. and Foskett, R.D. (1956–57). Ulrasonic Machining, Philips Technical Review, Vol. 18, pp. 225, 268.
[18] Pandey, P.C. and Shan, H.S. (1980). "*Modern Machining Processes*", Tata McGraw Hill, New Delhi.
[19] Shaw, M.C. (1956). "Ultrasonic Grinding", *Microtecnic*, Vol. 10, p. 257.
[20] Soundararajan, V. and Radhakrishnan, V. (1986). "An Experimental Investigation on the Basic Mechanism Involved in Ultrasonic Machining", *Int. J. Mach. Tool Des. and Res.*, Vol. 26(2), pp. 207–222.
[21] Sun, X.Q., Masuzava, T. and Fujino, M. (1996). Micro ultrasonic machining and its applications on MEMS, Sensors and Actuators A, Vol. 57, pp. 159–164.
[22] Yeo, S.H., Nagoi, B.K.A. and Chua, L.Y. (1997). Ultrasonic deburring, Int *J. Adv. Mfg. Tech.*, pp. 333–341.

SELF TEST QUESTIONS

Q1. Write True (T) or False (F).

1. In USM, for the same static load but higher tool diameter penetration rate is higher.
2. Audible frequency would give higher MRR than Ultrasonic frequency.
3. Magnetostrictive transducers are made up of laminated construction.
4. The ratio of MRR to tool wear rate can be as low as 0.5 in case of USM using WC as work material and alumina as abrasive material.

Q2. Multiple Choice Questions.

1. During calculation of MRR for USM (using Shaw's throwing model), the abrasive density by mistake has been used as 2 units in place of 3 units. Calculate the ratio of MRR with actual density and wrong density, and find out the correct answer from the following:
   (a) 1.71 (b) 1.36 (c) 0.74 (d) none
2. In USM, the frequency of oscillation is
   (a) 10–15 kHz (b) below 10 kHz (c) above 15 kHz (d) none of these
3. Lowering the temperature of the workpiece during USM would result in:
   (a) increased MRR_v (b) decreased in MRR_v
   (c) no effect on MRR_v (d) unpredictable
4. Performance of USM process is better in case of:
   (a) electrically conductive work material
   (b) work material having high thermal conductivity
   (c) soft material (d) no such constraint.
5. The number of impacts (in USM) on the workpiece is
   (a) directly proportional to the square of the diameter of the grit
   (b) inversely proportional to grit diameter
   (c) inversely proportional to the square of the grit diameter (d) none of these
6. Volume of material removal in USM is
   (a) directly proportional to frequency 'f' (b) inversely proportional to 'f'
   (c) directly proportional to $f^{2.5}$ (d) none of these two
7. Relationship between MRR_l and the particle size shows
   (a) monotonously increasing trend (b) monotonously decreasing trend
   (c) a maxima (d) a minima
8. During USM, abrasive mean diameter size is increased from 2 units to 3 units while workpiece hardness is increased to 2.25 times. What will be the ratio of new depth of penetration to the old depth of penetration?
   (a) 0.33 (b) 3.00 (c) 1.00 (d) none of these

9. Which of the following material(s) can be used for tool holder in USM?
   (a) Mild steel (b) Brass
   (c) Titanium (d) Monel
10. Which of the following is/are not true for USM?
    (a) It is a thermal type non-traditional machining process.
    (b) It can not be used for electrically non-conducting materials.
    (c) Tool and tool holder are designed in such a way that resonance can be avoided.
    (d) Due to the throwing & hammering action of the tool, very thin and fragile components can not be machined by this process.
11. Which of the following is/are true for magneto-strictive transducer?
    (a) It changes its length when subjected to a strong magnetic field.
    (b) These are generally made of quartz.
    (c) Its conversion efficiency (upto 95%) is much higher than piezoelectric transducer's efficiency.
    (d) No cooling is essential here as there is very less waste heat.
12. Which of the following is/are the desirable properties of tool holder in USM setup?
    (a) It should have good acoustic properties.
    (b) It should have high resistance to fatigue cracking.
    (c) It should have high electrical conductivity.
    (d) It should not have scratches or machining marks.
13. In USM, the material is removed by
    (a) Spark erosion (b) Etching (c) Anodic dissolution
    (d) Impact erosion (e) None of the above
14. If the amplitude of oscillation of the tool is doubled and frequency is made half, then the volumetric material removal rate in USM due to throwing mechanism (V_2) becomes,
    (a) $V_2 = 2V_1$ (b) $V_2 = 2^{2/4}V_1$ (c) $V_2 = 0.5V_1$
    (d) $V_2 = 2^{3/2}V_1$ (e) None
    (V_1 is initial MRR before changing oscillation and frequency)
15. In the previous question, the volumetric material removal rate in USM due to hammering mechanism (V_2) becomes,
    (a) $V_2 = 2V_1$ (b) $V_2 = 2^{2/4}V_1$ (c) $V_2 = 0.5V_1$
    (d) $V_2 = 2^{3/2}V_1$ (e) None
16. Which of the following is/are the desirable properties of tool in USM?
    (a) Ductile (b) high wear resistance
    (c) No scratches/machining marks (d) high electrical conductivity

Answers

1.*	2.*	3.*	4.*
5.*	6.*	7.*	8.*
9.c & d	10.all	11.a	12.a, b & d
13.D	14.c	15.e.	16.a, b & c

(*Find the solution)

Q3. Write short notes on the following.

(a) Transducers used in USM machine.
(b) Effect of amplitude of vibration, frequency of vibration, particle size, and % of abrasives concentration, on MRR in USM.
(c) Possible effects of use of audible frequency in USM.
(d) Functions of slurry, horn, transducer, and oscillator in USM.
(e) Types of abrasives used in USM.
(f) Conventional grinding and USM.
(g) Working principle of USM (show the necessary sketch).
(h) How USMM is different from USM?
(i) What do you understand by USMM?
(i) What kinds of problems you anticipate while using USMM for porous materials?

Q4. Solve the following questions.

(a) What do you understand by "Transducer" and "magnetostriction effects"?
(b) Explain the functions of a "horn" in USM.
(c) Draw the relationship observed during USM for the following cases:
  (i) frequency of tool vibration *vs* penetration rate,
  (ii) abrasive particle size *vs.* machining rate (mm/s),
  (iii) concentration of abrasive particles *vs.* machining rate,
  (iv) ratio of workpiece hardness to tool material hardness *vs* MRR.
(d) A cylindrical impression of 10 mm diameter and 1 mm deep is to be made on a WC specimen. Feed force is constant, and is equal to 5 N. Average diameter of grains in the slurry is 10 μm. Tool oscillates with the amplitude of 20 μm at 20 kHz. Abrasive and water ratio in the slurry is 1. Fracture hardness of WC workpiece may be taken as 7000 N/mm^2 and that of copper tool as 1500 N/mm^2. Calculate the time required to complete the job if only 20% of pulses are effective. Assume $K_1 = 0.3$, $K_2 = 1.8$ mm and $K_3 = 0.6$. Make the assumptions if necessary, and write them clearly.
(e) Write the four basic mechanisms by which material removal in USM can take place.
(f) Derive an equation suggested by Shaw to obtain volumetric material removal rate (consider both throwing and hammering mechanisms) during USM.

(g) USM is used for drilling a hole (under the same machining conditions) in aluminum and C.I. Which one will have higher depth of the drilled hole under the same machining conditions?

NOMENCLATURE

$a/2$	Amplitude of oscillation
A	Area
$d\ (=2R)$	Abrasive diameter
D_1, D_2	Crater diameter
E_m	Coefficient of magnetostrictor elongation
f	Frequency of vibration of tool
F	Force
h	Crater depth
h_h	Total penetration depth of an abrasive particle due to hammering mechanism
h_{tl}	Abrasive penetration depth in the tool
h_{th}	Depth of penetration of an abrasive particle due to throwing mechanism
h_w	Abrasive penetration depth in the workpiece
j	Ratio of hardness of workpiece material to the tool material
KE	Kinetic Energy
K_1, K_2	Constants
K_3	Probability of an abrasive particle under the tool being effective
L	Length of the magnetostrictor coil
l_1, l_2	Depth of penetration
m	Mass of a grit
N	Number of impacts on the workpiece by the grits in each cycle
r	Crater radius
t	Time
T	Cycle time
V_c	Amount of material removed per cycle
V_s	Volume of material removed per second
V_g	Volume of material removed per grit per cycle
V_h	Volumetric MRR from the workpiece due to hammering mechanism
V_{th}	MRR due to throwing mechanism
W_g	Work done by each grit
Y	Displacement of the tool
$\dot{Y}$	Velocity of the tool

GREEK SYMBOLS

ΔL	Change in length of the magnetostrictor coil
ΔT	Small time period
ρ_a	Density of abrasive particles
σ_w	Mean stress acting on the workpiece surface
σ_{tl}	Mean stress acting on the tool surface

SUFFIX

a	abrasive
avg.	average
h	hammering
max	maximum
th	throwing
tl	tool
w	workpiece

ACRONYMS

KE	Kinetic Energy
MRR	Material Removal Rate
TWR	Tool Wear Rate
USM	Ultrasonic Machining
USMM	Ultrasonic Micromachining
WP	Workpiece

CHAPTER 2
AT-A-GLANCE
ULTRASONIC MACHINING (USM)

What is Ultrasonic?	Vibratory Waves having Frequency Greater Than the Upper Audible Frequency Limit (~16 KHz) of the Human Ear.
Wave	Longitudinal Waves • Used in Ultrasonics • Easily Propagate in Solids, Liquids & Gases Shear Waves
Logitudinal Magnetostriction	If a Ferromagnetic Material is Placed in Continuously Changing Magnetic Field ↓ IT Results in a Change in Length of Transducer ↓ Magnetostriction Type Transducer
Ultrasonic Transducer	Converts any form of Energy into Ultrasonic Waves

Working of USM System

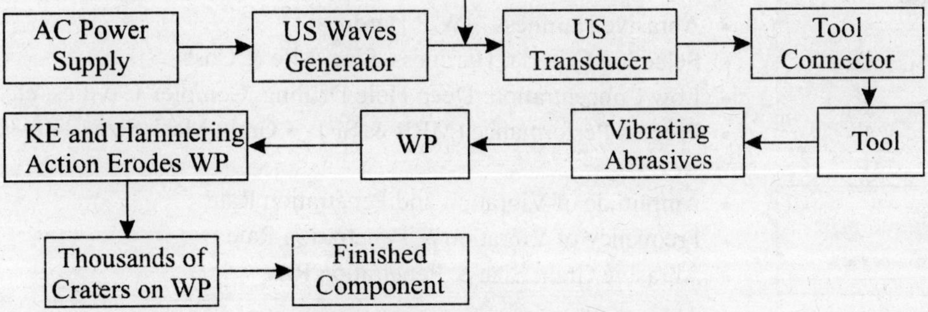

Some Elements of USM Systems

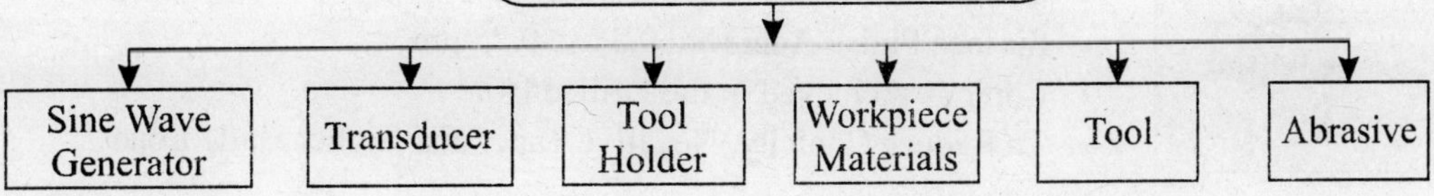

Sine Wave Generator	• Converts 60 Hz to 120 kHz
Transducer	• Piezoelectric Crystals (Quartz) → Up to 900 W & 95% Efficiency • Magnetostrictive Transducer – 2.4 KW or More; 20%–30% Efficiency – Cooling is Essential – Maximum Amplitude = 25 μm
Tool Holder and its Material	• Non-Amplifying • Transmits Energy • Amplitude of Vibration → 1 μm to 0.1 μm • Commonly Used: Monel, Ti, Stainless Steel • Good Acoustic Properties • High Resistance to Fatigue Cracking • Should Avoid Welding between Tool Holder & Transducer
Tool	• Material : Mild Steel, Stainless Steel, Brass, : Ductile, High Wear Resistance • Surface Condition: Good Surface Finish (SF), No Scratches/Machining Marks • Tool Design → Consideration for Overcut • Minimize Fatigue Problem → Brass (Tool & Toolholder)
Abrasive	• Al_2O_3, SiC, Baron Carbide • Abrasive Hardness > W/P Hardness • Selection Criteria: Hardness, Size, Life & Cost • Low Concentration: Deep Hole Drilling, Complex Cavities, etc. • Process Performance (MRR & SF) → Grain Mesh Size (240–800)
Parametric Analysis	• Amplitude of Vibration and Penetration Rate • Frequency of Vibration & Penetration Rate • Abrasive Grain Size & Penetration Rate
Process Capabilities	• Works Better for Workpiece Material Hardness > HRC 40 • Carbide, Ceramics, W, Glass etc. • Surface Finish Achieved 0.25 to 0.75 μm • Conicity Achieved in the Drilled Hole • To Reduce Conicity, Negative Taper and Higher Static Load

- Out-of-Roundness in Holes
- Tolerance: 7–25 μm
- Depth 51 mm. Even upto 152 mm (very special case)
- Aspect Ratio (Depth to Diameter) = 40:1

Process Applications

- Both Electrically Non-Conductive and Conductive
- Fragile Components
- Also for Multiple Holes
- Processing of Silicon Nitride Turbine Blades
- Glass, Ceramics, Titanium, Tungsten, etc.
- Drilling (Dentist Drilling Holes in Teeth), Grinding, Profiling
- USM Also Used In Conjunction with ECM, EDM.

Micromachining

- Same Working Principle of USM and USMM.
- Fabrication of Micro-Tool for USMM Operation Performed on the Same M/C.
- Micro-Tool Fabrication by EDMM, ECMM, Wire-EDMM etc.
- To Drill Holes as Small as 5–15 μm.
- Also Used for μ-Deburring.

Chapter Three

ABRASIVE JET MACHINING (AJM)

INTRODUCTION

A jet of air or inert gas consisting of very fine abrasive particles strikes the workpiece surface at high velocity (v) (usually, between 200 to 400 m/s) resulting in material removal through chipping/erosive action. This erosive action has been employed for cutting, cleaning, polishing and deburring. Kinetic energy ($KE = \frac{1}{2} mv^2$) of an abrasive particle is responsible for removal of material from the workpiece. Harder the workpiece material, higher should be the KE of the abrasive particle to remove material from the workpiece by erosive action. This method of material removal is quite effective on hard and/or brittle materials (*viz.*, glass, silicon, tungsten, ceramics, etc.) but not so effective on soft materials like aluminum, rubber, etc. It can produce fine and complicated details in the parts made of brittle materials [*Benedict*, 1987].

The essential parts of an Abrasive Jet Machining (AJM) setup, developed for laboratory purposes [*Verma and Lal*, 1984] are shown in a schematic diagram, Figure 3.1a.

ABRASIVE JET MACHINING SETUP

The main elements of the AJM setup include gas propulsion system, abrasive feeder, machining chamber, nozzle and abrasive particles.

Gas Propulsion System

The *gas propulsion system* supplies clean and dry inert carrier gas (air, nitrogen or CO_2) at high pressure to propel the abrasive particles. The gas may be supplied either by a compressor or a cylinder. In case of a compressor, air filter-cum-drier should be used to avoid water or oil contamination of the abrasive powder. The gas should be nontoxic, cheap, and easily available. It should not excessively spread when discharged from the nozzle into atmosphere.

Abrasive Feeder

Required quantity of abrasive particles is supplied by *abrasive feeder*. In this particular setup, abrasive quantity is controlled by inducing vibration of different magnitude and frequency to the feeder. The abrasive particles are propelled by carrier gas to the mixing chamber. The air-/gas-abrasive mixture moves further to the nozzle. The nozzle imparts high velocity to the mixture which is directed towards the workpiece surface. Material removal occurs due to erosive action of the jet of air-abrasive mixture impinging on the workpiece surface.

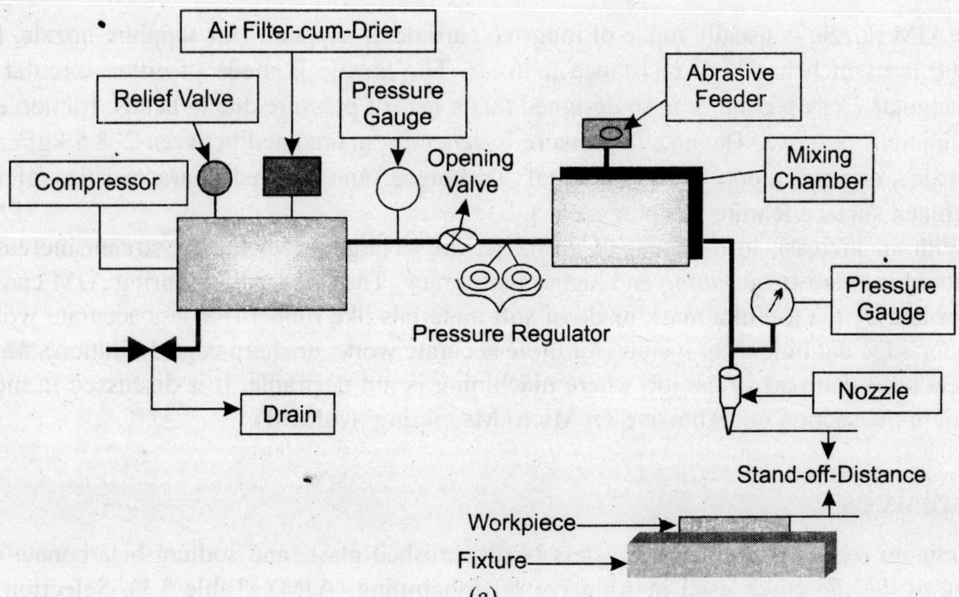

(a)

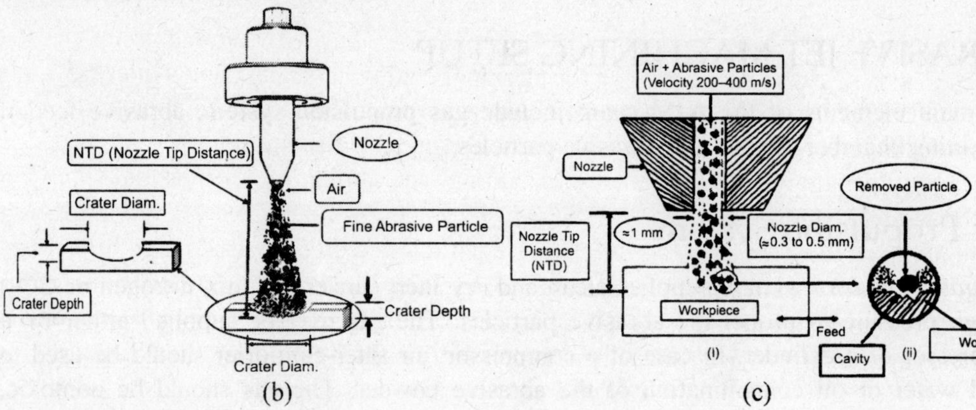

Fig. 3.1: (a) Schematic diagram of abrasive jet machining (AJM) setup, (b) Nozzle, air-abrasive jet and other details, (c) Air-abrasive jet and other details including mechanism of material removal.

Machining Chamber

The *machining chamber* is tightly closed so that the concentration of the abrasive particles around the working chamber or machining chamber does not reach to the harmful limit. Machining chamber is equipped with a *vacuum dust collector*. Special considerations should be given to the dust collection system if the toxic material (say, Beryllium) is being machined.

Nozzle

The AJM nozzle is usually made of tungsten carbide or sapphire (for sapphire nozzle, life ≃ 300 hr) which has high resistance to wear. The nozzle is made of either circular or rectangular cross-section. It is so designed that a loss of pressure due to bends, friction etc. is minimum possible. The nozzle pressure is generally maintained between 2–8.5 kgf/cm^2. Its value depends upon the material of workpiece, and desired characteristics of the machined surface/feature (accuracy, etc.).

With an increase in the wear of a nozzle, the divergence of the jet stream increases resulting in more *stray cutting* and higher inaccuracy. The stray cutting during AJM can be controlled by the use of a mask made of soft materials like rubber (for less accurate works or poor edge definition), or metals (for more accurate works or sharp edge definition). *Mask* covers only that part of the job where machining is not desirable. It is discussed in more detail in the section on 'Abrasive Jet Micro Machining' (AJMM).

Abrasives

Aluminum oxide, silicon carbide, glass beads, crushed glass, and sodium bicarbonate are some of the *abrasives* used in Abrasive Jet Machining (AJM) (Table 3.1). Selection of abrasive(s) depends upon the type of work material, desired Material Removal Rate (MRR), and accuracy. Al_2O_3 is good for cleaning, cutting, and deburring while SiC is also used for the similar applications but for harder work materials. For obtaining matte finish, glass beads are good while crushed glass performs better for giving sharper edges. However, operations such as cleaning, deburring, and cutting of soft materials are better performed by sodium bicarbonate. Table 3.1 shows abrasive particles size and hardness of some abrasive particles. The sizes of abrasive particles commonly used range from 10 μm to 50 μm. Small abrasive particles are used for cleaning and polishing (to make the surface smooth and shining) while large particles perform better during cutting. Fine abrasive particles are less irregular in shape hence their cutting ability is poor.

Table 3.1: Types of abrasives and corresponding types of machining operations performed using air + abrasive jet

S. No.	*Abrasive*	*Grain Size*	*Machining Operation*
1.	Aluminium oxide (Al_2O_3)	10, 20, 50 μm	Cutting and grooving
2.	Silicon carbide (SiC)	25, 40 μm	Cutting and grooving
3.	Sodium bicarbonate	27 μm	Light finishing below 50°C
4.	Dolomite	≈ 200 mesh	Etching + polishing
5.	Glass beads	0.635–1.27 μm	Light polishing and fine deburring

The abrasives should have sharp and irregular shape, and be fine enough to remain suspended in the carrier gas. Re-use of the abrasives is usually not recommended because of the following reasons. Firstly, abrasives get contaminated with eroded work material which may block the nozzle passage during their re-use. Secondly, cutting ability in terms of Material Removal Rate (MRR) of the used abrasive particles goes down. Further, cost of the commonly used abrasives is also comparatively low.

MECHANICS OF MATERIAL REMOVAL

In AJM, as discussed above, high velocity air-cum-abrasive jet strikes the work surface. The kinetic energy of the tiny abrasive particles results in micro-indentation in the workpiece. Separate models have been proposed for brittle and ductile workpiece materials. These proposed models have been critically reviewed in the literature [*Jain and Jain*, 2001]. In case of USM, slurry is used which consists of liquid (normally, water) as the carrier fluid. Further, in case of throwing model for USM, the velocity with which an abrasive particle hits the workpiece surface mainly depends on the momentum transferred to the abrasive particles through the tool. Similar situations exist in case of Abrasive Jet Machining (AJM). Here, the carrier fluid is normally air (or, some inert gas) and the velocity with which an abrasive particle hits the workpiece surface depends on the air/gas pressure at the nozzle exit end. Hence, the basic difference is that in case of USM, the tool is vibrating at ultrasonic frequency while in case of AJM the air-abrasive jet flow is in continuous mode. Keeping in view the similarities and differences in these two processes, the philosophy of throwing model proposed by Shaw can be applied to AJM also with certain modifications. The model is based on the following two assumptions:

(a) The abrasive particles are spherical in shape and their diameter is equal to the average particle size (Figure 3.2).

(b) Volume of material removed is equal to the volume of hemisphere of diameter equal to the chord length of indentation in case of brittle material (Figure 3.2). Here, an assumption is made that $r >> h$ ('r' is chord radius as shown in Figure 3.2). Following relationship (Eq. 3.1) can be derived exactly the same way as Eq. 2.2 has been derived in Chapter 2.

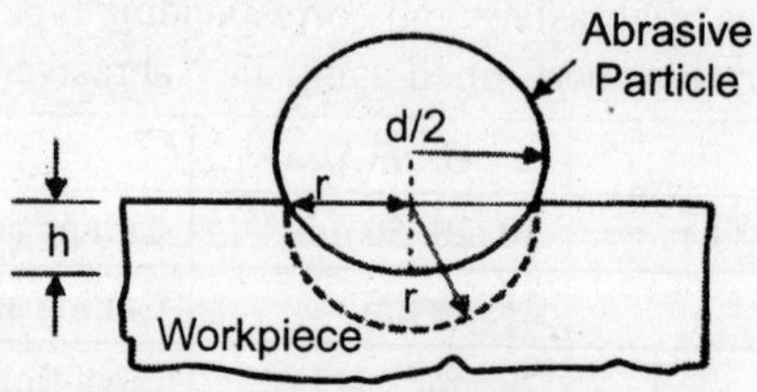

Fig. 3.2: Abrasive particle of diameter 'd' hitting a brittle material with depth of penetration as 'h'.

$$r = \sqrt{dh} \qquad \ldots (3.1)$$

Where, r is chord radius, h is depth of indentation and d is average diameter of the abrasive particle.

Volume of hemisphere (radius = r) is given by

$$\frac{1}{2}\left(\frac{4}{3}\pi r^3\right) = \frac{2}{3}\pi r^3 = \frac{2}{3}\pi(dh)^{3/2} \qquad \ldots (3.2)$$

Suppose, 'N' is number of Abrasive Particles (APs) that strike the work surface per unit time. Then, volumetric Material Removal Rate (MRR_{vb}) in case of brittle material can be evaluated from Eq. 3.3.

$$MRR_{vb} = \frac{2}{3}\pi(dh)^{3/2}.N \qquad \ldots (3.3)$$

When a particle strikes on the workpiece surface, Kinetic Energy (KE) (assuming no loss in the path) is utilized in indenting the workpiece, or KE equal to the work done on the work surface.

$$KE\ (= \frac{1}{2}mv^2) = \frac{1}{2} \times (\frac{4}{3}\pi r^3 \rho_a)v^2 = \frac{\pi}{12}d^3\rho_a v^2 \qquad \ldots (3.4)$$

Where, m is mass of an individual particle, v is velocity with which it hits the workpiece surface, ρ_a is density of the abrasive particle.

Let, H_w be the hardness of the workpiece material. Then, the work done on the workpiece in indenting a depth equal to 'h' is given by the following equation.

$WD_b = \frac{1}{2}Fh$ (F is the force acting on the work surface and varying in the right angle triangle form, see chapter 2, Figure 2.4c)

$$= \frac{1}{2}(\pi r^2 H_w)h \qquad \text{... (3.5)}$$

The work done by the particle should be equal to its KE. Hence,

$$\frac{\pi}{12}d^3\rho_a v^2 = \frac{1}{2}\pi(dhH_w)h \text{ (using Eqs. (3.1, 3.4 and 3.5))}$$

Or, $$h = \sqrt{\frac{\rho_a}{6}\left(\frac{d^2v^2}{H_w}\right)}$$

$$h = dv\sqrt{\frac{\rho_a}{6H_w}}. \qquad \text{... (3.6)}$$

From the abrasive (mass) flow rate ($\dot{m}_a$) and mass of each particle (m_p), the number of abrasive particles hitting the work surface per unit time can be evaluated:

$$N = \frac{\dot{m}_a}{m_p} = \left(\frac{\dot{m}_a 6}{\pi d^3 \rho_a}\right). \qquad \text{... (3.7)}$$

Therefore, by substituting the values of N (Eq. 3.7) and h (Eq. 3.6) in Eq. 3.3, the value of MRR_{vb} can be calculated as follows.

$$MRR_{vb} = \frac{2}{3}\pi\left\{d^2 v\sqrt{\frac{\rho_a}{6H_w}}\right\}^{3/2}\frac{6\dot{m}_a}{\pi d^3 \rho_a}$$

$$MRR_{vb} = \frac{\dot{m}_a v^{\frac{3}{2}}}{\rho_a^{\frac{-1}{4}} H_w^{\frac{3}{4}}} \cdot K_1 \qquad \text{... (3.8)}$$

where, K_1 is a constant.

Semi-empirical models have been proposed by different researchers as reported in [*Jain and Jain,* 2001]. Such models have limitations that they can be applied only under the machining conditions from which these models have been derived. Usually, extrapolation based on these models may lead to erroneous conclusions. One of the simple models that can be used for MRR_v calculation for brittle materials was proposed by *Pandey and Neema* [1977], and it was based on the assumption that each abrasive particle has a normal impact (usually, all abrasive particles do not hit the workpiece surface at 90°).

$$MRR_g(kg/s) = \frac{0.1156\rho_a M_a V_n^2}{10^3\sigma_{fw}} \qquad \text{... (3.9)}$$

Where, ρ_a is abrasive particles density (kg/mm^3), M_a is mass flow rate of abrasive particles (kg/s), V_n is component of particle velocity normal to the work surface and σ_{fw} is flow stress of workpiece material. For calculation of MRR_{vd} for ductile materials, Sheldon and Kanhere proposed the following model.

$$MRR_{vd}\ (\text{mm}^3/s) = \frac{K_1 d_m^3 V_a^3}{10^{4.5}}\left[\frac{\rho_a}{H_{vw}}\right]^{1.5} \qquad \text{... (3.10)}$$

Where, k_1 is constant of proportionality, d_m—mean diam. of abrasive particle (mm), V_a—volume of impacting particle (mm^3), ρ_a—abrasive particle density (kg/mm^3) and H_{vw}—Vicker's diamond pyramid hardness number of the target (workpiece) material.

Finnie [1960] showed that the volume of material (Q) eroded by impacting particles of mass M carried in a stream of air can be calculated as follows,

$$Q = \frac{Cf(\theta)\,M\,v^n}{\sigma}. \qquad \text{... (3.11)}$$

Where, C & n are constants, σ is minimum flow stress of work material, v is velocity of impacting particles, and $f(\theta)$ is a function depending on impingement angle.

Sarkar and *Pandey* [1976] suggested a model to calculate MRR_v ($\dot{Q}$) during AJM.

$$\dot{Q} = \chi Z d^3 v^{3/2}\left(\frac{\rho}{12H_w}\right)^{3/4}. \qquad \text{... (3.12)}$$

Where, Z is no. of particles impacting per unit time, d is mean diameter of abrasive particle, v is velocity of abrasive particle, ρ is density of abrasive particles, H_w is hardness (or the flow stress) of work material, and χ is a constant. There is a significant effect of change in mesh size (or abrasive particle diameter) and abrasive particle velocity on MRR_v. The Eq. 3.12 also indicates that harder the work material, smaller will be MRR_v for the same machining conditions.

PARAMETRIC ANALYSIS

Important parameters that affect material removal rate during AJM are stand-off-distance (*i.e.*, SOD, or sometimes also called as nozzle tip distance—NTD), type and size of abrasive particles, flow rate of abrasive particles, gas or air pressure, work material properties and feed rate. Figure 3.3 shows process parameters as a fish bone diagram. The effects of some of these parameters on the process performance are discussed in the following sub-sections.

Stand-off-Distance

Effects of a change in Stand-off-Distance (SOD) on volumetric material removal rate (MRR_v) as well as linear material removal rate (or penetration rate—MRR_l) are shown in Figure 3.4. Qualitatively, it also shows how the cross-section of the machined cavity varies with a change in SOD. Schematic of cross-sections of the actually machined profiles are shown in Figure 3.5a at different SODs. This change in cross-section happens due to the divergence of the abrasive-air jet. In a range of SOD which usually varies from 0.75 to 1.0 mm, the accuracy is maximum and the kerf width is minimum. A decrease in SOD improves accuracy, decreases kerf width, and reduces taper in the machined groove. Figure 3.5b shows a photograph of the actual machined cavities in the transparent workpiece material. However, light operations like cleaning, frosting, etc. are conducted at a large SOD (say, 12.5–75 mm).

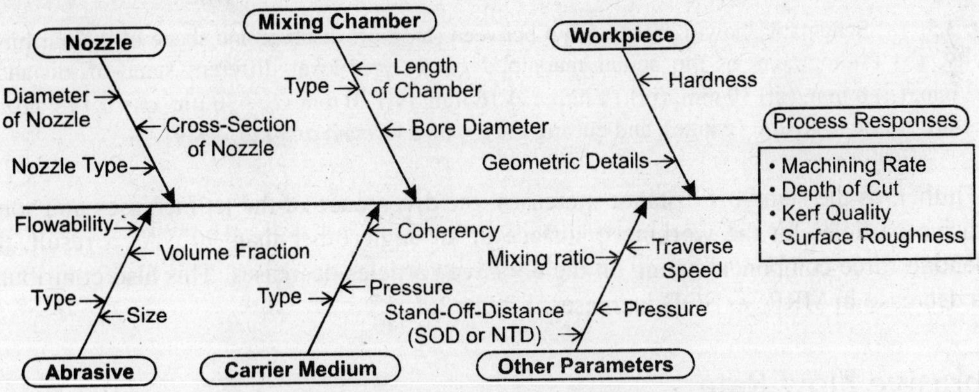

Fig. 3.3: Abrasive jet machining process parameters

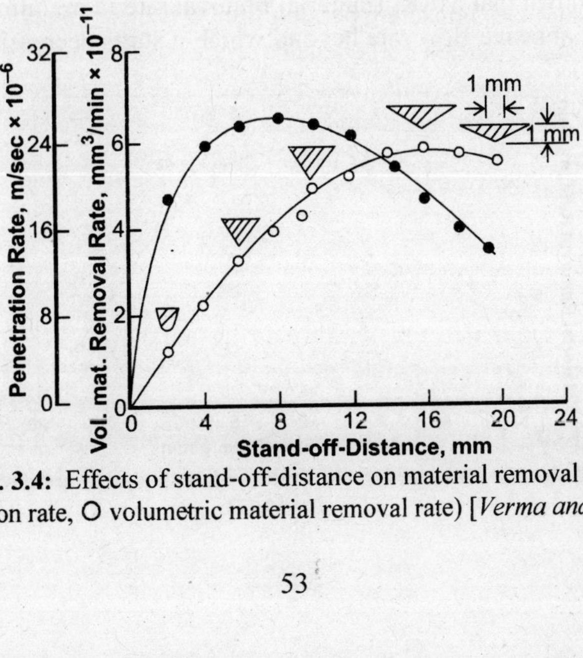

Fig. 3.4: Effects of stand-off-distance on material removal rate (● penetration rate, O volumetric material removal rate) [*Verma and Lal*, 1984].

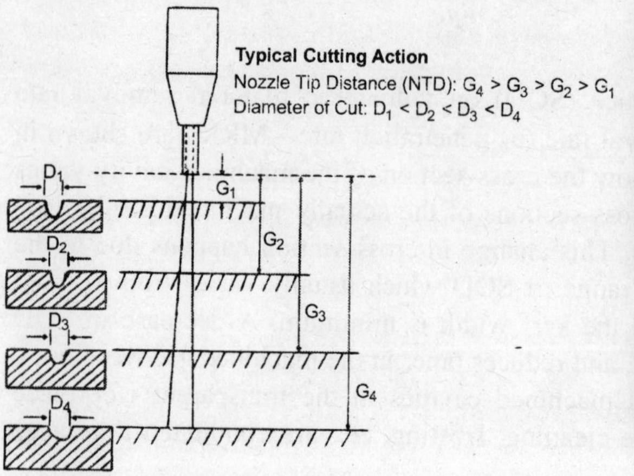

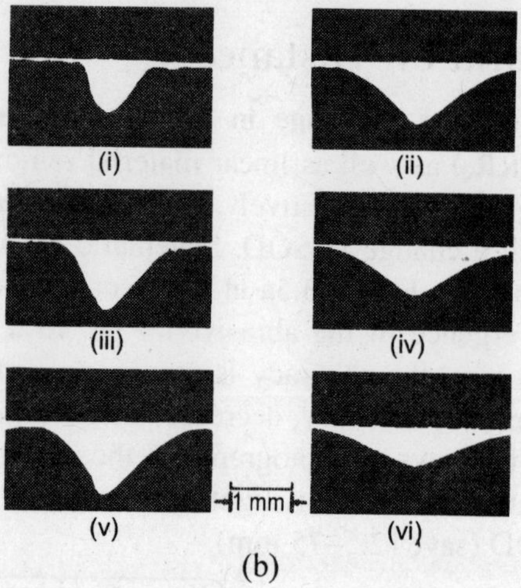

Fig. 3.5: (a) Schematic showing relationship between stand-off-distance and shape of the machined cavity, (b) Photographs of the actual machined cavity profile at different stand-off-distances (i) 2 mm, (ii) 6 mm, (iii) 10 mm, (iv) 14 mm, (v) 16 mm, (vi) 20 mm, $d_p = 30$ μm, $\alpha = 0.148$, nozzle pressure = 147.15 kN/m^2 (gauge), and cutting time = 60 s [*Verma and Lal*, 1984].

Further, as the stand-off-distance increases, the divergence of the jet increases and some abrasive particles hit the workpiece surface at an angle other than 90°. As a result, the indenting force component acting on the abrasive particles decreases. This also contributes to a decrease in MRR_v as SOD increases, Figure 3.4.

Abrasive Flow Rate

Ingulli [1967] has shown that MRR_g (material removal rate in mg/min) increases only up to a certain value of abrasive flow rate beyond which it starts decreasing (Figure 3.6). As

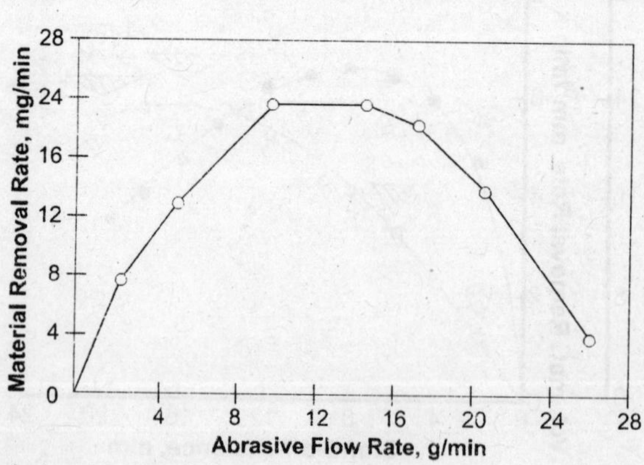

Fig. 3.6: Variation in material removal rate with a change in abrasive flow rate [*Ingulli*, 1967]

abrasive flow rate increases, the number of abrasive-particles hitting/cutting the workpiece also increases thereby increasing MRR_g. However, with a further increase in abrasive flow rate (other parameters remaining unchanged), the abrasive flow velocity goes down, or KE of each abrasive particle decreases. It is so because air/gas pressure does not change but the abrasive particles quantity increases hence abrasive flow velocity decreases, resulting in decreased KE of each abrasive particle. This decrease in abrasive flow velocity (or KE) causes a reduction in MRR_g.

Nozzle Pressure

Effect of nozzle pressure on MRR_v is shown in Figure 3.7. Kinetic energy of the abrasive particles is responsible for removal of material by erosion process. Abrasives must impinge on the work surface with a certain minimum velocity so that the erosion can be initiated. This minimum velocity for initiating machining process in glass by SiC particles (Size: 25 μm) is found to be around 150 m/s [*Sheldon and Finnie*, 1966]. For the case shown in Figure 3.7, the minimum pressure to achieve minimum velocity of abrasive to initiate erosive action is shown in the figure by point B (= 10).

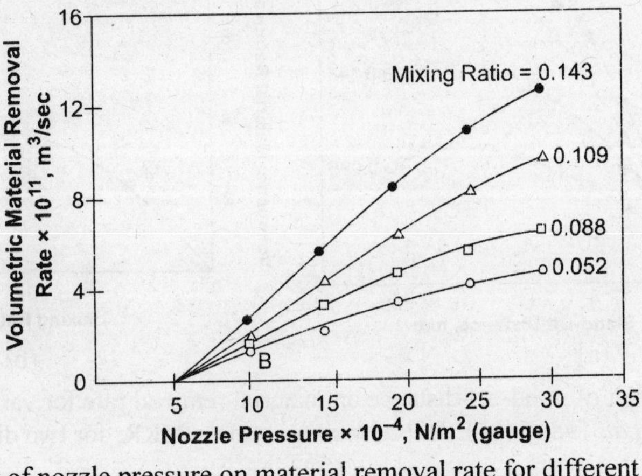

Fig. 3.7: Effect of nozzle pressure on material removal rate for different mixing ratios [*Verma and Lal*, 1984]

Mixing Ratio

Mixing ratio (M) also influences $\mathbf{MRR_v}$. M can be calculated using Eq. 3.13 as follows:

$$M = \frac{\text{Volume flow rate of abrasve particles}}{\text{Volume flow rate of carrier gas}} = \frac{\dot{V}_a}{\dot{V}_g} \qquad \dots (3.13)$$

An increase in the value of 'M' increases MRR_v but a large value of 'M' may decrease jet velocity and sometimes may block the nozzle also. Thus, an optimum value of mixing ratio has been observed that gives maximum MRR_v. Beyond the maximum value of MMR_v, if SOD is further increased, MRR_v starts decreasing. Figure 3.8a shows the effect of stand-off-distance on MRR_v for different mixing ratios and Figure 3.8b shows the effect of mixing ratio on MRR_v for two values of SODs. In place of '*M*', the mass ratio (α, Eq. 3.2) is easier to determine experimentally.

$$\alpha = \frac{\dot{M}_a}{\dot{M}_{a+c}}, \qquad \text{... (3.14)}$$

where, $\dot{M}_a$ is abrasive mass flow rate, and $\dot{M}_{a+c}$ is abrasive and carrier gas combined mass flow rate. *Verma and Lal* [1984, 1985] have studied its effect on optimum volumetric material removal rate.

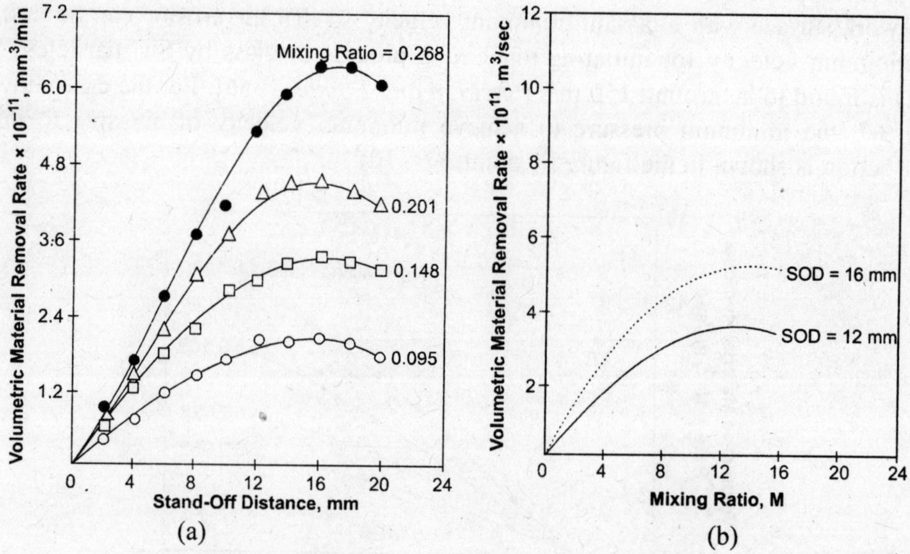

Fig. 3.8: (a) Effect of stand-off-distance on material removal rate for various mixing ratios [*Verma and Lal*, 1984], (b) Effect of mixing ratio on MRR_v for two different SODs.

PROCESS CAPABILITIES

Although AJM gives low MRR_v (approx. 0.015 cm^3/min) but it can easily produce intricate details in hard and brittle materials. Production of narrow slots (0.12–0.25 mm), low tolerances (± 0.12 mm), good surface finish (Ra = 0.25–1.25 μm), and sharp radius (0.2 mm) are some of the characteristics of the AJM process. The quality of the machined component depends on the machining conditions and workpiece material. Steels up to 1.5 mm thick and glass up to 6.3 mm thick have been possible to cut by AJM but at very low MRR and large taper. The process has a special application of machining thin sectioned brittle materials particularly in the areas which are inaccessible for conventional machining methods. Since the heat generation is very small, the resulting surface damage is also insignificant.

APPLICATIONS

AJM is useful [*Butler*, 1980; *Dombrowski*, 1983] in the manufacture of electronic devices, deburring of plastics, nylon & teflon parts, marking on electronic products, permanent marking on rubber stencils, deflashing small castings, cutting titanium foil, and drilling glass wafers. It is also used for engraving say, on the toughened glass used for car windows. This process is also used for frosting glass surfaces and cutting thin sectioned fragile components.

Deburring is one of the bottlenecks in manufacturing. AJM has been gaining increasing acceptability for deburring applications. Detailed investigations have not been carried out about the shape of the abrasive jet machined surfaces. The effects of various process parameters on the shape produced during AJM have been reported. An interesting application in deburring of cross-drilled holes by secondary erosion of an abrasive jet has been discussed [*Balasubramaniam et al.*, 1998, 1999, 2000, 2002]. An empirical linear model has been presented to evaluate the edge radius generated during AJM (Figure 3.9 and Eq. 3.15).

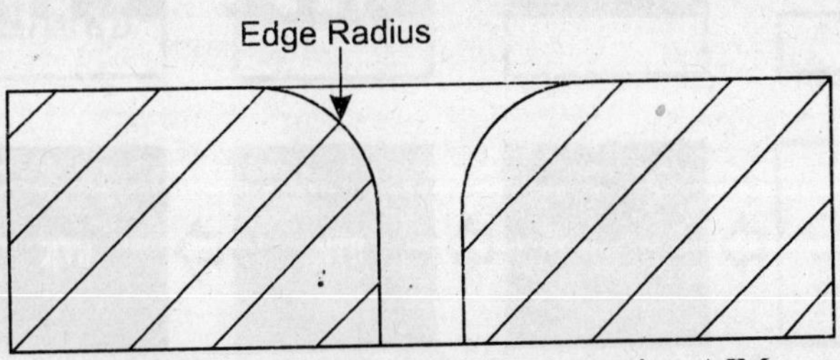

Fig. 3.9: Deburring and radiusing using AJM

$$\text{Edge radius} = \oint\begin{pmatrix}\text{Abrasive grit size, Mixing ratio,}\\ \text{Nozzle diam., SOD, Plate thickness}\end{pmatrix}$$

$$= C_o + \sum_{i=1}^{5} C_i X_i \qquad \ldots (3.15)$$

Where, C_o is constant, C_i are coefficients and X_i are variables.

ABRASIVE JET MICROMACHINING (AJMM)

Introduction

Abrasive Jet Micromachining (AJMM) is one of the mechanical type micromachining methods, also known as non-lithography based micro and nano-machining process. The basic working principle remains the same as that of AJM. In this process, a jet of mixture of gas (usually, air) and very fine abrasive particles (≤ 100 μm diameter) [*Ko*, 2014] is used. The (air + abrasive) jet impinges on the workpiece surface to remove material by erosive action or by plastic deformation, depending on the properties of workpiece material and angle of impact/impingement of abrasive particles on the workpiece surface. The erosion rate mainly depends on the velocity and mass of the abrasive particles hitting the workpiece surface. The angle at which an abrasive particle hits the workpiece also affects the erosion rate and mode of material removal. For achieving high level of precision in micro size features, the area from where no machining is desirable, should be protected by erosion resistant material, known as '*mask*'. This method can be used for making micro-patterns, holes, channels, grooves, etc. on the workpiece.

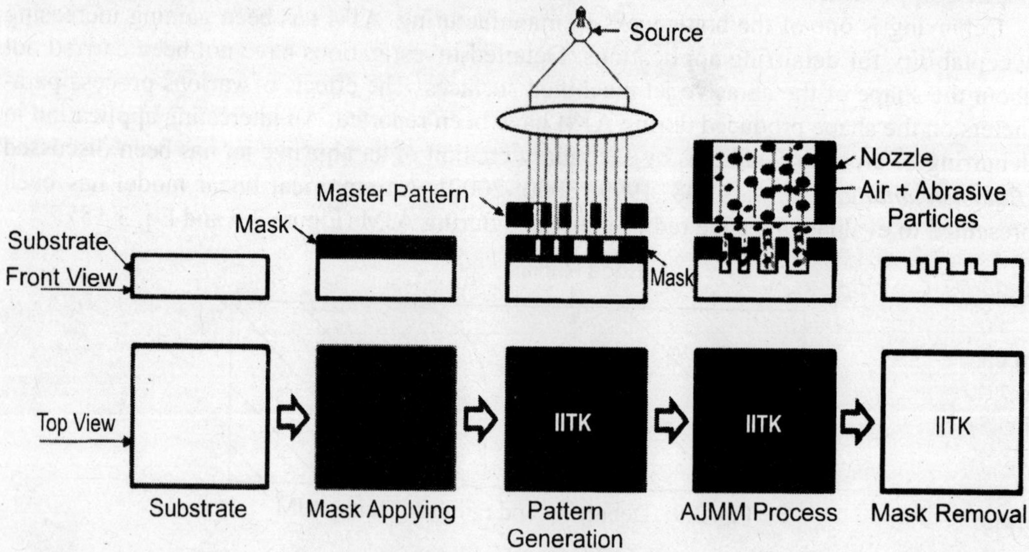

Fig. 3.10: Abrasive jet machining using mask

Figure 3.10 shows different steps followed in AJMM process. The workpiece or substrate is covered with a thin layer of mask (say, a photoresist material), ductile material or elastic material having a pattern to be produced on the substrate. If it is a Photoresist material, then it is subjected to the UV-light passing through the master pattern, to create the desired pattern on the mask. Then, the workpiece/substrate is exposed to the abrasive-laden air jet with a pressure ranging between 0.2 to 0.8 MPa. The area on the workpiece surface which is uncovered by the mask gets eroded while rest of the area wherever the abrasive particles hit, the mask gets eroded. In case of large workpiece area, the workpiece has to be scanned by giving relative motion between the workpiece and air-abrasive-particles jet/beam. Once the whole workpiece surface has been scanned and the desired pattern/feature has been generated, the last step is to remove the mask and clean the substrate/workpiece surface (see, Figure 3.10).

As is obvious from the above discussion, the erosion of the mask will take place hence the mask should have low erosion rate or high wear resistance. Apart from the machining conditions, the accuracy of the feature transferred on the mask will decide the accuracy of the feature created on the workpiece.

Mask Types

As mentioned elsewhere, the mask can be a metal sheet, elastomer, or photoresist. Ductile metal mask is always preferred because of lower erosion rate in AJMM. It can be in the form of a thin sheet of stainless steel, copper or some other suitable metal. If the mask material is magnetic in nature, then fixing the mask on the workpiece can be done with the help of a magnet otherwise, it has to be fixed by adhesive layer. However, with this type of mask, feature size ($\geq$ 50 μm) limitation is there. To overcome this limitation, copper mask can be prepared by electroplating. Elastomers do not wear by brittle fracture due to lateral cracks or plastic deformation just like in metals. In case of elastomers, the erosion of elastomers is based on fatigue mechanism. The correlation of erosion with rebound resilience agrees with the dependence of polymer fatigue life on mechanical damping in the material [*Arnold and Hutching*, 1992; 1993]. However, the performance of elastomers as a mask depends on the temperature, rate of deformation and particle velocity. Such mask materials are available in the form of foil with self-adhesive properties.

Another type of mask is Photoresist. They are of two types—positive and negative. In case of positive photoresist, the area from where the material is to be removed is exposed to Ultraviolet (UV) light. The chemical reaction takes place and becomes more soluble in the developer, and it is washed away by the developer solution. It leaves a copy of the pattern which is to remain on the substrate/workpiece. The negative resist behaves just in a reverse manner. SU-8 is an epoxy- based good negative photo resist [*Saragih and Ko*, 2009].

Abrasives and Nozzle

One of the problems of AJMM process is handling of fine abrasive particles (10 μm–25 μm). Its flowability and compactability depend on particles' shape, size, moisture content and surface texture [*Li et al.*, 2004]. Various approaches are in practice to control powder feeding, namely, pressurized powder feed system, fluidized bed powder spray system, and vibratory and screw feeding system. Powder compaction, amount of powder in the reservoir, powder stratification and powder humidity are some of the factors which affect powder feeding quality and efficiency both [*Ko*, 2014].

Another important element of the system is nozzle configuration. There are two configurations of the AJMM nozzles discussed in the literature [*Ko*, 2014]: blasting nozzle and jet stream nozzle. This AJMM process has been applied in the fabrication of micro features and micro parts, for example, micro transformer, capillary electrophoresis chips [*Schlautmann et al.*, 2001], spring like mixer [*Saragih and Ko*, 2009], etc.

PROBLEMS

Problem 1

During AJM, the mixing ratio used is 0.2. Calculate mass ratio if the ratio of density of abrasive and density of carrier gas is equal to 20.

Solution

We know, mixing ratio $= \dfrac{\dot{V}_a}{\dot{V}_g}$

Mass ratio, $\alpha = \dfrac{\dot{M}_a}{\dot{M}_{a+g}} = \dfrac{\rho_a \dot{V}_a}{\rho_a \dot{V}_a + \rho_g \dot{V}_g}$

Or,

$$\frac{1}{\alpha} = \frac{\rho_a \dot{V}_a + \rho_g \dot{V}_g}{\rho_a \dot{V}_a}$$

$$= 1 + \frac{\rho_g}{\rho_a} \times \frac{\dot{V}_g}{\dot{V}_a}$$

$$= 1 + \frac{1}{20} \times \frac{1}{0.2}$$

$$= 1 + 0.25$$

$$= 1.25$$

or, $\alpha = \dfrac{1}{1.25} = 0.80$

$\boxed{\alpha = 0.80.}$

Problem 2

Diameter of the nozzle is 1.0 mm and jet velocity is 200 m/s. Find the volumetric flow rate (cm³/s) of the carrier gas and abrasive mixture, $\dot{V}_{a+g}$.

Solution

Cross-sectional area of the nozzle $= \pi \times (0.5)^2 \times 10^{-2}$ cm²

$$= \pi \times 25 \times 10^{-4} \text{ cm}^2$$

$$\therefore \dot{V}_{a+g} = (\pi \times 25 \times 10^{-4}) \times 200 \times 10^2 \text{ cm}^3/\text{s}$$

$$= \pi \times 25 \times 2 \text{ cm}^3/\text{s}$$

$$\boxed{\dot{V}_{a+g} = 50\pi \text{ cm}^3/\text{s}}$$

BIBLIOGRAPHY

[1] Arnold, J.C. and Hutching, I.M. (1992). A model for erosive wear of rubber at oblique impact angle, *J. Phy. D: App. Phys.,* Vol. 25, pp. A222–A229.

[2] Arnold, J.C. and Hutching, I.M. (1993). Erosive wear of rubber by solid particles at normal incidence, *Wear,* Vol. 161, pp. 213–221.

[3] Balasubramaniam, R., Krishnan, J. and Ramakrishnan, N. (1998). Investigation on AJM for deburring, *J. of Material Processing Tech.,* Vol. 79, pp. 52.

[4] Balasubramaniam, R., Krishnan, J. and Ramakrishnan, N. (1999). An experimental study on abrasive jet deburring of cross-drilled holes, *J. of Material Processing Tech.,* Vol. 91, pp. 178.

[5] Balasubramaniam, R., Krishnan, J. and Ramakrishnan, N. (2000). An empirical study on the generation of edge radius in abrasive jet external deburring (AJED), *J. of Material Processing Tech.,* Vol. 99, pp. 49–53.

[6] Balasubramaniam, R., Krishnan, J. and Ramakrishnan, N. (2002). A study on the shape of the surface generated by abrasive jet machining, *J. of Material Processing Tech.,* Vol. 121, pp. 102–106.

[7] Benedict, G.F. (1987). Non-traditional Manufacturing Processes, Marcel Dekker Inc., New York.

[8] Bitter, J.G.A. (1962–63). A study of Erosion Phenomenon-Part I, Wear, Vol. 5–6, p. 5.

[9] Bitter, J.G.A. (1962–63). A study of Erosion Phenomenon - Part II, Wear, Vol. 5–6, pp. 169–190.

[10] Butler, K. (Sept/Oct 1980). Blast Your Problems Away, *Tool Carbide*, pp. 12–15.

[11] Dombrowski, T.R. (Feb 1983). The How and Why of Abrasive Jet Machining, *Modern Machine Shop,* pp. 76–79.

[12] Finnie, I. (1960). Erosion of Surface by Solid Particles, Wear, Vol. 3, p. 87.

[13] Finnie, I. (1966). The Mechanism of Metal Removal in the Erosive Cutting of Brittle Materials, *Trans ASME, Series B*, Vol. 88, p. 393.

[14] Ingulli, C.N. (1967). Abrasive Jet Machining, *Tool and Manuf. Engrs.,* Vol. 59, p. 28.

[15] Jain, N.K. and Jain, V.K. (2001). Modeling of material removal in mechanical type advanced machining process: A State-of-Art review, *Int. J. Mach. Tool Manuf.,* Vol. 41, pp. 1573–1635.

[16] Ko Tae, Jo (2014). Abrasive Jet Micromachining; in Introduction to Micromachining Processes (Ed. V.K. Jain), (2nd Edition) Narosa Publishing House, Delhi.

[17] Li, Q., Rudolph, V., Weigl, B. and Earl, A. (2004). Interparticle van der Waals force in powder flowability and compactability, *Int. J. Pharm.*, Vol. 208, pp. 77–93.

[18] McGeough, J.A. (1988). Advanced Methods of Machining, Chapman and Hall, London.

[19] Neema, M.L. and Pandey, P.C. (1977). Erosion of glass when acted upon by abrasive jet, in: *Proceedings of International Conference on Wear of Material,* pp. 387–392.

[20] Pandey, P.C. and Shan, H.S. (1980). Modern Machining Processes, Tata McGraw Hill Publishing Co. Ltd. , New Delhi, p. 39.

[21] Saragin, Agung Shamsuddin and Ko, Tae Jo (2009). Fabrication of passive glass mixture with third dimensional feature by employing SU8 mask on micro abrasive jet machining, *Int. J. Adv. Manuf. Tech.*, Vol. 42, pp. 474–481.

[22] Saragih, A.S. and Ko, T.J. (2009). A thick SU-8 mask for micro abrasive jet machining on glass, *J. Adv. Manuf. Tech.,* Vol. 41, pp. 734–740.

[23] Sarkar, P.K. and Pandey, P.C. (1976). Some Investigations in Abrasive Jet Machining , *J. Inst. of Engrs* (I), Vol. 56, Pt. ME-5, p. 284.

[24] Schlautmann, S., Wensink, H., Schasfoot, R., Elwenspolk, M. and Berg, A. Van Der (2001). Powder blasting technology as an alternative tool for microfabrication of electophoresis chips with integrated conductivity sensor, *J. Micromech. Microeng.,* Vol. 11, pp. 386–389.

[25] Sheldon, G.L. and Finnie, I. (1966). The Mechanics of Material Removal in the Erosive Cutting of Brittle Materials, *Trans ASME*, Sr. B, Vol. 88, p. 393.

[26] Sheldon, G.L. and Kanhere, A. (1972). An investigation of impingement erosion using single particle, *Wear,* Vol. 21, pp. 195–209.

[27] Snoeyes, R., Staelens, F. and Dekeyser, W. (1986). Current Trends in Non-conventional Material Removal Processes, *Annals of the CIRP,* Vol. 35(2), p. 467.

[28] Venkatesh, V.C. (1984). Parametric Studies on AJM, *Annals of the CIRP,* Vol. 33, No. 1, p. 109.

[29] Verma, A.P. and Lal, G.K. (1985). Basic Mechanics of Abrasive Jet Machining, The Institution of Engineers (I), *Production Engg. Div.,* Vol. 66, pp. 74–81.

[30] Verma, A.P. and Lal, G.K. (1984). An Experimental Study of Abrasive Jet Machining, *Int. J. Mach. Tool Des. Res.,* Vol. 24, No. 1, pp. 19–29.

SELF TEST QUESTIONS

Q1. Write all the correct answers.

1. Material used for making a nozzle employed in AJM process may be
   (a) copper (b) stainless steel
   (c) sapphire (d) WC
2. In AJM, mixing ratio is governed by amplitude and frequency of vibration of sieve. The frequency of vibration of sieve is
   (a) 50–60 Hz (b) 10–15 kHz
   (c) above 15 kHz
3. In case of AJM, higher carrier gas pressure would yield
   (a) higher MRR and higher nozzle-life (b) lower MRR and lower nozzle life
   (c) higher nozzle life but lower MRR (d) None of these
4. In AJM, with the increase in stand-off distance, the width of cut
   (a) decreases (b) increases
   (c) remains constant
5. During AJM, increase in mass flow rate of abrasive particles would
   (a) decrease the value of mixing ratio (b) increase the value of mixing ratio
   (c) no definite effect on mixing ratio
6. With an increase in abrasive particle size in AJM,
   (a) MRR as well as surface roughness (Ra value) increase
   (b) MRR decreases but surface roughness (Ra value) increases
   (c) MRR increases but surface roughness (Ra value) decreases
7. The range of size of abrasive particles used in AJM is
   (a) 0.001–0.05 mm (b) 0.1–0.5 mm (c) 1–5 mm
8. AJM is best suited for machining
   (a) Al (b) Glass (c) M.S.
9. In AJM, MRR increases with
   (a) increase in NTD (b) decrease in NTD
   (c) no effect of NTD (d) none of these
10. AJM can be recommended to machine
    (a) M.S. (b) C.I. (c) WC
11. AJMM without a mask, would produce the feature with
    (a) higher accuracy (b) poorer accuracy
    (c) no effect of mask on accuracy
12. Which mask material would produce a feature with comparatively more sharp edge
    (a) copper (b) perspex (c) rubber

Q2. Examine whether the following statements are true (T) or false (F).

1. The surface damage during AJM is negligible.
2. AJM is not a good process for cleaning metallic mould cavities.
3. Glass is machined by AJM process using SiC abrasive. The minimum jet velocity required has been found to be 150 m/s.
4. AJM process can be employed to machine materials irrespective of whether they are insulator or conductor of electricity.
5. AJM and sand blasting are similar kind of processes from application point of view.
6. Carrier gas used in AJM is either air or O_2.
7. Shape of the abrasive particles has no effect on MRR in AJM.
8. One obtains uniform diameter hole along its length in AJM process.
9. The size of the workpiece is 2 × 10 × 20 mm and a hole of size 0.80 mm is drilled using AJM process. It does not come under the category of AJMM.

Q3. Multiple Choice Type Questions.

1. In AJM, the material is removed by
   (a) Spark erosion (b) Etching
   (c) Anodic dissolution (d) Erosive/Chipping action
2. For which of the following material(s), AJM is not very effective?
   (a) Glass wafers (b) Aluminum
   (c) Tungsten (d) Germanium (fragile material)
3. Which of the following is/are the desirable properties for the gas used in AJM?
   (a) it should be clean and dry
   (b) it should spread excessively when discharged from the nozzle
   (c) the gas should be non-toxic
   (d) it should be cheap and easily available
4. Which of the following is/are not true for AJM?
   (a) Larger size abrasive particles are better for cleaning and polishing rather than cutting.
   (b) Gases like CO_2, O_2 can be used as the carrier gas.
   (c) Re-use of abrasives is recommended.
   (d) The velocity with which the abrasive particles strike the workpiece is about 10 m/s.
5. Which of the following abrasives will you suggest for stainless steel & ceramics?
   (a) Al_2O_3 (b) SiC
   (c) $NaHCO_3$ (d) Glass beads
6. A decrease in stand-off distance
   (a) Improves accuracy (b) Decreases kerf width
   (c) Reduces taper in machined groove (d) Increases nozzle wear

7. What is the mixing ratio used in AJM, if the ratio of density of abrasive and density of carrier gas is 10 and the mass ratio is 0.8?
   (a) 0.2 (b) 0.4
   (c) 0.8 (d) 2.5 (e) 5
8. Which of the following material(s) can not be used for AJM nozzle?
   (a) Al_2O_3 (b) Sapphire
   (c) Tungsten carbide (d) Boron carbide
9. Which of the following is/are true for AJM?
   (a) It can easily produce intricate details in hard & brittle materials.
   (b) MRR is very high.
   (c) Due to very low heat generation, surface damage is insignificant.
   (d) Abrasive grain size has no effect on MRR_v.
10. Which of the following abrasives is/are good for matte finish?
    (a) Al_2O_3 (b) SiC
    (c) $NaHCO_3$ (d) Glass beads
11. Abrasive-air jet pressure in AJMM is:
    (a) 0.2–0.8 MPa (b) 2–8 MPa
    (c) 10–20 MPa (d) None of these.
12. SU-8 is epoxy based Photoresist used for AJMM. Whatever area of this Photoresist is exposed to UV light, it is washed away by developer solution.
    (a) True (b) False

Answers

1. (d)	2. (b)	3. (a), (c), (d)	4. (a), (b), (c), (d)
5. (b)	6. (a), (b), (c)	7. (b)	8. (a)
9. (a), (c)	10. (d)	11. *	12. *

(Find solution your self)*

Q4. Answer in brief.

1. Draw a schematic diagram of AJM system and label it.
2. Explain the working principle of AJM process.
3. With the help of sketches, show the effects of stand-off-distance on (a) width of cut, (b) material removal rate.
4. "AJM is not recommended to machine ductile materials". Comment in brief.
5. Show the effect of carrier gas pressure on MRR during AJM.
6. Write five important variables of AJM process. Draw a sketch showing the effect of one of these variables on MRR.
7. Write the applications of different types of abrasives used in AJM.

8. Show (sketch only) the effect of jet pressure on MRR_v in AJM, for different abrasive particles sizes.
9. (a) In a AJM system, the nozzle diameter is 0.9 mm and jet velocity is 210 m/s. Find volumetric flow rate of the carrier gas and abrasive mixture.

   (b) In the above problem, take mixing ratio = 0.25. Calculate mass ratio if the density ratio $\rho_a/\rho_g = 25$.

   (c) Over a long period of use, the nozzle wears and its average diameter becomes 1.1 mm from 0.9 mm (Q. 9(a)). Calculate new volumetric flow rate (cm^3/s) of the carrier gas $(\bar{V}_a)$, and carrier gas and abrasive mixture $(\bar{V}_{a+g})$. Assume that the jet velocity decreases to 195 m/s. What will be its effects on (a) MRR_v, (b) penetration rate, (c) MRRg? Give justification.
10. Explain different steps to be followed during AJM process using copper as a mask material. AJMM is to be used for engraving on a glass sheet.
11. Which mask material will have maximum effect on its performance if the temperature of the surrounding rises from 10°C to 35°C?
12. Why abrasive powder used in AJMM should be dry? Explain in brief.

NOMENCLATURE

C, C_o, n, χ	Constants
C_i	Coefficient
d, d_m	Mean diameter (or average diameter) of the abrasive particles
F	Force
h	Depth of indentation
H_w	Flow stress (or hardness) of the work material
K_I	Constant
m	Impacting particle mass
M	Mixing ratio
MRR_g	Material Removal Rate (g/s)
MRR_ℓ	Linear Material Removal Rate (mm/s)
MRR_{vb}	Volumetric Material Removal Rate (mm^3/s) (brittle material)
$\dot{M}_a$	Mass flow rate of abrasive
$\dot{M}_{a+c}$	Mass flow rate of abrasive and carrier gas combined
Q	Volume of eroded material
$\dot{Q}$	Volumetric material removal rate
r	Chrod radius
$\dot{V}_a$	Volume flow rate of abrasive
$\dot{V}_g$	Volume flow rate of carrier gas
V_n	Normal component of abrasive particle velocity
X_i	Variables
Z	No. of particles impacting per unit time

GREEK SYMBOLS

α	Mass ratio
σ	Flow stress of work material
θ	Impingement angle
v	Impacting particle (or abrasive particle) velocity
ρ_a	Density of abrasive particles
ρ_g	Density of carrier gas

SUFFIX

a	Abrasive
b	Brittle material
c	Carrier gas
f	Flow stress
g	Gas
l	Linear
m	Mean
p	Particles
v	Volumetric or Vicker's diamond pyramid hardness number
w	Workpiece

ACRONYMS

AJM	Abrasive Jet Machining
AJMM	Abrasive Jet Micromachining
KE	Kinetic Energy
MRR	Material Removal Rate
NTD	Nozzle-Tip-Distance
SOD	Stand-Off-Distance
USM	Ultrasonic Machining

CHAPTER 3
AT-A-GLANCE
ABRASIVE JET MACHINING (AJM)

How does it Work?

↓

Air-Abrasive Jet (Inert Gas + Fine Abrasives)→Srikes Workpiece (WP) at Very High Velocity (200–400 m/s)

↓

Material Removal by Erosive/Chipping Action. In Some Cases by Plastic Deformation

What Kinds of Operations it can Perform?

↓

* Cutting, * Cleaning, * Polishing, * Duburring, etc.

- Effective on Hard and/or Brittle Materials
- Can Produce Fine & Complicated Details

Elements of AJM Setup

Gas Propulsion System
→ Supplies Clean & Dry Gas to Propel Abrasive Particles
→ Compressor or Cylinder Filled with Gas
→ Gas → Non-Toxic, Cheap, Easily Available, Small Spreading

Abrasive Feeder
→ Quantity Control → By Vibration

Machining Chamber
→ Well Closed → Abrasive Particles Content should be below Harmful Limit
→ Vacuum Dust Collector

Nozzles
→ WC/Sapphire (= 300 hr)
→ Circular/Rectangular X-Section
→ Nozzle Pressure = 2–8.5 kgf/cm^2 (∮) (WP Material, MRR Requirements).
→ Nozzle Wear → Stray Cutting & Higher Inaccuracy

↓

Controlled by Mask

↓

* Less Accurate – Rubber
* More Accurate – Metal

Abrasives
→ Alumina, SiC, Glass Beads, Sodium Bicarbonate, etc.
→ Selection (∮) (WP Material, MRR, Accuracy)
→ Al_2O_3, Sodium Bicarbonate, SiC (For Hard Materials)
→ Cleaning, Cutting, Deburring, etc.
→ Glass Beads → Good for Matte Finish
→ Size → 10–50 μm
Small Size → Cleaning & Polishing
Large Size → Cutting
Re-Use → Not Recommended
: Contamination with Chips → Blocks Nozzle Passage
: Lower Cutting Ability
: Low Cost

Parmeteric Analysis

Stand-off-Distance	• 0.75–1.0 mm → Maximum Accuracy • Low SOD → High Accuracy → Low Kerf Width → Low Taper
Abrasive Flow Rate	• Maxima Observed. • Higher Abrasive Flow Rate → More # of Active Abraisve Particles (APs) → Higher MRR • Lower Abrasive Flow Rate → Lower MRR.
Nozzle Pressure	• Affects MRR • Removes Material → Erosive Action • Certain Minimum Velocity → 150 m/s for SiC → 25 μm
Mixing Ratio	• Volume Flow Rate of Abrasive/(Volume Flow Rate of Carrier Gas) • Maxima Observed Between 'M & MRR'

Process Capabilities

- Mass Ratio $= \dot{M}_a/\dot{M}_{a+c}$
- $MRR_v = \left(\dot{Q}\right) = c\, f(\theta) m\, V^n/\sigma$
- Other Models Also
- Low MRR
- Intricate Details can be Produced
- Narrow Slots (0.12 to 0.25 mm)
- Low Tolerances (+0.12 to –0.12 mm)
- Minimization of Taper → Angle of Nozzle WRT WP
- Thin Sectioned, Brittle Material, Inaccessible Areas
- Almost no Surface Damage

Applications

- Manufacture of Electronic Devices
- Deburring of Plastics, Nylon, Teflon Parts
- Deflashing Small Castings
- Drilling Glass Wafers

Micromachining

- Micro Sized (One Dimension < 1000 μm) Features
- Machining by Erosive Action (KE is Used)
- Masks: Metal, Elastomers, Photoresist
- Self Fixing Mask
- Mask Performance Depends: Temperature, Particle, Velocity, Mask Material Properties etc.
- Positive Photoresist and Negative Photoresist
- Abrasive Types, Particles Size, Abrasive Concentration
- Powder Feeding System
- Nozzle Configuration

Chapter Four

ABRASIVE FINISHING PROCESSES

The need for high accuracy and high efficiency machining of difficult-to-machine materials is making the application of abrasive finishing technologies increasingly important. The most labour intensive, uncontrollable area in the manufacture of precision parts involves final machining (or finishing) operations. The cost of surface finish increases sharply for a roughness value of less than one micron. The result of high quality finish on the parts is improved performance, and considerable increase in the length of life of the component.

The basic idea of abrasive fine finishing processes is to use a large number of random cutting edges with indefinite orientation and geometry for effective removal of material with chip sizes smaller than those obtained during machining using cutting tools with defined edges. Because of extremely thin chips produced in abrasive machining, it allows better surface finish, closer tolerances, generation of more intricate surface features, and machining of harder and difficult-to-machine materials [*Subramanian, 1994*]. The capabilities of four important fine finishing process, are compared in Table-4.1 [*Jain and Jain, 1998*]. Lapping and honing are the two important but well known processes while the other two—magnetic abrasive finishing (MAF) and abrasive flow machining (AFM)—are the new ones. This chapter deals with AFM and MAF processes only.

Table–4.1 Comparison of four finishing processes [*Jain and Jain, 1998*]

Sl.No.	Process features	Lapping	Honing	MAF	AFM
1.	Surface finish (μ m)	0.025–0.1	0.025–0.5	0.04–1.0	0.05–1.0
2.	Dimensional tolerance (μ m)	0.5	0.5–1.25	0.5	5.0
3.	Material removal (mm)	< 0.0025	0.061–0.183	0.002–0.007	0.008–0.010
4.	Pressure	0.01–0.2 N/sq. mm	1–3 N/sq.mm	0–0.007 kPa	0.69–22.0 MPa
5.	Abrasive product type	Abrasive grain entrained in a liquid vehicle	Bonded abrasives	Magnetic abrasives composed of ferromagnetic particles and conventional abrasive grits	Semisolid abrasive media composed of viscoelastic carrier and abrasive grits
6.	Work surface configuration	Flat, cylindrical, and spherical surfaces	Cylindrical surfaces	Flat and cylindrical surface	Inaccessible areas and complex internal passages

(A) ABRASIVE FLOW FINISHING (AFF)

WORKING PRINCIPLE

Developments in the area of materials science are taking place at a fast rate; at the same time demand for better quality and lower cost products is also increasing. There is a consistent demand for a decreased lead time from design to production. Further, finishing operations usually cost approximately 15% of the total machining cost [*Rhoades, 1988*] in a production cycle. In view of this, a need of automated finishing operations to substitute manual finishing operations is felt. To meet such requirements a non-traditional finishing process named as **abrasive flow machining (AFM)** has been developed. It seems to have a potential to offer better accuracy, and higher efficiency, economy and consistency.

Abrasive flow machining is a kind of finishing process in which a small quantity of material is removed by flowing a semisolid abrasive laden putty over the surface to be finished. The media has such a high viscosity that it can be held between fingers (Fig. 4.1) like a rubber ball which can be deformed by applying a little pressure. Two vertically opposed cylinders (Fig. 4.2) extrude abrasive media back and forth through passage(s) formed either by workpiece and tooling, or by

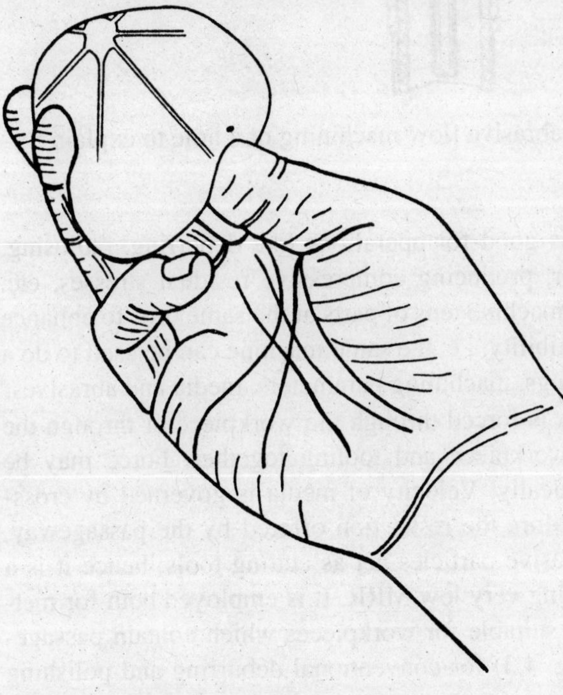

Fig. 4.1 Media held between fingers.

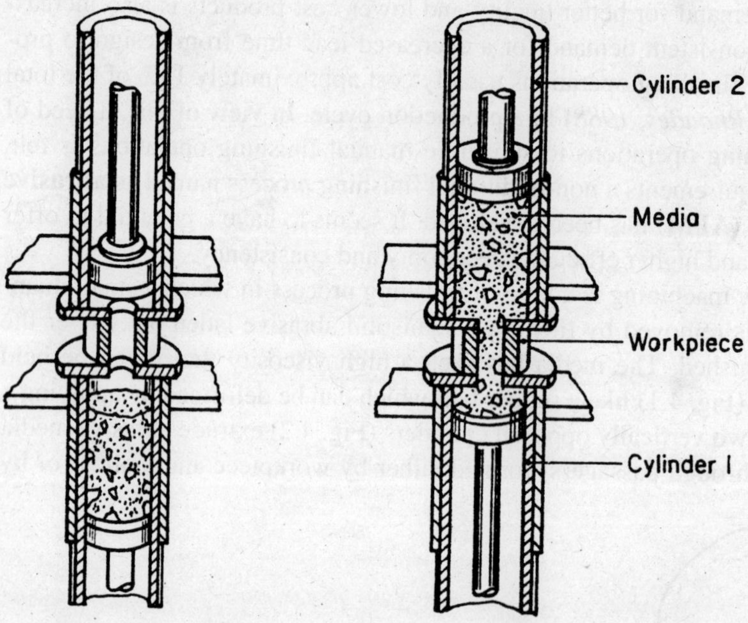

Fig. 4.2 Schematic diagram of abrasive flow machining of a hole to explain its working principle

workpiece alone. This process is good for operations like deburring, radiusing, polishing, removing recast layer, producing compressive residual stresses, etc. The process can be employed to machine tens of parts at the same time to enhance productivity. This has a high flexibility, i e, the same machine can be used to do a variety of jobs by changing toolings, machining parameters, media and abrasives.

The semisolid abrasive media is forced through the workpiece or through the restrictive passage formed by workpiece and tooling together. Force may be applied hydraulically or mechanically. Velocity of media is governed by cross-sectional area of passageways. More the restriction offered by the passageway, larger is the force required. Abrasive particles act as cutting tools; hence it is a **multi-point cutting process** giving very low MRR. It is employed both for metals and non-metals. It is equally suitable for workpieces which contain passageways that are *not accessible* (Fig. 4.3) for conventional deburring and polishing tools.

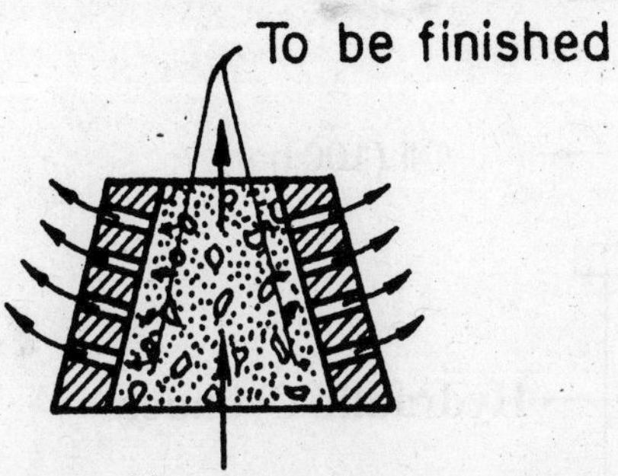

Fig. 4.3 Deburring/finishing of inaccessible holes using AFM process.

ABRASIVE FLOW MACHINING (AFM) SYSTEM

It consists of three **elements**, viz., machine, tooling, and media.

(i) Machine

Machine having two cylinders and abrasive media controls the *extrusion pressure* and flow volume. The media is extruded (or forced) back and forth from one cylinder to another with the help of a hydraulic ram. These cylinders/chambers are clamped together with the workpiece sandwiched between them. The enclosed workpiece area through which the media is forced, is called the **extrusion passage**. The media extruding pressure ranges from 0.70 to 22 MPa. To maintain a constant media viscosity, in some applications, coolers for lowering the temperature of the media are also used. Manual or computer control machines are available [*Extrude Hone*]. *The controllable variables are volume of media, no. of stroke, number of cycles and pressure*. Fig. 4.4 shows a schematic diagram and labelled photograph of an AFM system [*Sunil Jha*, 1998].

(ii) Tooling

Tooling is that element of an AFM system which is used to confine and direct the flow of media to the appropriate areas (Fig. 4.5(d)). Basic principle of tooling design in AFM system is to selectively permit or block the flow of media into or out of workpiece passages where deburring, radiusing, and surface improvements are desired. *Tooling* for AFM machine is designed with two **aims**:

(i) to hold the parts in position, and
(ii) to contain the media and direct its flow.

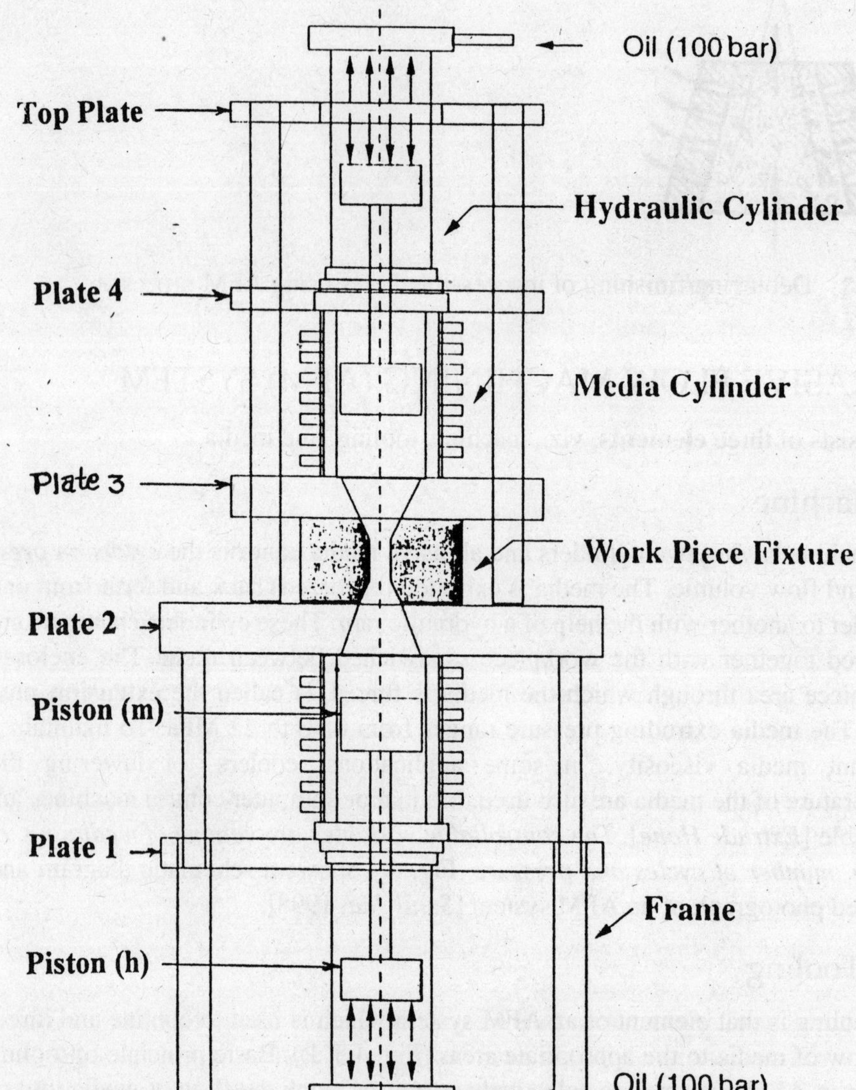

Fig. 4.4a Schematic diagram of abrasive flow machining setup [Sunil Jha, 1998].

Fig. 4.4b Photograph showing abrasive flow machining setup [Sunil Jha, 1998].

Maximum machining takes place wherever there is a maximum restriction to the flow of abrasive. While machining **internal surfaces** like bore of a shaft, configuration of the bore itself decides the extent of restriction present to the flow of media. Figs. 4.5a, 4.5b and 4.5c show finishing of internal surfaces. In case of **external surfaces**, the designer of tooling decides the extent of restriction to be imposed (Figs. 4.5d and 4.5e). In case of polishing of spur gear teeth, diameter of a cylinder placed around the gear teeth determines the extent of restriction present for the flow of media (Fig. 4.5d).

Passages of similar nature can be *processed in parallel* (Fig. 4.5a). In case of a non-uniform cross-section, MRR of narrowest section will be maximum and that of widest section will be minimum (Fig. 4.5b). **Replaceable inserts** made of nylon, teflon, or similar other materials are used for restricting flow of media to induce abrasive action. Once these inserts are worn out, they are replaced by new ones. Life of such inserts may be in terms of thousands of parts if they are properly designed.

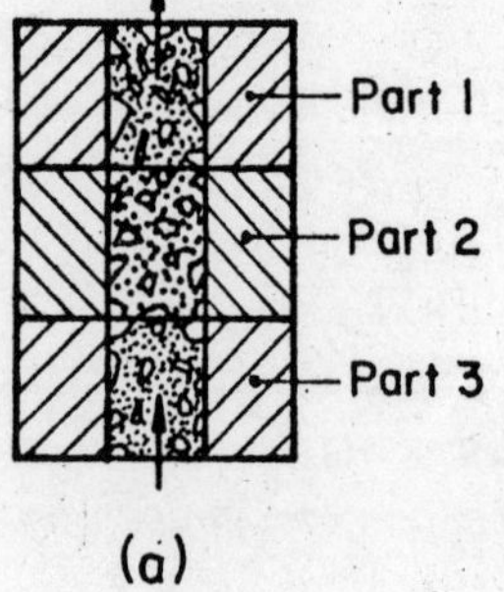

Fig. 4.5a Finishing of multiple parts having the same configuration.

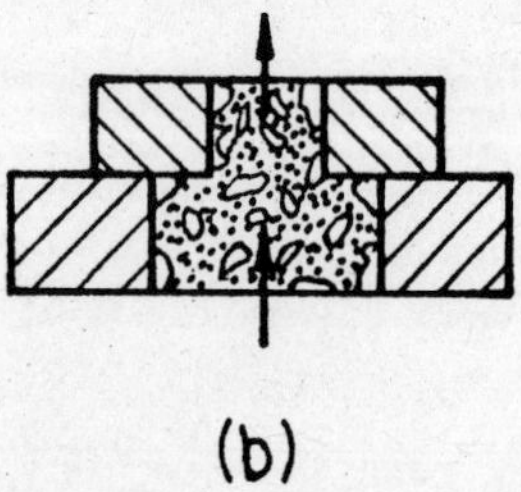

Fig. 4.5b Finishing of two parts but with different configurations.

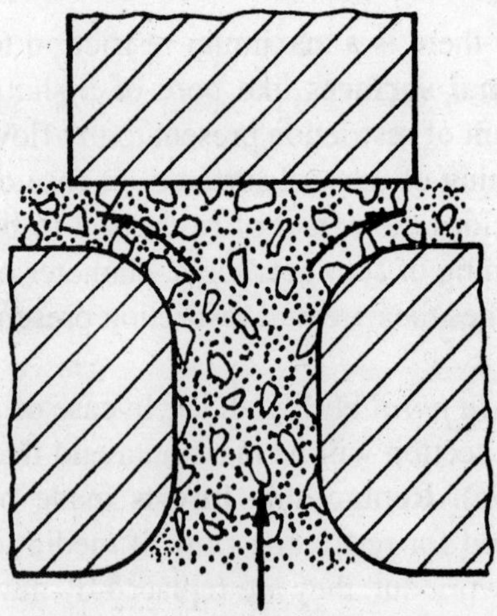

Fig. 4.5c Finishing and radiusing of an internal hole.

(iii) Media

Characteristics of media determine aggressiveness of action of abrasives during AFM process. AFM media is pliable (easily moulded, Fig. 4.1) material which is resilient enough to act as a self *deforming grinding stone* (Fig. 4.6) when forced through a passageway [*Rhoades, 1988*]. It consists of base material and abrasive grits. The base material (**viscoplastic/viscoelastic material**) is made up of an organic polymer, and hydrocarbon gel. Composition of the base material determines its degree of stiffness. The *stiffest media* is used for largest hole, while *soft media* is used for small holes. High stiffness of the media results in a kind of pure extrusion while soft media will lead to a faster flow in the centre than along walls. It is reported that a more stiff media finishes a passageway more uniformly while a less stiff media results in a greater radius at the passage opening [*Williams*, 1989].

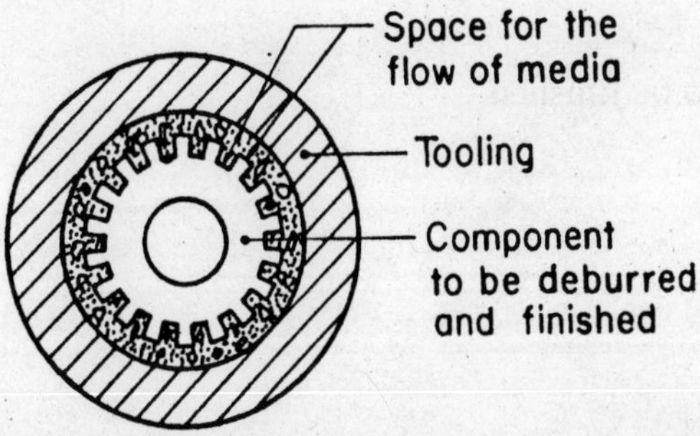

Fig. 4.5d Tooling for deburring and finishing of a gear using AFM process.

When the abrasive grains laden media comes in contact with the workpiece surface to be machined, the abrasive grains are held tightly in place and the media acts as a **deformable grinding stone**. Fig. 4.6 shows that how the same media takes different shapes depending upon configuration of the workpiece to be finished. Hundreds of holes in a combustion liner can be sized to a tolerance of hundredth of a millimetre. **Types of abrasives** used are Al_2O_3, SiC, cubic boron nitride (CBN) and diamond (written in the order of their increasing hardness and cost). These abrasives are available in different grit sizes. The abrasives have limited life. As a thumb rule, when the media has machined an amount equal to 10% of its weight, it must be discarded. To assure a *proper mixing* of a new batch of media and abrasive, it should be cycled 20-50 times through a scrap part. Machined parts should be properly cleaned before use, by air or vacuum.

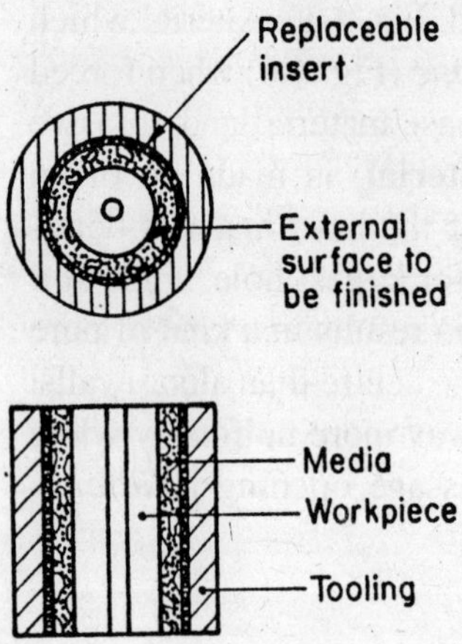

Fig. 4.5e Tooling for external surface to be finished.

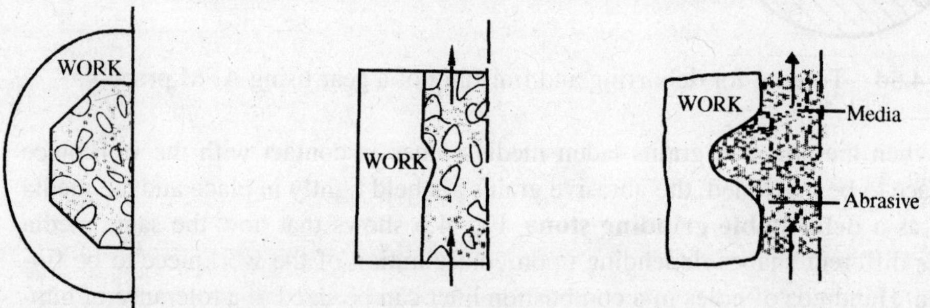

Fig. 4.6 Media acting as a '*self-deformable stone*'.

PROCESS VARIABLES

Abrasive flow machining process is in its *infancy*. Mechanism of the process, parametric relationships, surface integrity, and other related issues are yet to be effectively addressed. The discussion under this section is based on very limited experimental results reported by researchers [*Williams*, (1989); *Williams and Rajurkar* (1989); *and Williams, Rajurkar and Rhoades* (1989)].

Important factors that affect performance of the process and the product quality (viz, accuracy, MRR, and surface integrity), are shown in Fig. 4.7. They are workpiece material (hardness and composition), machine and tooling (fixture design, cylinder size, clamping pressure, etc), adjustable parameters (pressure, number of cycles, etc), geometry of the component(s) (passage shape, length to diameter ratio, length of flow path, etc), and media (viscosity and its change during the process, volume flow rate, and abrasive-type, -size, and -concentration).

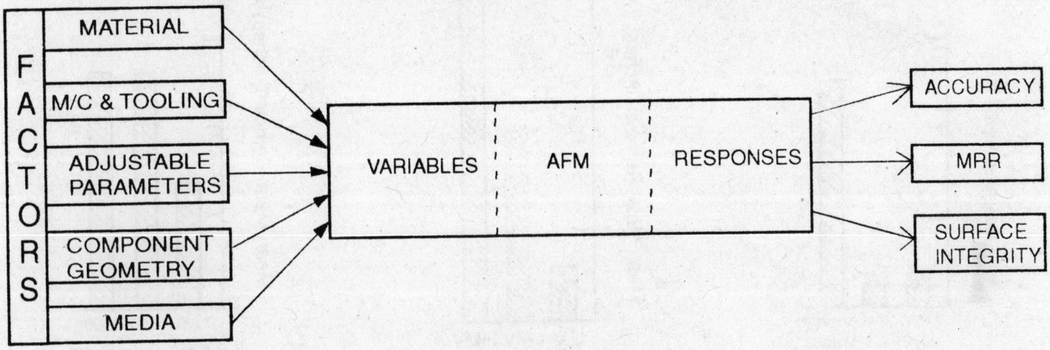

Fig. 4.7 Variables and responses of AFM process.

Williams and others [1989] have conducted experiments to evaluate the effects of viscosity, pressure and flow rate of media on MRR, surface finish, and surface integrity. They have also attempted to stochastically model and analyze the machined surface profile for understanding the mechanics of surface generation during AFM. Experiments were carried out on sleeve type parts made of AISI 4140. Some experiments were planned based on *full factorial design* while others were done as *one variable at a time* approach.

In the absence of actual values of viscosity, three viscosity levels were chosen

[*Williams and co-workers*, 1989] as low (LV), medium (MV), and medium high (MHV) viscosity. Pressure levels chosen were 500, 1000 and 1500 psi. Workpieces were run for 10 cycles with a media flow volume of 70 in^3/stroke. Concentration of abrasives in the media was 66%. Linear **material removal** was measured. To minimize the effect of time error variability, the trials were randomized. Fig. 4.8a shows the *effect of workpiece material* on material removal. Fig. 4.8b shows the effect of viscosity on linear material removal (in inches). It was concluded based on statistical tests that there was a significant difference between LV, and MV as well as LV, and MHV but insignificant difference between MV and MHV. Fig. 4.8c shows the *effect of pressure* on mean metal removal (in inches). At 1500 psi, there is hardly any improvement in MR as compared to 1000 psi.

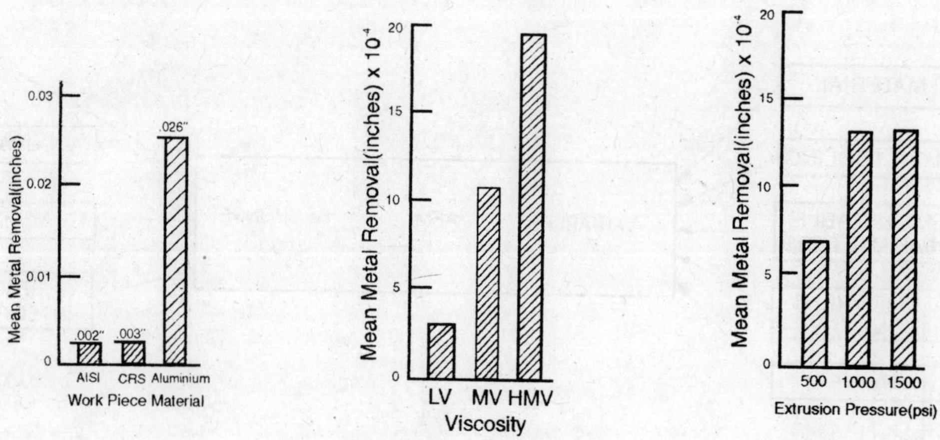

Fig. 4.8(a) *Fig. 4.8(b)* *Fig. 4.8(c)*

Fig. 4.8 Effect of (a) workpiece material, (b) viscosity, (c) pressure, on mean material removal during AFM. [*Williams and others*, 1989].

Figs. 4.9a and 4.9b show the *effect of viscosity and pressure*, respectively on mean final **surface roughness**. It was also concluded that the *effect of flow rate* on material removal is insignificant.

Scanning electron micrographs (*SEM*) of some specimens after abrasive flow finishing were taken. Some of the photographs taken during experimentation at 1600 × magnification are shown in Fig. 4.10. Fig. 4.10a shows the photograph or *original specimen* having rough surface with many tears present. It is evident from Figs. 4.10b-4.10d [*Williams and others, 1989*] *that major changes in surface finish* take place just after a few cycles. Number of cycles required to achieve the desired surface finish depends on the values of the variables set, and the difference between the initial and final surface finish.

ANALYSIS AND MODELLING OF ABRASIVE FLOW MACHINED (AFM'd) SURFACES

Profiles of AFM'd surfaces are unidirectional and random in nature, and have been modeled by stochastic modeling and analysis technique (called data dependent system - DDS) [*Williams and Rajurkar, 1992*; *Pandit and Wu, 1983*]. It gives differential/difference equation model using the raw data. This fits **ARMA (n, n-1) models** using non-linear least square method till the residual sum of squares of random disturbances is significantly reduced as determined by F-test criterion. Surface roughness profiles of AFM'd surfaces are digitized and analyzed using data acquisition software. A typical ARMA (2,1) model for the surface profiles generated during AFM (Fig. 4.11), is as follows:

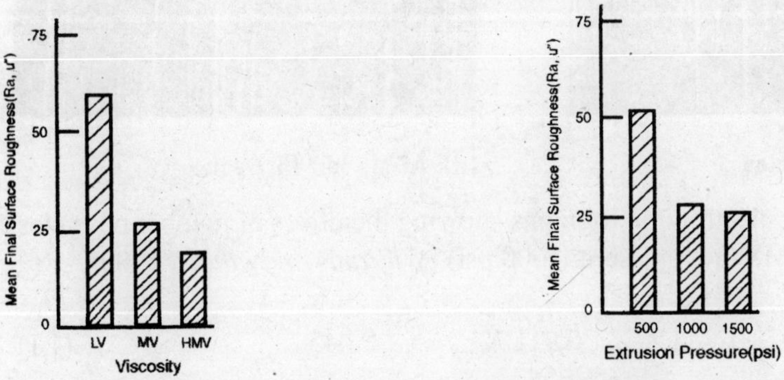

Fig. 4.9(a) *Fig. 4.9(b)*

Fig. 4.9 Effect of (a) viscosity, (b) pressure, on mean final surface roughness [*Williams and others*, 1989].

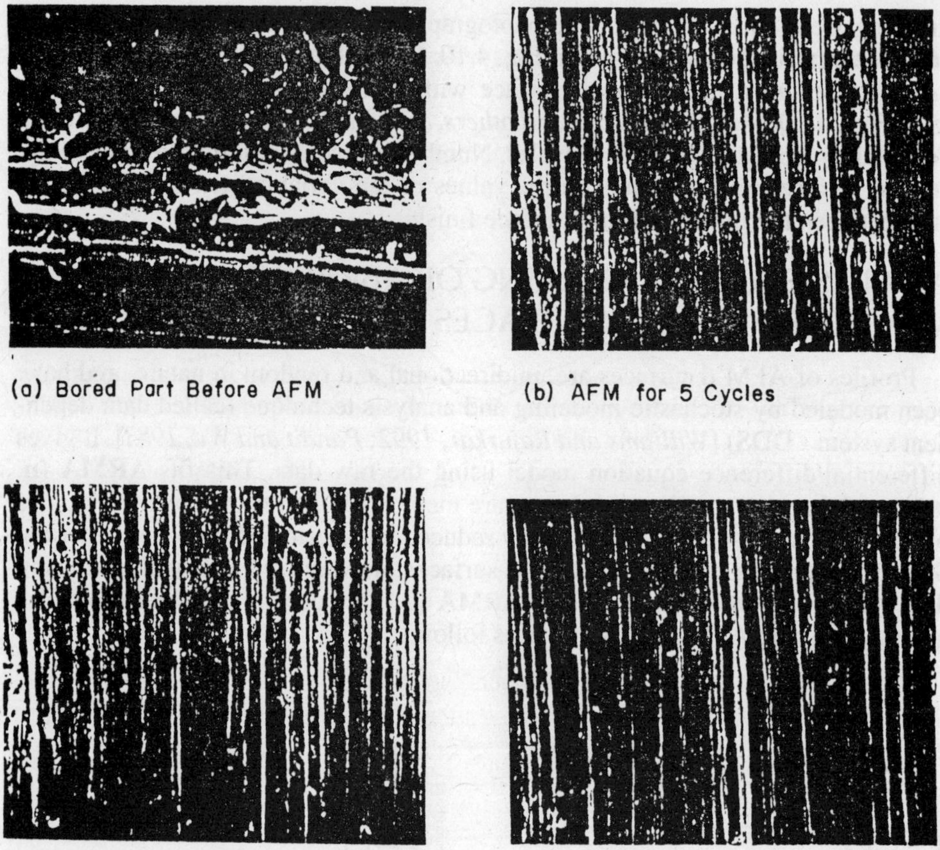

(a) Bored Part Before AFM

(b) AFM for 5 Cycles

(c) AFM for 10 Cycles

(d) AFM for 15 Cycles

Fig. 4.10 Scanning electron micrographs showing the effect of number of cycles (media: MV-36, pressure: 1000 psi) [*Williams and others*, 1989].

$$X_t - 1.883\, X_{t-1} + 0.911\, X_{t-2} = a_t + 0.045\, a_{t-1} \qquad \ldots(4.1)$$

where X_t, X_{t-1} and X_{t-2} are heights of the machined surface profiles, a_t and a_{t-1} are random disturbances (or noise), and t, t−1 and t−2 (subscripts) indicate distances.

This technique has also been used to compare peak to valley profile height (R_{max}) with center line average (CLA) roughness (R_a). The computed and experimentally obtained values from profilometer are found to compare favourably.

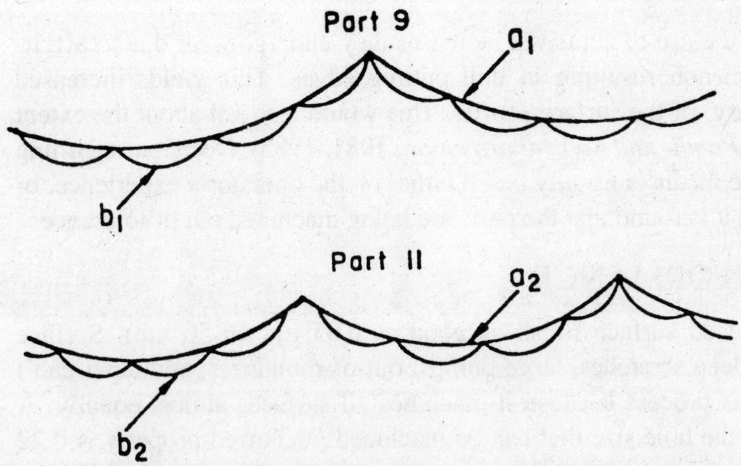

Fig. 4.11 A typical ARMA model (2, 1) for the surface profiles generated during AFM [*Williams and Rajurkar*, 1992].

Wavelength decomposition results have been explained based on the viscosity of the media. Higher viscosity restricts the freedom of movement and orientation of the grains. As a result, even the dull grains continue to cut the material and hence lower surface roughness. On the other hand, lower viscosity allows a continuous rapid change of position and orientation. It results in higher pseudo frequency and higher surface roughness.

Number of Active Grains

For consistency and accuracy of the process, number of **dynamically active grains (C_d)** is required. Using the frequency decomposition data, C_d can be evaluated as

$$C_d = \frac{\text{Frequency} \times \text{Time for one stroke}}{\text{Cross-sectional area of extrusion passage}} \qquad \ldots(4.2)$$

With the passage of time more number of grains would fracture resulting in larger number of grains but smaller in size. It is conjectured that after the critical number of strokes, the value of C_d will go down sharply. It is analogous to wheel loading in grinding process demanding dressing.

Wear of Abrasive Grains

As the time for the use of abrasives increases they undergo wear due to **attritious wear** phenomenon resulting in dull cutting edges. This yields increased wavelength (primary) of the surface profile. This would also tell about the extent of wear of grains [*Pandit and Sathyanarayanan*, 1981, 1985]. Decision regarding when to replace the media is usually based either on the operator's experience, or at the instant when it is found that the parts are being machined out of tolerance.

PROCESS PERFORMANCE

AFM has produced surface finish as good as 0.05 μm (= 50 nm). Surface irregularities like deep scratches, large bumps, out-of-roundness, and taper can't be corrected by this process because it machines all surfaces almost equally. A minimum limit on the hole size that can be machined / deburred properly is 0.22 mm and the largest size (or diameter) that has been machined is around 1000 mm. It can produce dimensional tolerance as good as ± 0.005 mm (± 5 μm).

APPLICATIONS

AFM is *suitable to* **automate** finishing operations that ask for high cost and which are labour intensive, and can be performed manually. It is very useful for finishing of extrusion dies, nozzle of flame cutting torch, and airfoil surfaces of impellers; deburring of aircraft valve bodies and spools; removing recast layer after EDM, etc. It is also employed for finishing operations, specially in the industries related to the manufacture of aerospace, automotive, semiconductor, and medical components. Some of the specific examples are given in the following:

Aerospace:

1. Conditions of the airfoil surface (compressors and turbines) are improved. Such surfaces are made by ECM, EDM, casting (say, investment castings), or milling. For finishing of such surfaces, this process *easily works even in inaccessible areas.*
2. Removal of thermal recast layers left by EDM, or laser beam machining.
3. Accessory parts like fuel spray nozzles, fuel control bodies, and bearing components are finished.

4. Resistance offered to the flow of air by blades, nozzles, diffusers, etc can be adjusted or tuned accurately by modifying surface using AFM process.
5. AFM *improves* the mechanical *fatigue* strength of blades, disks, hubs, shafts, etc.
6. AFM is also employed for removing coke and carbon deposits, and to improve surface integrity.
7. It can also be employed to remove left out light machining marks. The surface finish can be improved from say 1.75 μm to 0.4 μm.
8. Both radiusing and deburring of cooling turbine blades are done in one pass by AFM.

Dies and Moulds:

1. Multiple passages can be processed at one time.
2. Surface finish can be improved to a large extent with least change in dimension.
3. It saves considerable time when compared with finishing by skilled hands. Jobs demanding hours for polishing can be completed in minutes automatically.
4. Finishing of two-stroke cylinders and four-stroke engine heads is done using AFM for improved air flow and better performance. Stainless steel impeller made by investment casting is polished to 0.37 μm using this process.

Precise edge finishing of gears, and deburring, radiusing and polishing of fuel injector components at the rate as high as 30 components per minute in a single fixture can be done by AFM. Uniform finishing of threaded holes is also one of the common applications of this process.

BIBLIOGRAPHY

1. Benedict G.F. (1987), *Nontraditional Manufacturing Processes*, Marcel Dekker Inc., New York, pp. 53-65.
2. Extrude Hone Technical Pamphlet, U.S.A.
3. Jha Sunil (1998), *On the Abrasive Flow Machining Process Performance*, M. Tech. Thesis, I.I.T. Kanpur, India.

4. Kohut T. (1984), Automatic Die Polishing: Extrusion Productivity Through Automation, *The Aluminium Association*, Washington D.C., Vol. 1, pp. 193-202.
5. Kohut T. (1988), Surface Finishing with Abrasive Flow Machining, Proc. 4th *Int. Aluminium Extrusion Technology Seminar*, Vol. 2, The Aluminium Association, Washington D.C.
6. Przyklen K. (1986), Abrasive Flow Machining—A Process for Surface Finishing and Deburring of Workpieces, with a Complicated Shape by Means of Abrasive Laden Medium, In '*Advances in Nontraditional Machining*', *PED*, Vol. 22., ASME, New York, pp. 101–110 (Ed.: Rajurkar K.P.).
7. Pandit S.M. and Sathyanarayanan G. (1981), A New Approach to the Analysis of Wheel Workpiece Interaction in Surface Grinding, *Proc.* 9th *North American Manufacturing Research Conference*, pp. 311–320.
8. Pandit S.M. and Wu S.M. (1983). *Time Series and System Analysis with Applications*, John Wiley, New York.
9. Rhoades L.J. (1988), Abrasive Flow Machining, *Manuf. Eng*., pp. 75-78.
10. Sathyanarayanan G. and Pandit S.M. (1985), Fracture and Attritious Wear Rates in Grinding by Data Dependent System, *Proc. 13th North American Manufacturing Research Conference*, pp. 314–320.
11. Williams R.E. (1989), *Metal Removal and Surface Finish Characteristics in Abrasive Flow Machining*, M.S. Thesis, University of Nebraska at Lincoln. Nebraska, (U.S.A.).
12. Williams R.E. and Rajurkar K.P. (1989), Mechanics of Deburring and Surface Finishing Processes, paper presented at Winter Annual Meeting of ASME, held at San Francisco, Dec. 10–15, 1989, *PED*, Vol. 38, pp. 93–106.
13. Williams R.E., Rajurkar K.P. and Rboades L.J. (1989), Performance Characteristics of Abrasive Flow Machining, SME Technical Paper, FC 8980-610 to F8980-616.
14. Williams R.E. and Rajurkar K.P. (1992), Stochastic Modeling and Analysis of Abrasive Flow Machining, *Trans ASME, J. Engg. for Industry*, Vol. 114, pp. 74-80.

REVIEW QUESTIONS

1. With the help of a neat diagram, explain the working principle of abrasive flow machining process.

2. Justify the statement: "AFM has high flexibility".
3. How the restriction offered by passageway governs MRR? Make necessary sketches.
4. Write independent and dependent variables of AFM process.
5. What is the basic principle of tooling design for AFM process? Suggest the tooling for finishing of 200 gears of one type made of the same material. Would you consider parallel processing? Justify your answer.
6. Write the salient features of AFM system.
7. Detail the important process variables and responses.
8. Write five applications (component names) of AFM process.
9. In which cases of the following, you will recommend the use of AFM process:
   (a) A component has been machined by EDM process. It is found that the top layer (≈ 7 μm) has to be removed before it can be put to use.
   (b) ECM'd component is found weak in fatigue strength. It is decided to remove a layer of material ($\approx 1 \times 10^{-5}$ m) to improve its fatigue strength.
   (c) During gear cutting, burrs are formed. Gears can be used only after removing these burrs.
   (d) A cylinder has 100 small sized (0.5 mm diam.) holes to be finished. Every day 50 such cylinders are to be finished.
10. A hollow cylinder (inner diam. = 4 mm, outer diam. = 14 mm) requires improved surface finish on its outer and inner curved surfaces. Would you recommend AFM? Note that workpiece material is hardened steel and permissible dimensional change in its diameter is 20 μm. Draw the suggested tooling, if any.
11. Comment on the re-use of abrasives in AFM process.

SELF-TEST QUESTIONS

12. (A) Write true (**T**) or false (**F**).
    i. AFM can be used to reduce the diameter of a mild steel rod from 14 mm to 12 mm.
    ii. Stiff media is used for radiusing.
    iii. At the intersection of two holes drilled at an angle (< 90°), burrs are produced. Can you use AFM for deburring?

iv. After AFM, you obtain HAZ of approximately 0.5 μm.
v. Workpiece material has no effect on the time required for AFM.
vi. Major improvement in surface finish takes place in the initial five cycles or so.
vii. Centre line average roughness value (Ra) changes from 3 to 1. The surface roughness has deteriorated.
viii. Self deformable stone means the workpiece deforms to suit the media.
ix. The life of media in AFM is very high (approaches infinity).
x AFM process gives good dimensional tolerances.

NOMENCLATURE

X	Height of the machined surface profile.
a	Random disturbance (or noise)
t	Distance
C_d	Number of dynamically active grains

(B) MAGNETIC ABRASIVE FINISHING (MAF)

INTRODUCTION

Although the magnetic abrasive finishing (MAF) process was originated in U.S., it was in the former U.S.S.R and Bulgaria that much of the **developments** took place. The researchers have shown that the technique can be applied to a wide range of products. Japanese researchers followed this work and conducted research for finish polishing applications. Mainly two configurations of the MAF system have been in use, viz. one for cylindrical workpieces (Fig. 4.12a for external surfaces and Fig. 4.12b for internal surfaces), and another for flat workpieces (Fig. 4.12c). *Shinmura et al.* [1984–1985, 1990] and *Fox et al.* [1994] proposed designs for various equipments to study internal finishing of tubes, external

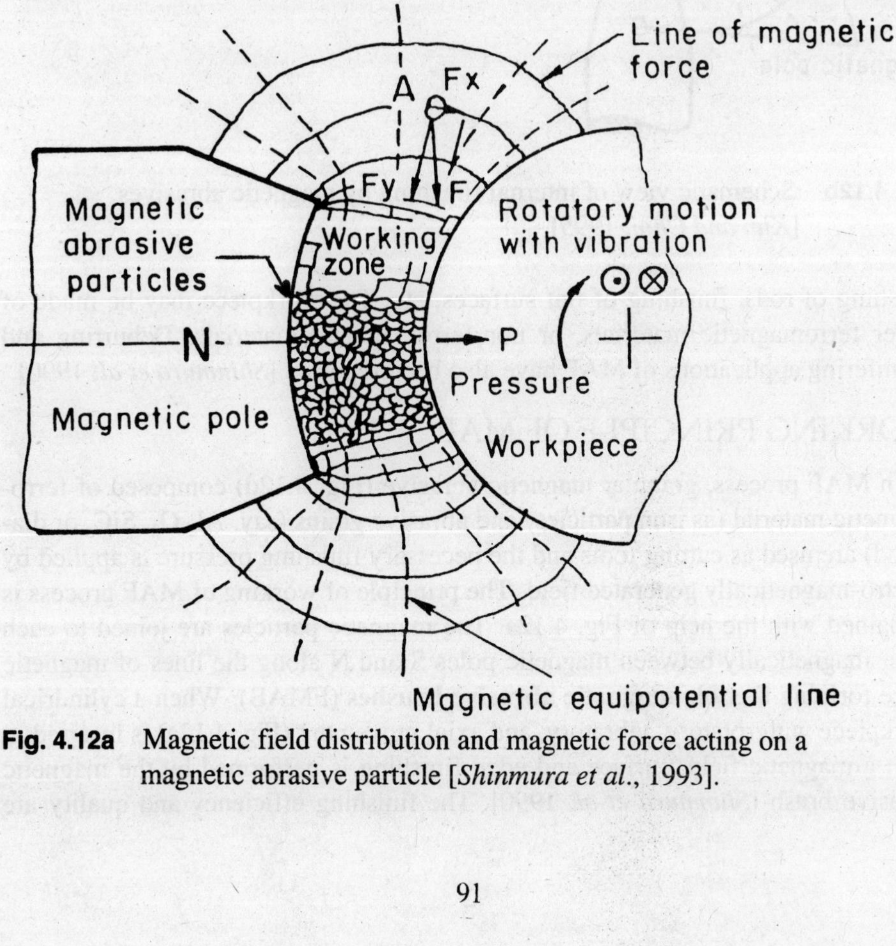

Fig. 4.12a Magnetic field distribution and magnetic force acting on a magnetic abrasive particle [*Shinmura et al.*, 1993].

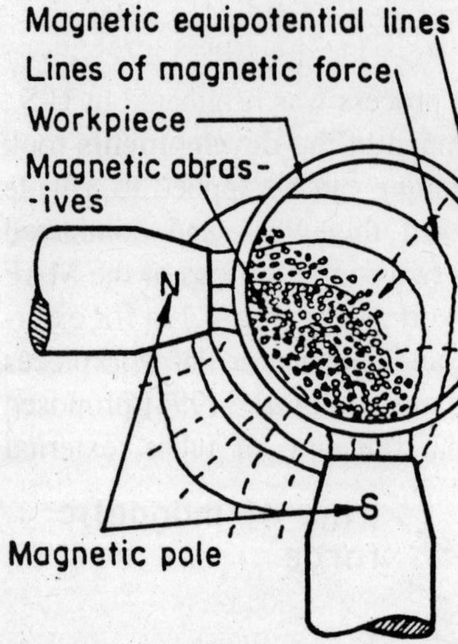

Fig. 4.12b Schematic view of internal finishing by magnetic abrasives [*Kim and Choi*, 1995].

finishing of rods, finishing of flat surfaces, etc. The workpiece may be made of either ferromagnetic materials, or non-ferromagnetic materials. Deburring and chamfering applications of MAF have also been reported [*Shinmura et al*, 1990].

WORKING PRINCIPLE OF MAF

In MAF process, granular magnetic abrasive (Fig. 4.12d) composed of ferromagnetic material (as iron particles) and abrasive grains (say, $Al_2 O_3$, SiC, or diamond) are used as cutting tools and the necessary finishing pressure is applied by electro-magnetically generated field. The principle of working of MAF process is explained with the help of Fig. 4.12a. The magnetic particles are joined to each other magnetically between magnetic poles S and N along the lines of magnetic force forming **flexible magnetic abrasive brushes** (FMAB). When a cylindrical workpiece with rotatory, vibratory, and axial movement (Fig. 4.12a) is inserted in such a magnetic field, surface and edge finishing is performed by the magnetic abrasive brush [*Shinmura et al*, 1990]. The finishing efficiency and quality are greatly influenced by the rigidity of the magnetic abrasive brush. If the workpiece is of non-magnetic material, the lines of magnetic field go around it (i.e. through the magnetic abrasives), and if it is of magnetic material then they pass through the workpiece.

Magnitude of *magnetic force* prevailing between the two poles is also affected by the material, shape, and size of the workpiece, and shape and size of the

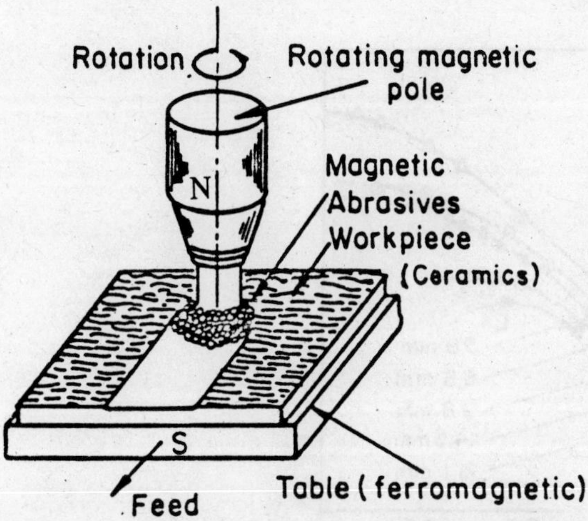

Fig. 4.12c Schematic view of plane (flat surface) magnetic abrasive finishing [*Shinmura et al.*, 1984].

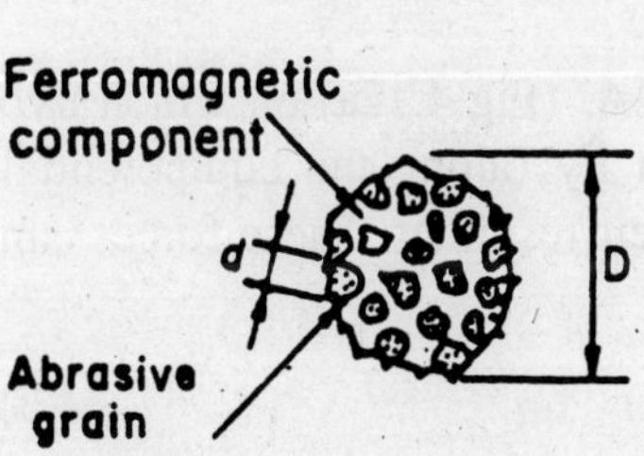

Fig. 4.12d Magnetic abrasive particlc [*Shinmura et al.*, 1990].

magnetic poles. These factors therefore, should be considered during the design of the system [*Shinmura et al.*, 1985]. It is experimentally observed that the pressure exerted by the magnetic abrasives is decreased as the clearance (i e, gap between the magnetic pole and the workpiece) is increased provided the filling density, ρ (g/cm^3) of the abrasive grains in the gap remains constant. Fig. 4.13 indicates that the magnetic abrasive pressure P acting on the work surface increases as the flux density on the magnetic abrasive grains increases for a given value of clearance.

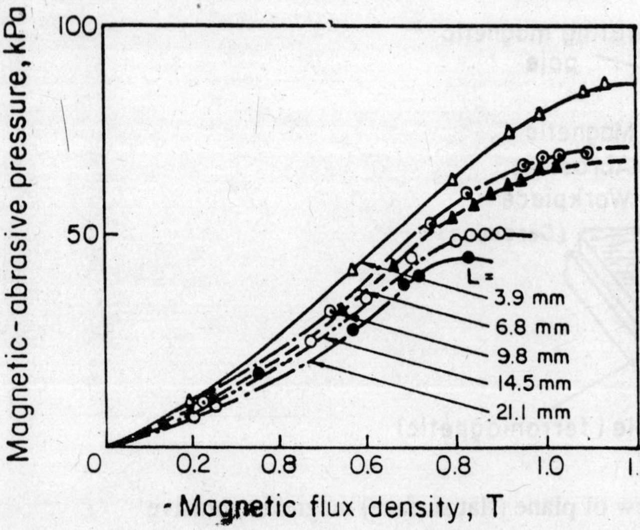

Fig. 4.13 Relationship between magnetic abrasive pressure and magnetic flux density for various values of clearance [*Shinmura et al*, 1985].

Consider a diamond magnetic abrasive particle 'A' (Fig 4.12a) on which two forces Fx (along the lines of magnetic force) and Fy (along the equipotential magnetic lines) act simultaneously, and F shows their resultant. These forces can be evaluated using Eqs. (4.3) and (4.4),

$$Fx = (\pi D^3/6)\, k\, H\, (\partial H/\partial x). \quad \ldots(4.3)$$

$$Fy = (\pi D^3/6)\, k\, H\, (\partial H/\partial y). \quad \ldots(4.4)$$

where, D and k are diameter (Fig. 4.12d) and susceptibility of magnetic abrasive particle, respectively and H is the magnetic field strength.

The **magnetic force Fy** is responsible for actuating the abrasive particles such that they take part in finishing the workpiece. The force Fx acts on the abrasive grains along the line of magnetic force, and it is mainly responsible for cutting action (or penetration in the workpiece). Any change in the strength of the magnetic field in the direction of the line of magnetic force near the workpiece surface, will actuate the magnetic abrasive particles. In addition to improving the surface finish, it also enhances the surface integrity by introducing residual compressive stresses.

The effective way of changing the force (or finishing pressure) and rigidity of the **magnetic abrasive brush** is through the change in diameter D of magnetic abrasive particle. Hence, ferromagnetic particles of several times the diameter of diamond abrasive (d) are mixed to form the magnetic abrasive brush. This configuration results [Shinmura et al, 1984] in higher rigidity of the brush and permits to use only one kind of small diameter (d) diamond abrasives (which are responsible for actual finishing/material removal) with the different sized ferromagnetic particles. A **'trade-off'** between 'd' and 'D' is required because the two have oppositing effects on surface finish and material removal rate, as discussed later. Further, the mixing proportion of these two basic constituents determines the number of cutting edges.

Some **variables** of the MAF process are type and size of magnetic abrasives, mixing ratio of abrasive grains with ferromagnetic particles, working clearance, rotational speed, vibration (amplitude and frequency both), properties of workpiece material, magnetic flux density and relative speed of magnetic abrasive to the workpiece surface. The effects of these variables on the various responses (viz material (or stock) removal, machined depth, surface finish, out of roundness, and surface integrity) have been studied by various researchers. Some of them are discussed in what follows.

MATERIAL REMOVAL (OR STOCK REMOVAL) AND SURFACE FINISH

Type and Size of Grains

In the plane finishing process, magnetic force on the surface and rigidity of the magnetic abrasive brush are governed mainly by the grain size of magnetic abra-

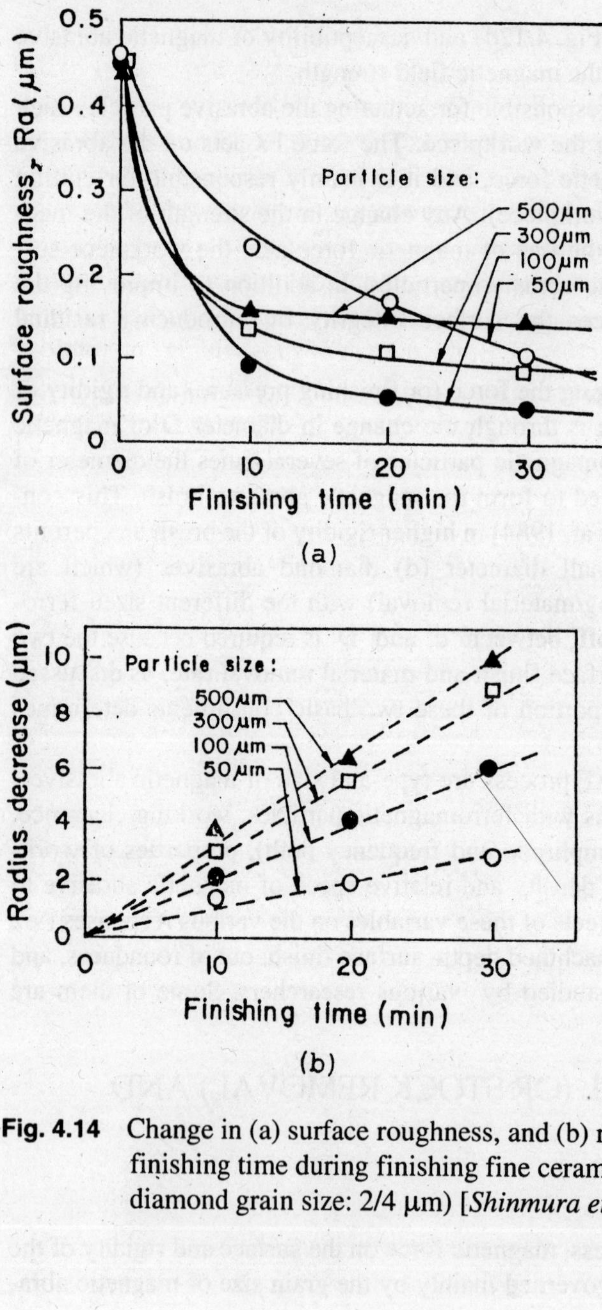

Fig. 4.14 Change in (a) surface roughness, and (b) radius decrease with finishing time during finishing fine ceramics (Workpiece: Si_3N_4; diamond grain size: 2/4 μm) [*Shinmura et al.*, 1990]

sives. Finishing tests were carried out using silicon nitride fine ceramic bar as workpiece. As shown in Fig. 4.14a, surface roughness rapidly improves in the beginning, then it levels off to a constant value. Larger is the particle size, poorer is the finished surface (except for the case of 50 μm particles) but higher is the stock removal (or radius decrease), which increases linearly with finishing time (Fig. 4.14b) [*Shinmura et al.*, 1990]. The process performance further improves if irregular shaped iron particles are used.

Finishing efficiency can be remarkably improved by mixing small sized diamond abrasive with irregular shaped large sized ferromagnetic particles (iron particles) [*Shinmura et al.*, 1984]. Both the size and mixing proportion of ferromagnetic particles influence characteristics of the finished surface. In some cases, there is an optimum value of mixing weight percentage of ferromagnetic particles for obtaining the best finish and the largest machined depth (Fig. 4.15). Diamond magnetic abrasives, composed of cast iron balls (diameter ≈ 100 μm)

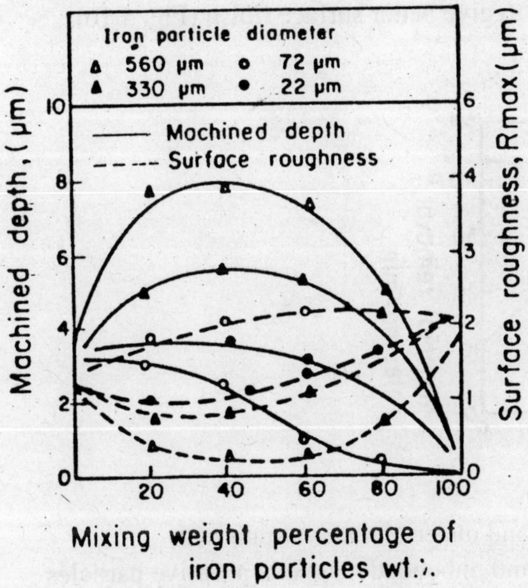

Fig. 4.15 Change in machined depth and surface roughness with mixing weight percentage of iron particles [*Shinmura et al.*, 1984].

and diamond micron sized powders (diameter ≈ 2/4 μm) were used for finishing Si_3N_4 fine ceramics. Surface roughness of 0.45 μm was efficiently reduced to 0.04 μm [*Shinmura et al.*, 1990]. Using Cr_2O_3 and Fe (Chromium oxide abrasives of 3 μm size and iron particles of 75, 330 and 510 μm size) mixing type magnetic abrasives, the surface finish of silicon nitride fine ceramic plate could be improved from 0.8 μm to 0.05 μm [*Shinmura et al.*, 1994]. Use of iron particles alone is not recommended because it is not capable to improve the surface finish appreciably.

Bonded and Unbonded Magnetic Abrasives

Magnetic abrasives have been used in the form of either a mixture (unbonded) of abrasive and ferromagnetic particles, or as magnetic abrasive conglomerate, i e, abrasives held in a ferromagnetic matrix (bonded) formed by sintering or other techniques. *Fox et al.* [1994] conducted experiments with bonded as well as unbonded magnetic abrasive particles for finishing stainless steel rollers. It was found that unbonded magnetic abrasives yield higher removal rates whereas bonded magnetic abrasives are found to give better surface finish (Fig. 4.16).

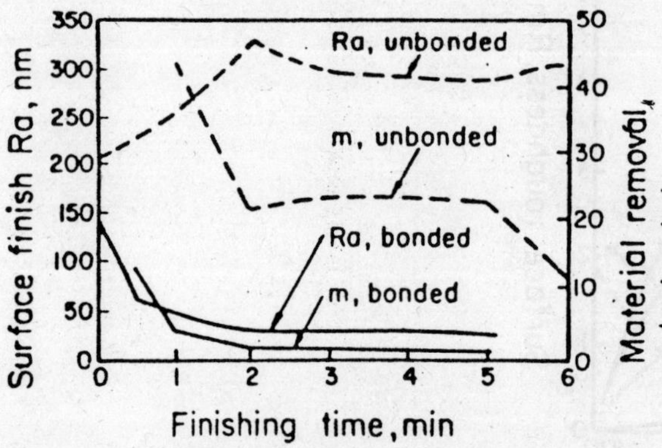

Fig. 4.16 Variation in surface finish and material removal rate with finishing time for bonded and unbonded magnetic abrasive particles (Surface speed: 1.3 m/s; flux density: 0.37 T; lubricant: oil (SAE 30)) [*Fox et al.*, 1994].

Machining Fluid

Addition of machining fluid with magnetic abrasives during MAF can be recommended [*Shinmura*, 1986], because it has revealed improvements in stock removal by the use of various types of commonly used grinding fluids. The best surface finish that could be achieved was 0.5 μm (Rmax). However, a better surface finish has been reported in the absence of cutting fluid.

Magnetic Flux Density

An increase in the magnetic flux density increases the tangential finishing force for various % of grinding fluid [*Shinmura*, 1986]. Machined depth increases with increase in magnetic flux density (T) and iron particle size (Fig. 4.17a), and decreases with increase in working clearance, (Fig. 4.17b). It has been reported [*Shinmura et al.*, 1985], in plane finishing operations, that surface roughness improves with the increased magnetic flux density, and finishing time (Fig. 4.18). Further, the grinding burrs are more efficiently removed than turning burrs [*Shinmura et al.*, 1985]. This is attributed to the fact that magnetic flux concentrates on the edges of ferromagnetic substance.

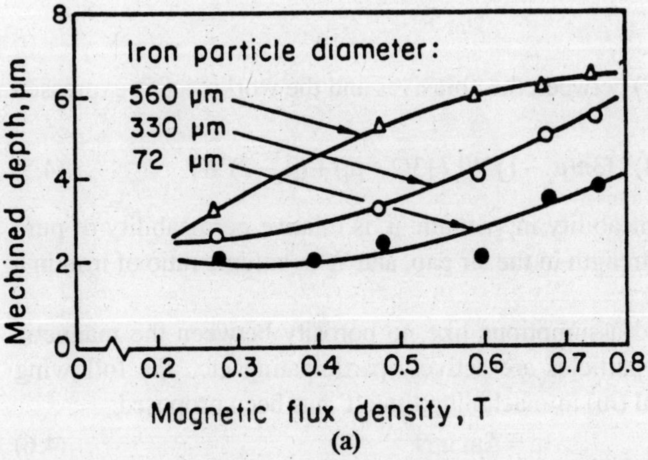

Fig. 4.17a Change in machined depth with magnetic flux density [*Shinmura et al.*, 1984].

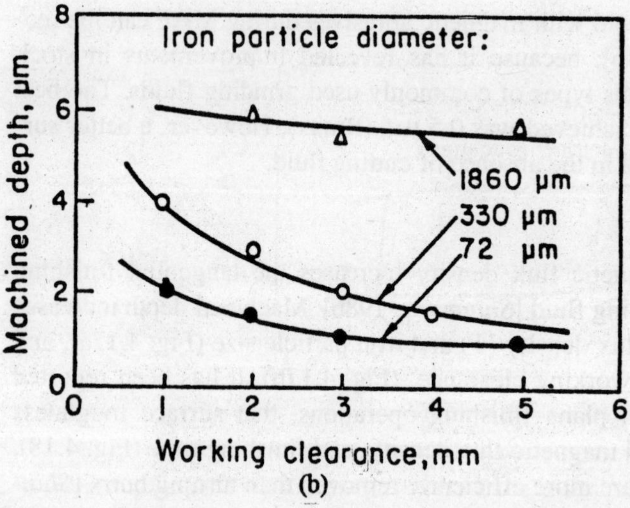

Fig. 4.17b Change in machined depth with working clearance (Magnetic flux density: 0.6 T) [*Shinmura et al.*, 1984].

ANALYSIS

The machining pressure (P) between the abrasives and the workpiece is expressed [*Kim and Choi*, 1995] as:

$$P = \mu_o (H_a^2/4) \cdot [3\pi(\mu_r - 1) W] / [3(2 + \mu_r) + (\mu_r - 1) W] \qquad \ldots(4.5)$$

where, μ_o is magnetic permeability in vacuum, μ_r is relative permeability of pure iron, H_a is magnetic field strength in the air gap, and W is volume ratio of iron in a magnetic abrasive particle.

Based on the simplified assumptions like no porosity between the magnetic abrasive particles, all the particles are actively participating etc, the following equation for **stock removal** (m) in machining time 't' has been proposed.

$$m = \Delta m\, n\, N \qquad \ldots(4.6)$$

where, n = number of edges of a magnetic abrasive particle simultaneously acting on the surface, N = number of active magnetic abrasive particles, and Δm = volume of material removed by an edge in time 't', and

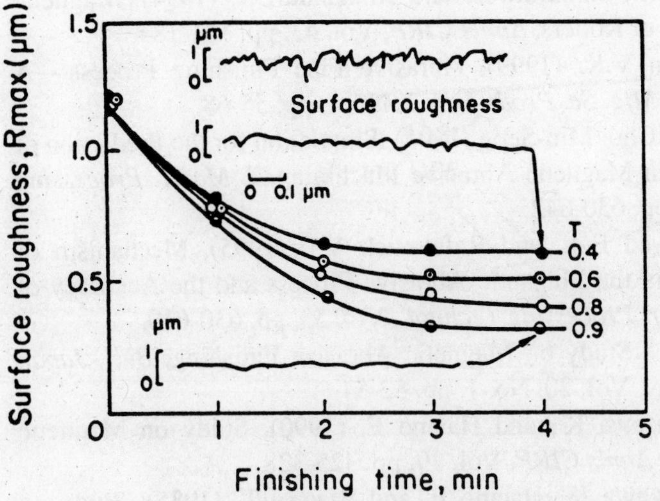

Fig. 4.18 Relationship between surface roughness and finishing time for various magnetic flux densities (Working clearance: 3 mm; peripheral speed of rotating pole: 166 m/min) [*Shinmura et al*, 1985].

$$\Delta m = C\,(\Delta f / H_w \pi \tan\theta)\,(1 - R_a/R_{ao})\; v\, t \qquad \ldots(4.7)$$

where, Δf is force acting on a grain edge, H_w is workpiece hardness, θ is mean angle of asperity of abrasive cutting edges, R_a is final surface roughness, R_{ao} is initial surface roughness, v is speed of magnetic abrasives, and t is machining time.

Using the relationship between the surface roughness and stock removal, the following equation for evaluation of surface roughness is derived.

$$R_a = R_{ao} - C^1\,(R_{ao})^{-1/8}\,(l_w)^{-5/4}\,(nN\Delta f\, v\, t / H_w\, \pi\, l_w \tan\theta)^{1/4} \qquad \ldots(4.8)$$

where, l_w is machined length and C^1 is a constant.

Using above models, surface roughness and stock removal in internal surfaces have been computed [*Kim and Choi*, 1995; *Keremen et al*, 1995], and comparison of computed and experimental results are found satisfactory.

BIBLIOGRAPHY

1. Fox M., Agrawal K., Shinmura T. and Komanduri R. (1994), Magnetic Abrasive Finishing of Rollers, *Annls CIRP*, Vol. 43, pp. 181-184.
2. Jain R.K. and Jain V.K. (1999), Abrasive Fine Finishing Process—A Review, *The Int. J. Mfg. Sc. Prod.*, Vol. 2, No. 1, pp. 55-68.
3. Kim Jeong-Du and Choi Min-Seog (1995), Simulation for the Prediction of Surface Accuracy in Magnetic Abrasive Machining, *J. Mater. Processing Technol.*, Vol. 53, pp. 630-642.
4. Kermen G.Z., Elsyed E.A. and Rafalovich V.I. (1995), Mechanism of Material Removal in the Magnetic Abrasive Process and the Accuracy of Machining, *J. Mater. Processing Technol.*, Vol. 53, pp. 630-640.
5. Shinmura T. (1986), Study on Magnetic Abrasive Finishing, *Bull. Japan Soc. Precision Engg.*, Vol. 20, No. 1, pp. 52-54.
6. Shinmura T., Takazawa K., and Hatano E. (1990), Study on Magnetic Abrasive Finishing, *Annls CIRP*, Vol. 39, pp. 325-328.
7. Shinmura T., Takazawa K., Hatano E. and Aizawa T. (1985), Study on Magnetic Abrasive Process—Process Principle and Finishing Possibility, *Bull. Japan Soc. Precision Engg.*, Vol. 19, No. 1, pp. 54-55.
8. Shinmura T., Wang F.H. and Aizawa T. (1984), Study on a New Finishing Process of Fine Ceramics by Magnetic Abrasive Machining, *Int. J. Japan Soc. Precision Engg.*, Vol. 28, pp. 99-104.
9. Shinmura T. and Wang F.H. (1994), A New Process for Precision Finishing of Silicon Nitride Fine Ceramics by the Application of Magnetic Abrasive Machining Using Chromium Oxide Abrasives Mixed with Iron Particles, *Int. J. Japan Soc. Precision Engg.*, Vol. 28, pp. 229-230.
10. Subramanian K. (1994), Finishing Methods Using Multipoint or Random Cutting Edges, *Surf. Engg., ASM*, Vol. 5, pp. 91-109.

SELF-TEST QUESTIONS

1. Write true (**T**) or false (**F**).
   i. MAF is suitable for finishing fine holes.
   ii. Stainless steel powder and Al_2O_3 abrasive particles will form a strong magnetic abrasive brush during MAF.

iii. Brass rod cannot be finished by MAF because it is non-magnetic in nature.

iv. Due to low magnetic abrasive filling density, strength of the magnetic abrasive brush is low, hence MRR is also low.

v. Diamond particles are smaller in size but material removal is done by them, and not by ferromagnetic particles.

vi. Unbonded abrasives give better surface finish compared to bonded abrasives.

vii. With a change in iron particle size, machined depth during MAF is not affected.

viii. During MAF, major % improvement in surface finish takes place during first few minutes only.

REVIEW QUESTIONS

1. Explain the working principle of MAF with the help of a neat sketch. Clearly show lines of magnetic force, magnetic equipotential lines, direction of pressure acting on the workpiece, direction of rotatory motion, and a semi-magnetic abrasive particle.
2. Make a schematic diagram of MAF of internal surface of a cylinder showing the details mentioned in Q1.
3. Can you use MAF process to finish a ceramic thin plate (3 mm thick)? If yes, sketch a schematic diagram for the same, else justify your answer.
4. During experimentation, it is found that material removal decreases with the increase in clearance (ie., gap between the w/p and the magnet). Explain the reason(s) for the observation.
5. What is gap filling density? Explain with the help of a figure.
6. Write the type (material type) and size of the abrasive and magnetic particles used in MAF.

NOMENCLATURE

C^1 Constant

D Mean diameter of the magnetic abrasive particle

$FMAB_s$ Flexible magnetic abrasive brushes

F_X	Magnetic force acting on a magnetic abrasive particle along the magnetic lines of force
F_Y	Magnetic force acting on a magnetic abrasive particle along the equipotential magnetic lines
H	Magnetic field strengthd
H_a	Magnetic field strength in the air gap
H_w	Workpiece hardness
k	Susceptibility of the magnetic abrasive particle
l_w	Machined length
m	Stock removal (volumetric)
MAF	Magnetic abrasive finishing
N	Number of active magnetic abrasive particles
n	Number of edges of a magnetic abrasive particle simultaneously acting
P	Machining pressure
R_a	Final surface roughness
R_{ao}	Initial surface roughness
t	Machining time
v	Speed of magnetic abrasives
W	Volume ratio of iron in a magnetic abrasive particle
Δf	Force acting on a grain edge
μ_0	Magnetic permeability in vacuum
μ_r	Relative permeability of pure iron
Δm	Volume of material removed by a grain edge
θ	Mean angle of asperity of abrasive cutting edges

CHAPTER 4A
AT-A-GLANCE
ABRASIVE FLOW MACHINING (AFM)

ABRASIVE FLOW MACHINING

- FINISHING PROCESS : DEBURRING, RADIUSING, POLISHING, PRODUCING COMPRESSIVE RESIDUAL STRESSES, ETC.
- SEMISOLID ABRASIVE LADEN PUTTY → DEFORMABLE GRINDING WHEEL → REMOVES SMALL AMOUNT OF MATERIAL
- FORCE: HYDRAULIC OR MECHANICAL
- MORE RESTRICTION → LARGER FORCE & VELOCITY
- FOR BOTH METALS & NON-METALS
- AREAS NOT ACCESSIBLE FOR CONVENTIONAL METHODS → MAY BE FINISHED

ELEMENTS OF ABRASIVE FLOW MACHINING

↓ MACHINE ↓ TOOLING ↓ MEDIA ↓ WORKPIECE

MACHINE

HOW AFM WORKS?

- MEDIA CONTAINING TOOLS (ABRASIVE PARTICLES) MOVES OVER THE W/P SURFACE AND REMOVES MATERIAL → JUST LIKE GRINDING
- PRESSURE → 0.70 TO 22 MPa
- FOR MAINTAINING CONSTANT VISCOSITY → COOLERS TO LOWER TEMPERATURE OF MEDIA
- CONTROLLABLE VARIABLES : VOLUME / STROKE, NO. OF CYCLES AND PRESSURE

TOOLING

- HOLDS MEDIA
- DIRECTS THE FLOW OF MEDIA
- MACHINING RATE → FUNCTION OF THE RESTRICTION TO THE FLOW OF MEDIA

------> INTERNAL SURFACES → ITSELF DECIDE THE EXTENT OF RESTRICTION
------> EXTERNAL SURFACES → DESIGNER DECIDES THE EXTENT OF RESTRICTION
------> NARROWEST SECTION : MAXIMUM MRR
------> WIDEST SECTION : MINIMUM MRR

- INSERTS FOR RESTRICTING AREAS → RESIST ABRASIVE ACTION → NYLON, TEFLON, ETC.
  - * CHARACTERISTICS : PLIABLE (EASILY MOULDED) MATERIAL ACTS AS A SELF-DEFORMING GRINDING STONE
  - * CONSTITUENTS : BASE (ORGANIC POLYMER & HYDROCARBON GEL) + ABRASIVE GRITS
  - * DEGREE OF STIFFNESS : DETERMINED BY COMPOSITION
    : FOR LARGE HOLE - STIFF MEDIA
    : FOR SMALL HOLE - SOFT MEDIA
  - * ABRASIVES : Al_2O_3, SiC, CBN, DIAMOND
    : REPLACE MEDIA IF MACHINED AMOUNT = 10% OF ITS WEIGHT
  - * MIXING OF A NEW MEDIA : CYCLE 20-25 TIMES THROUGH A SCRAP PART
  - * CLEANING OF A MACHINED PART : AIR / VACUUM / CHEMICAL

PROCESS VARIABLES

* NO. OF CYCLES
* PRESSURE
* MEDIA
* WORKPIECE CONFIGURATION

PROCESS PERFORMANCE

* SURFACE FINISH → 0.05 μm
* DIMENSIONAL TOLERANCE → ± 0.05 μm
* SURFACE IRREGULARITY (SCRATCHES, BUMPS, OUT-OF-ROUNDNESS) CANNOT BE CORRECTED

WHY?

=> MACHINES ALL SURFACES ALMOST EQUALLY

APPLICATIONS

* FINISHING AIRFOIL SURFACES OF IMPELLERS & EXTRUSION DIES
* NOZZLE OF TORCH FLAME, ETC.

CHAPTER 4B
AT-A-GLANCE
MAGNETIC ABRASIVE FINISHING (MAF)

INTRODUCTION

HISTORY : US / USSR / BULGARIA / JAPAN <- DEVELOPMENTS

SETUP CONFIGURATIONS:
- FOR FLAT SURFACES
- FOR CYLINDRICAL SURFACES
  - + EXTERNAL
  - – INTERNAL

W/P MATERIAL : FERROMAGNETIC / NON-FERROMAGNETIC

TYPES OF OPERATIONS : FINISHING, RADIUSING, DEBURRING, INTRODUCING RESIDUAL COMPRESSIVE STRESSES, ETC

WORKING PRINCIPLE

MAGNETIC ABRASIVE PARTICLES (MAPs) $\rightarrow$
- FERROMAGNETIC PARTICLES + ABRASIVE PARTICLES (FIG.4.12d)
- ACT AS CUTTING TOOLS

FLEXIBLE MAGNETIC ABRASIVE BRUSH (FMAB) $\rightarrow$
- MAPs JOIN EACH OTHER UNDER THE INFLUENCE OF MAGNETIC FIELD

WORKPIECE $\rightarrow$
- ROTATES, VIBRATES AND MOVES AXIALLY
- FMAB DOES FINISHING BY SHEARING PEAKS ON W/P SURFACE

FACTORS $\rightarrow$ MAGNETIC FLUX DENSITY $\rightarrow$ ϕ (W/P MATERIAL (TYPE, SHAPE, AND SIZE), MAGNETIC POLE <- SHAPE & SIZE)

MAGNETIC ABRASIVE PRESSURE ON W/P $\rightarrow$ ϕ (FLUX DENSITY, CLEARANCE)

FORCES ACTING ON A PARTICLE $\rightarrow$ F_X, F_Y (Eqs. 4.3, 4.4)

HOW TO CHANGE RIGIDITY OF MAB $\rightarrow$ EASY WAY $\rightarrow$ CHANGE 'D' OF MAPs AND FINISHING PRESSURE

'TRADE OFF' BETWEEN 'D' & 'd' $\rightarrow$ HAVE OPPOSING EFFECT ON MRR & SF

VARIABLES

- MAGNETIC ABRASIVES $\rightarrow$ + TYPE & SIZE + MIXING RATIO
- WORKING CLEARANCE
- ROTATIONAL SPEED, VIBRATION (FREQUENCY & AMPLITUDE), AND AXIAL MOVEMENT
- PROPERTIES OF W/P MATERIAL
- MAGNETIC FLUX DENSITY

RESPONSES

- MATERIAL REMOVAL
- MACHINED DEPTH
- SURFACE FINISH

- OUT-OF-ROUNDNESS
- SURFAE INTEGRITY

MATERIAL REMOVAL AND SURFACE FINISH

TYPE AND SIZE OF GRAINS

- WITH FINISHING TIME, SURFACE ROUGHNESS RAPIDLY IMPROVES; LATER ON LEVELS OFF TO A CONSTANT VALUE
- LARGER PARTICLE SIZE → POOR SURFACE FINISH (FIG.4.14a)
  → HIGHER STOCK REMOVAL
  ↓
  INCREASES LINEARLY WITH FINISHING TIME (FIG 4.14b)
- IRREGULAR SHAPED IRON PARTICLES → BETTER PERFORMANCE
- MIXING WEIGHT PERCENTAGE → OPTIMUM MACHINED DEPTH (FIG 4.15a)
  → OPTIMUM SURFACE ROUGHNESS (FIG 4.15b)
- ONLY FERROMAGNETIC PARTICLES → NOT RECOMMENDED. DOES NOT IMPROVE SURFACE FINISH APPRECIABLY

BONDED AND UNBONDED MAGNETIC ABRASIVES

BONDED MAGNETIC ABRASIVES → BETTER SURFACE FINISH
UNBONDED MAGNETIC ABRASIVES → HIGHER MRR

MACHINING FLUID

- USE OF MACHINING FLUID → HIGHER STOCK REMOVAL
  → POORER SURFACE FINISH

MAGNETIC FLUX DENSITY

MACHINED DEPTH INCREASES BY → INCREASED MAGNETIC FLUX DENSITY
→ INCREASED PARTICLE SIZE
→ DECREASED WORKING CLEARANCE

SURFACE FINISH IMPROVED BY → INCREASED FLUX DENSITY
→ INCREASED FINISHING TIME
→ HIGHER RELATIVE SPEED
→ SMALLER WORKING CLEARANCE

ANALYSIS

STOCK REMOVAL 'm' IN MACHINING TIME 't', m = Δ m • n×n

$$\Delta m = C \times (\Delta f/(H_w \pi \tan\theta)) \times (1 - R_a/R_{ao}) \times Vt$$

SURFACE ROUGHNESS, $R_a = R_{ao} - C^1 (R_{ao})^{-1/8} (1_w)^{-5/4} \left(\dfrac{nN\Delta f vt}{H_w \pi \tan\theta 1_w} \right)^{1/4}$

Chapter Five

WATER JET CUTTING (WJC)

INTRODUCTION

Water Jet Cutting (WJC) process is good for cutting and slitting of porous non-metals like wood, leather, foam, etc. It is also used for cutting composites, wire stripping, and deburring. This process works on the principle of erosion effects of a high velocity, small diameter jet of water. Quality of the machined edge obtained during this process is usually superior to the other conventional cutting processes. In Water Jet Machining (WJM), water jet is a *cutting tool* which never dulls or breaks. This process does not generate airborne dust hence health hazard during machining of fiber composites and asbestos, are minimum. The upper limit of jet velocity may be as high as 900 m/s. Basic cutting action is performed by tight water jet core. Interaction with air diverges the shroud around the core.

WATER JET STRUCTURE

Figure 5.1 shows a schematic diagram of a pure water jet after it travels into atmosphere. It has been divided into three different zones namely, initial zone, transition zone and droplet zone. In the first zone, the jet is having diameter almost the same as that of the orifice. In this zone, velocity profile, stagnation pressure, and density are approximately constant across its cross-section. In the next zone, the transition from laminar to turbulent flow takes place. In this zone across the cross-section of the jet, the jet properties change, namely, velocity profile, stagnation pressure and density. In the third zone that is the droplet zone, the jet breaks into the droplets due to drag forces [*Hoyt and Taylor*, 1979]. The length of each zone along the axis depends on Reynolds's number, orifice diameter, type of flow (Laminar or turbulent), and nozzle geometry. The initial zone jet is a coherent jet with uniform energy distribution across x-section. It is achieved at low stand-off-distance. This jet is good for cutting while the jet of droplet zone (obtained at high stand-off-distance) is good for comparatively lighter work. The abrasive water jet of the first zone is suitable specially for micro-machining while the abrasive-water-droplet jet (zone III) is good for hydro blasting, hydro demolition and rough cutting (discussed in chapter 6).

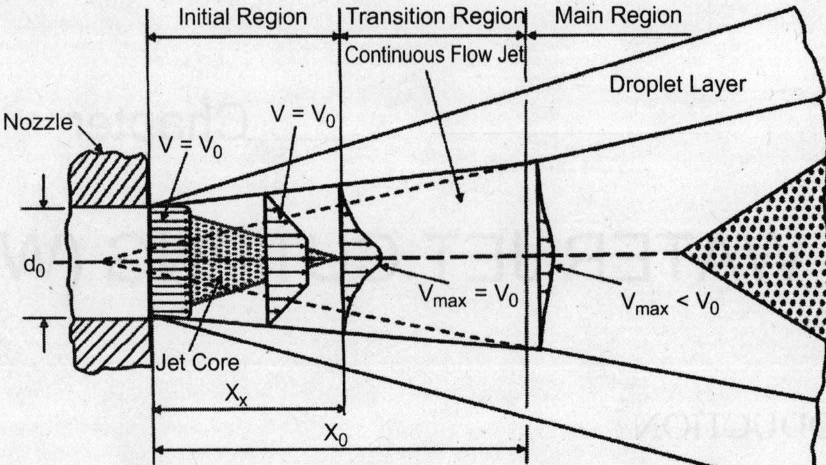

Fig. 5.1: Schematic diagram of different zones of a water jet

WATER JET ANALYSIS

Figure 5.2 shows a pipe with an orifice used to produce a water jet exiting with high velocity in the atmosphere. Suppose a stream of water at steady state pressure P_1, density

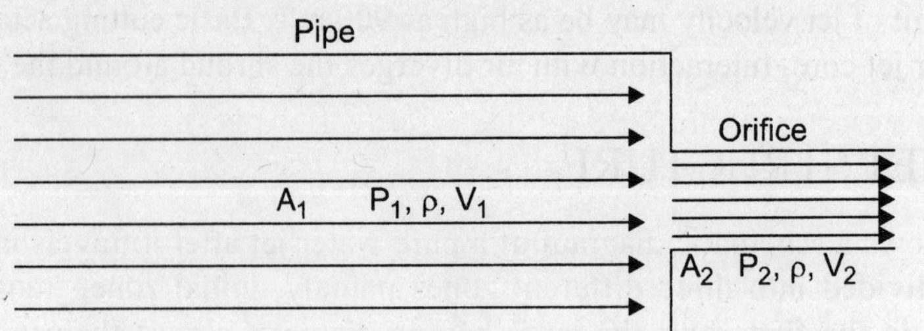

Fig. 5.2

ρ, velocity V_1 and pipe cross-sectional area A_1, passes through the orifice. Then, its pressure drops to P_2, cross-sectional area changes to A_2 and velocity increases to V_2 while assuming that the density ρ remains unchanged. In this situation, the principle of conservation of energy (Bernoulli's theorem) Eq. 5.1, can be applied to evaluate the exit velocity of the water jet.

$$P_1 + \frac{\rho V_1^2}{2} = P_2 + \frac{\rho V_2^2}{2} = P_s. \qquad \text{... (5.1)}$$

Where, $\frac{\rho V_1^2}{2}$ represents the dynamic pressure of the fluid and P_s is total pressure or stagnation pressure. For high velocity water jet, $P_1 >>> P_2$ (P_2 is usually atmospheric pressure) and $V_2 >>> V_1$. Therefore, P_2 and V_1 in Eq. 5.1 can be neglected. Therefore,

$$P_1 = \frac{\rho V_2^2}{2}$$

$$\text{Or, } V_2 \simeq \sqrt{\frac{2P_1}{\rho}} \qquad \text{... (5.2)}$$

In actual practice, there are frictional losses, and losses in the bends of the pipe, hence the theoretical velocity calculated from Eq. 5.2 is always higher than the actual exit velocity. To correct the theoretical velocity that is to make it very close to the experimental velocity, Eq. 5.2 can be modified as,

$$V_2 = K\sqrt{\frac{2P_1}{\rho}} \qquad \text{... (5.3)}$$

Where, K is a constant of proportionality and it is different for different setups. Some researchers [*Momber and Kovacevic*, 1998] have suggested its value as 0.8–0.98.

Depending upon the stand-off-distance, in water jet cutting, material is sheared by high speed water jet or water droplets. Due to the impact of high speed jet or droplets, it momentarily generates very high pressure called '*water hammer pressure*'. If this pressure is high enough, it induces large scale plastic deformation due to shearing in ductile materials or cracks in brittle materials.

WJC MACHINE

Figure 5.3 shows a schematic diagram of *WJC* system. The pumping *unit* (*i.e.* Oil pump) is driven by an electric motor. The oil drawn from a reservoir is pumped to an intensifier which uses low pressure oil to produce very high pressure water. The intensefier that acts as a high pressure pump produces water pressure as high as 40 times that of the oil. Water pressure (P_w) can be determined from the following equation:

$$P_w = (p_o \times A_o)/A_w \qquad \text{... (5.4)}$$

where, p_o is oil pressure, A_o is oil piston area, and A_w is water piston area.

For generating pressure upto 400 MPa and flow rate upto 10 lit/min, direct pressurization of water using triplex direction drive pump is adequate. For producing ultra high pressure upto 1000 MPa, an intensifier pump is used. To minimize pulsation in water flow, a high pressure *accumulator* (*i.e.*, a pressure vessel to store high-pressure water and to give smooth outflow) is used. High pressure water is transported to nozzle through the rigid high pressure tubing and rigid connectors (not shown in Figure 5.3). Off/On control valve of water flow can be operated manually or electronically.

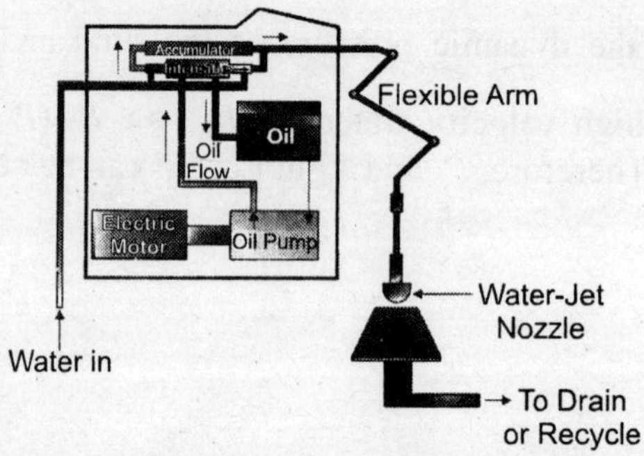

Fig. 5.3: Schematic diagram of water jet machining system
(*Source: Norwood and Johnston,* 1984]

Proper *design of nozzles* has made it possible to discontinue the use of *long-chain polymers* to lower down friction in fluid flow. Internal diameter of nozzle usually ranges from 0.07 to 0.50 mm. These nozzles are made of synthetic sapphire which is wear resistant but easily machinable. Presence of foreign particles (say, dirt) in water results in failure of nozzle by chipping. Sometimes, constriction of nozzle by mineral deposits also results in nozzle failure. Life of a sapphire nozzle is usually 250–500 hr.

Water jet out side the nozzle, travels at a very high speed (usually more than sound velocity). To minimize the exposed length of a jet, from safety point of view and also to minimize the process noise, a *catcher* (a slot type or tube type) is used. It is attached to a draining hose. The catcher uses hard and replaceable inserts to break the jet quickly and completely before it reaches bottom of the tube. Slot type catcher is comparatively less efficient in reducing noise level (Details of the types of catchers are given in Abrasive Water Jet Machining chapter (chapter 6)).

PROCESS CHARACTERISTICS

Important process parameters include water jet exit velocity, nozzle exit diameter, stand-off-distance, traverse speed, angle at which water jet hits the workpiece, coherency of water jet and properties of workpiece material. However, *Pressure, diameter of nozzle, traverse rate, and stand-off-distance* are four more important *variables* that affect the performance of *WJC* process to a larger extent. Stand-off-distance is comparatively less effective variable and its value normally lies between 3–25 mm. Cutting at higher water jet pressure is faster. Thick materials can be machined without difficulty at high water jet velocity and/or low traverse rate.

PROCESS PERFORMANCE

This process is useful for cutting materials which are porous, fibrous, granular, and/or soft. It includes the materials like corrugated board (Cutting speed (relative speed between water jet and the WP surface normal to the jet axis) = 3 m/s), granite (0.25 m/s), rubber (0.13 m/s), plywood (0.025 m/s), glass (0.025 m/s), aluminum (0.0025 m/s), etc. This process *does not require predrilled hole* to start cutting in any direction and location provided the location is accessible for the water jet. Too thick materials can be cut in *more than one pass*. Second and subsequent passes are used to make the cut deeper rather than wider. Energy consumption per unit length in *multi pass cutting* is less than single pass cutting of the same workpiece material but with a more powerful jet. *Machined surfaces* have no burrs, no thermal damage, and a good surface finish. Surface finish, tolerance, and straightness of the cut edges also depend on material thickness, cutting speed, and other machining parameters.

APPLICATIONS

Water jet at lower pressure (69–200 MPa) has been used to cut *insulation of the cables* without damaging the underlying metallic cable. The water jet nozzle rotates around the cable to cut all around. It takes about 5–10 s/cable depending upon the thickness of the insulation.

In aerospace and other similar industries, cutting of asbestos is done by WJ to minimize airborne dust in the atmosphere. Carbide grit safety walks (used on cargo aircraft loading ramps) are also cut by WJC. It is also used to cut fiber glass and polyethylene automotive parts. This technique is also used for high speed cutting of the corrugated box, Figure 5.4.

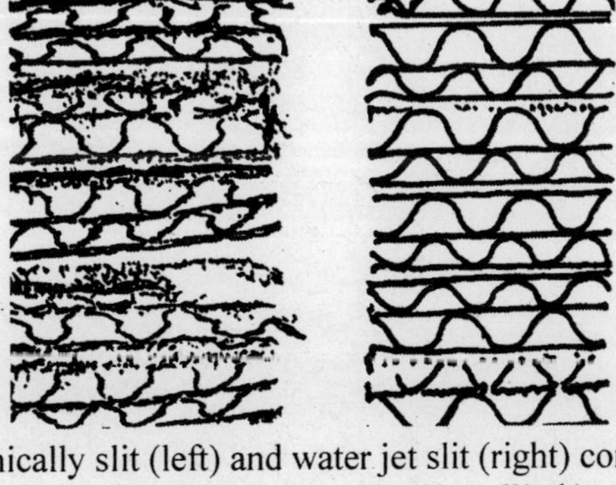

Fig. 5.4: Mechanically slit (left) and water jet slit (right) corrugated cardboard
(Source: Flow System, Inc., Kent, Wash)

WATER JET MICRO-MACHINING (WJMM)

Water Jet Micro-Machining (WJMM) has been applied for both drilling (with stationary water jet and workpiece) and profiling (with relative motion between Water Jet (WJ) and workpiece). This process of WJMM is used for soft and thin materials such as aluminum, copper, plastics etc. Top machined edges are usually rough and have burrs due to shearing and tearing mode of material removal. Fiber reinforced composites have been micro-machined without delamination [Miller, 2013].

BIBLIOGRAPHY

[1] Atkey, M. (1983). "Water Jet Cutting: A Production Tool", *Mach. Prod. Eng.*, Vol. 141(3631), pp. 18–19.

[2] Benedict, G.F. (1987). "Nontraditional Manufacturing Processes", Marcel Dekker Inc., New York.

[3] De Meis, R. (1986). "Cutting Metal with Water", *Aerosp Am*, Vol. 24(3), pp. 22–23.

[4] Engemann, B.K. (1981). "Water Jet Jutting of Fibre Reinforced Composite Materials", *Ind. Prod. Engg.*, Vol. 3, pp. 162–164.

[5] Hoyt, J.W. and Toylor, J.J. (1979). "Effect of nozzle shape and polymer additives on water jet appearance", *J. fluids Engg.*, Vol. 101(3), pp. 304–308.

[6] Konig, W., Wulf, Ch., Groβ, P. and Willerscheid, H. (1985). "Machining of Fiber Reinforced Plastics", *Annals CIRP*, Vol. 34(2), pp. 537-548.

[7] Labus, T.J. (1978). "Cutting and Drilling of Composites using High Pressure Water Jets", Paper G2 of the 4th *Int. Symp. on Jet Cutting Technology*, Univ. of Kental, Canterbury, pp. G2 9–18.

[8] McGeough, J.A. (1988). "Advanced Methods of Machining", Chapman & Hall, New York.

[9] Miller, D.S. (2013). "Micro-machinig with abrasive waterjets" (MAWs), In Introduction to Micro-machining (Ed. V.K. Jain), Published by Narosa Publishing House, New Delhi.

[10] Momber, A.W. and Kovacevic, R. (1998). Principles of Abrasive Water Jet Machining, Springer-Verlag London.

[11] Pandey, P.C. and Shan, H.S. (1980). "Modern Machining Processes", Tata McGraw Hill, New Delhi.

SELF TEST QUESTIONS

Q1. Write the most appropriate answer(s).

1. Best suited cutting process under explosive environment is:
   (a) ECM (b) WJC (c) AJM (d) EDM
2. Polymers are added in water jet in WJC to:
   (a) increase MRR (b) improve surface finish
   (c) improve dimensional accuracy (d) none of these
3. WJC can be applied only for:
   (a) metals having high thermal and high electrical conductivity
   (b) non-metals (c) metals as well as non-metals
   (d) none of these
4. The source of pulsation in water flow is:
   (a) intensifier (b) type of pump
   (c) nozzle design (d) long chain polymers
5. When a thick material is to be cut by WJC, one should prefer:
   (a) very powerful jet and single pass cutting
   (b) multipass cutting (c) low stand-off-distance
   (d) all of these
6. Water used in WJM:
   (a) can be reused directly (b) can not be reused
   (c) can be reused after post process treatment

Q2. Answer in brief.

(a) Give four applications of WJC process.
(b) What do you mean by pulsation? How do you minimize pulsation in water flow?
(c) What is the basic function of a catcher? What problems do you expect in the absence of a catcher during WJC?
(d) Explain the way you will cut the insulation of a large diameter underground electric power supply cable.
(e) Write the factors that affect the performance of WJM process.
(f) "Cutting tool" never dulls or breaks in WJM. Explain.
(g) Write the unique applications of WJM process (other than those given in Q.2(a)).
(h) For which types of materials, WJMM can be successfully employed?
(i) Why WJ cut edges are usually rough and have burrs?

NOMENCLATURE

A_o	Oil piston cross-sectional area
A_w	Water piston cross-sectional area
K	Constant
P_1	Water pressure in pipe
P_2	Water pressure in orifice (or environment)
V_1	Water flow velocity in pipe
V_2	Water flow velocity in orifice/in environment
P_o	Oil pressure
P_s	Total pressure
P_w	Water pressure

GREEK SYMBOL

ρ	Density

ACRONYMS

WJ	Water Jet
WJC	Water Jet Cutting
WJM	Water Jet Machining
WJMM	Water Jet Micro-Machining
WP	Workpiece

CHAPTER 5
AT-A-GLANCE
WATER JET CUTTING (WJC)

- Utilizes erosion effects of a high velocity small diameter jet of water.
- Water jet → Cutting tool → Upper limit of velocity ≈ 900 m/s.
- Cutting by tight water jet core.
- Shroud around the core → Diverges due to interaction with air.
- Cutting and slitting → Basically porous non-metals, composites, etc.
- Wire stripping and deburring also.

WJC Machine

- Intensifier increases water pressure.
  $P(w) = \{P(O). A(O)\}/A(w)$.
- To minimize pulsation in water flow → A high pressure accumulator.

Nozzle

- Internal diameter of nozzle → 0.07 mm to 0.5 mm
- Material: Synthetic sapphire → Wear resistant and easily machinable.
- Failure → Presence of foreign particles in water.
  → Constriction by mineral deposits.
- Life → 250–500 hr.

Catcher

- Jet speed → Around sound velocity.
- To minimize noise and exposed length of the jet → A catcher is used.
- Completely breaks the jet before it reaches bottom of the tube.
- Uses a hard, replaceable insert to break the jet quickly.

Process Characteristics

- Variables
  ↓
- Pressure → Higher value → Can cut thicker materials.
- Nozzle Diameter.

- Traverse rate → Decreased value for thicker material.
- Stand-off-Distance: 3 mm–25 mm.

Process Performance

- Material Cut
  - Porous/fibrous/grannular/soft.
  - Corrugated board, rubber, plywood, glass, aluminium, asbestos etc.
- No predrilled hole is required → any direction and location → accessible for the water jet.
- Too thick part → cut in more than one pass → energy consumption/unit length is less.
- Machined surface: no burrs, no thermal damage and good surface finish.
  - Tolerance, straight cut edges and finish = ϕ (WP thickness and cutting speed)
- To cut insulation of cables → 5–10 s/cable.

Applications

- Cutting of asbestos (minimize airborne dust).
- Carbide grit safety walks.
- Fibre glass and polyethylene automotive parts.
- High speed cutting of corrugated box.

Chapter Six

ABRASIVE WATER JET MACHINING (AWJM)

INTRODUCTION

Abrasive Jet Machining (AJM) and Ultrasonic Machining (USM) are the processes which use abrasive particles for removal of material from the workpiece. In AJM, air driven abrasive jet strikes the workpiece and removes the material while in USM, abrasive particles suspended in liquid slurry strike the workpiece surface at ultrasonic frequency and cut the material at low Material Removal Rate (MRR). Recent developments have witnessed improvements in jet cutting technology by using abrasive water jets where water is used as a carrier fluid. In principle, this process is similar to abrasive jet machining except that in this case water is used as a carrier fluid in place of gas or air in AJM. These processes offer advantages of being able to cut both electrically conductive and electrically non-conductive materials, as well as difficult-to-machine materials comparatively more rapidly and efficiently than other processes. Other advantages claimed for this process can be listed as practically no dust, high cutting speed, multidirectional cutting capability, no fire hazards, no thermal or deformation stresses, high quality of machined edges, easy adaptation for remote control, recycling of abrasives, low power requirements, and almost no delamination in case of machining of composite materials.

A water jet and a stream of abrasive particles both coming from two different directions, mix up and pass through the abrasive water jet nozzle. Here, a part of the momentum of water (= (1/2) mv^2) jet is transferred to the abrasives. As a result, velocity of the abrasive particles rises rapidly. Thus, a high velocity stream of mixture of abrasives and water impinges on the Workpiece (WP) and removes material. Depending upon the type of the WP material being cut, material removal may occur due to erosion, shear, or failure under rapidly changing localized stress field.

In Abrasive Water Jet Cutting (AWJC), an erosive action of an abrasive laden water jet is employed for cutting, drilling, and cleaning of hard materials [*Benedict*, 1987]. The pressure at which water jet operates is about 400 MPa which is sufficient to produce a jet speed as high as about 900 m/s. Water-abrasive mixture jet exiting from the nozzle at such a high velocity is fully capable to cut ceramics, composites, rocks, metals etc. Removal of material from upper most position of a kerf is governed by erosive action while that at depth it is governed by deformation wear [*Hashish,* April 1989].

ABRASIVE WATER JET MACHINING MACHINE

Abrasive Water Jet Machining (AWJM) set-up is made up of four important elements, *viz.* pumping system, abrasive feed system, abrasive water jet nozzle, and catcher (Figure 6.1).

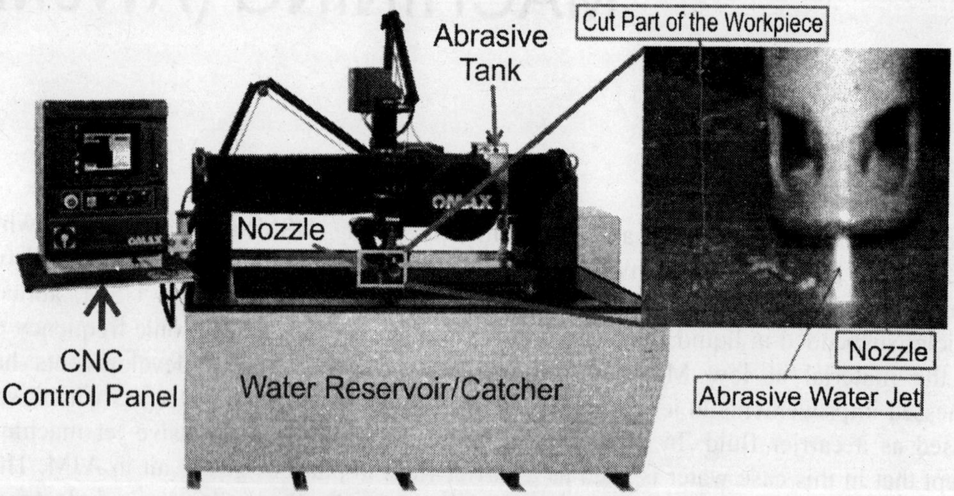

Courtesy: http://www.omax.com/abrasivejet-machining.php
http://www.omax.com/articles.php?c=3

Fig. 6.1: Abrasive water jet machining machine

Pumping System

It produces a high velocity water jet by pressurizing water to as high as 400 MPa by means of an intensifier. To acquire such a high pressure, 75 HP motor may be required. Water flow requirements up to approximately ten liters per minute are quite common. The water is recycled.

Highly pressurized water is passed through a nozzle of diameter ranging from 0.075 to 0.635 mm to obtain the desired velocity of the jet. For longer life (say, 250–500 hr) of the nozzle, it should be made of sapphire.

Abrasive Feed System

How abrasive particles are introduced to the water jet also affects to some extent, the performance of AWJC process. It must deliver a controlled flow of abrasive particles to the jet nozzle. Present abrasive feed systems deliver a stream of dry abrasives to the nozzle. Flow of water jet in a mixing tube is responsible to create enough suction for the flow of abrasives. Flow rate of abrasives can be varied by changing the diameter of the control orifice. These systems have limitation that they cannot supply the abrasives over a long distance. To overcome this drawback, researchers are developing a system in which it is possible to directly use slurry (mixture of abrasives and water) instead of mixing them in a nozzle. This System would make it possible to feed slurry over a long distance. However, it would require more power than the power required by the dry abrasive feed system.

Abrasive Water Jet Nozzle

It performs two functions: (i) mixing of water jet and abrasive, and (ii) to form a high velocity water abrasive jet. It should give a coherent, and focussed abrasive water stream at the exit from the nozzle which is made of sapphire, Tungsten Carbide (WC), or boron carbide. Internal details of abrasive water jet nozzle are shown in Figure 6.2(a). Three *kinds of abrasive water jet nozzles* (*viz*., annular jet central feed nozzle, single jet side feed nozzle and multiple jet central feed nozzle.) are shown in (Figure 6.2). In the annulus water jet central feed nozzle, the abrasive feed system is located in the center. The abrasive jet is surrounded by annulus water jet, Figure 6.2(b). Mixing of abrasives and water jet takes place outside the nozzle. In this case, the cuts are wider and machining is less accurate. In a single jet side feed nozzle (Figure 6.2(c)), abrasives fed from the side mix with water jet in the mixing chamber in the nozzle. This nozzle is less expensive, simple to make but does not provide an optimal mixing efficiency, and experiences a rapid wear at the exit part of the nozzle. The multiple jets central feed nozzle (Figure 6.2(d)) consists of a centrally located abrasive feed system surrounded by multiple water jets. In this nozzle, converging annulus of water is produced. It gives higher nozzle life and better mixing of abrasives into the *water jet*. However, it is difficult and costly to fabricate such nozzles because of the angle of convergence.

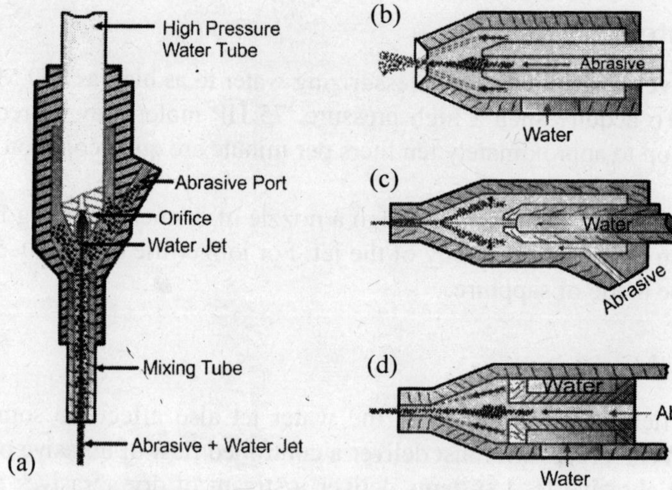

Fig. 6.2: (a) Details of construction of abrasive water jet nozzle, (b) Annulus water jet central feed nozzle, (c) Single jet side feed nozzle, (d) Multiple jets central feed nozzle [*Hashish*, 1982].

Catcher

Another element of the system is a *catcher.* There are two types of catchers used to avoid any possible accident to take place. In the '*long narrow tube type catcher*', a long narrow tube is placed under the point of cut to capture the used jet, Figure 6.3(a). In another case, when the workpiece remains stationary and the nozzle moves, a water filled settling tank is placed directly underneath the workpiece, Figure 6.3(b). The used jet dies out in the water of this tank. High pressure water from the pump is transmitted to the nozzle by a high pressure flexible hose (if pressure ≤ 24 MPa) or by a rigid tubing (if pressure > 24 MPa).

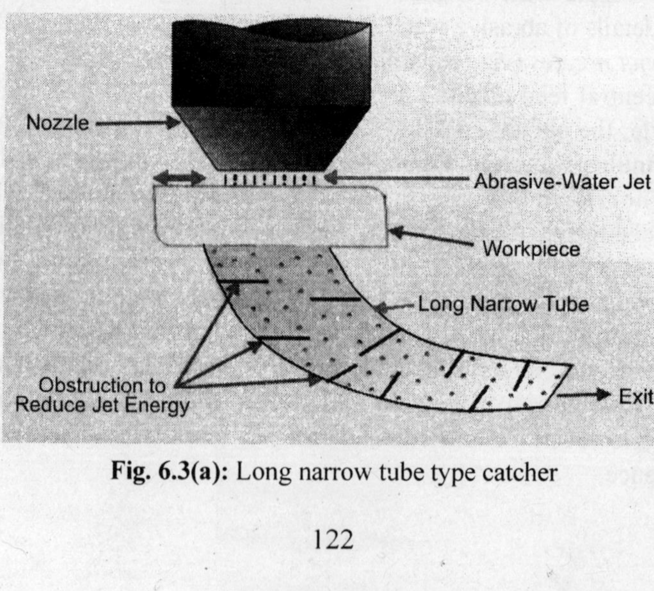

Fig. 6.3(a): Long narrow tube type catcher

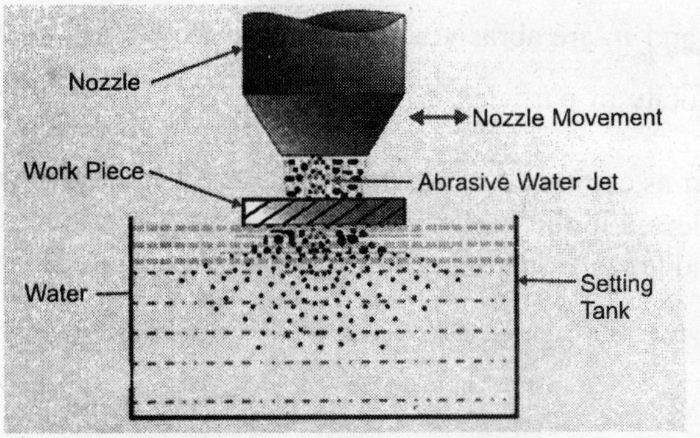

Fig. 6.3(b): Settling tank type catcher

MECHANICS OF MATERIAL REMOVAL IN AWJC

The main mechanism of material removal in AWJC is erosion caused by abrasive particles' impact. In case of brittle workpiece material, the removal of material is by brittle fracture while in case of ductile material it is by plastic deformation. Fatigue also plays its role in material removal. These mechanisms act simultaneously and which is dominating one, depends on the angle of impact of abrasive particles, particle KE, workpiece material properties, particle shape and size, and surrounding environmental conditions.

As discussed in chapter 4, the water jet theoretical velocity (V_{oth}) can be calculated by the following equation.

$$V_{oth} = \sqrt{\frac{2P}{\rho_w}} \qquad \text{... (6.1)}$$

In actual experiments, it can be modified as follow:

$$V_{o\exp} = \mu\sqrt{\frac{2P}{\rho_w}} \qquad \text{... (6.2)}$$

Where, μ is efficiency coefficient that characterizes momentum loss due to wall friction, fluid flow disturbances, and compressibility of water. $V_{0\exp}$ is actual or experimental velocity, P is water jet pressure and ρ_w is density of water.

The value of μ can be evaluated [*Momber et al.*, 1998] as follow:

$$\mu = \frac{V_{o\exp}}{V_{oth}} \qquad \text{... (6.3)}$$

In AWJM process, there is momentum transfer from the high velocity water jet to the abrasive particles, which are injected by abrasive feeder at a comparatively low velocity, and the air sucked in. The impulse balance gives the following equation,

$$\dot{m}_a V_{po} + \dot{m}_w V_o + \dot{m}_f V_f = (\dot{m}_a + \dot{m}_w + \dot{m}_f) V_p \qquad \text{... (6.4)}$$

Where, $\dot{m}_a$, $\dot{m}_w$ and $\dot{m}_f$ are abrasive, water and air flow rates, respectively. V_{po}, V_o and V_f are the flow velocity of particles, water and air, respectively. V_p is abrasive-water jet velocity at the exit.

The amount of air as compared to the mass of water and abrasive is very small (≤ 3%), hence it can be neglected in the further analysis. Secondly, it can also be assumed that the velocity of water and abrasive particles at the exit of the nozzle is same. Hence, it can be written as,

$$V_p = \eta_T \frac{V_0}{1+\frac{\dot{m}_a}{\dot{m}_w}} \quad \text{... (6.5)}$$

Where, η_T is known as momentum transfer parameter. It can be evaluated from the following equation.

$$\eta_T = \frac{(\dot{m}_w + \dot{m}_a)V_p}{\dot{m}_w V_0} \quad \text{... (6.6)}$$

From the literature survey, it has been concluded that the mechanism of AWJM has been reduced to the workpiece-particle interaction, and several theoretical models have been reported on this basis. Only two of them are given as follows:

Hashish [1987] expanded Finnie's model to calculate the volume removed (δ_v) during AWJM. Some researchers feel that it gives better results for cutting wear mode (at low impact angle).

$$\delta_v = \frac{7\text{m}}{\pi\rho_p}\left(\frac{V_p}{C_k}\right)^{2.5} \sin 2\alpha\sqrt{\sin\alpha} \quad \text{... (6.7)}$$

$$C_k = \sqrt{\frac{3\sigma R_f^{3/5}}{\rho_p}} \quad \text{... (6.8)}$$

Where, m = mass of particle, ρ_p = density of particle, V_p = velocity of particle, α = impact angle, C_k = characteristic velocity, σ = material flow stress, and R_f = particle roughness factor.

Bitter [1963] proposed a model that works well for 'deformation wear' mode at large impact angle for calculation of material removed (volumetric) due to deformation wear.

$$W_d = \frac{1}{2}m\frac{(V\sin\alpha - V_e)^2}{E_d} \quad \text{... (6.9)}$$

Where, V_e = maximum particle velocity at which the collision is still purely elastic, E_d = energy required to remove unit volume of WP material by deformation wear and V is the particle (jet) velocity with which it is hitting the workpiece surface.

PROCESS VARIABLES AND PARAMETRIC ANALYSIS

Parameters which affect performance of AWJM process are water (flow rate, and pressure), abrasives (type, shape, size, and flow rate), water nozzle and abrasive jet nozzle (design), cutting parameters (feed rate in vertical direction if any, traverse speed, and stand-off-distance), and workpiece material properties (Figure 6.4). Other cutting parameters can be listed as mixing tube diameter and its length, and angle of cutting [*Hashish*, April 1989]. Number of passes is also identified as one of the important variables that affects cutting performance during AWJC. Some of these variables and their effects on the performance of the process are discussed below.

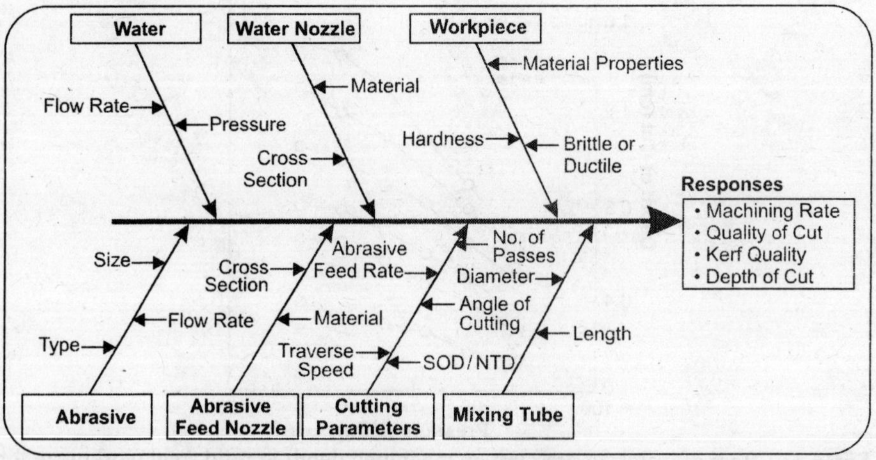

Fig. 6.4: Process variables and responses of abrasive water jet machining process

WATER

Water Jet Pressure during Cutting

Relationships between water jet pressure and depth of cut (or machined depth) for different abrasive flow rates (kg/min) are shown in Figure 6.5 [*Hashish*, 1986]. There is a minimum pressure (*i.e.* critical pressure or threshold pressure, P_c) below which no machining would take place [*Hashish*, 1983]. This *critical pressure* (P_c) exists because a minimum abrasive particle velocity (or KE) is required to cut a particular material. Each material has different mechanical properties hence it requires different pressure to be applied to cut it. P_c is therefore obviously different for different workpiece materials. The machined depth tends to stabilize, beyond a certain value of water jet pressure. The machined depth versus pressure relationship becomes steeper as the abrasive flow rate increases. An increase in pressure also increases rate of nozzle wear and cost of pump maintenance. AWJs are more suitable for cutting as compared to gas abrasive jets because earlier ones are more coherent.

Volumetric water flow rate ($\dot{Q}_w$) is proportional to the square root of pressure ($\dot{Q}_w \alpha \sqrt{P}$) and square of diameter of the nozzle ($\dot{Q}_w \,\alpha\, d_n^2$). It is concluded from the experimental study that a percentage increase in depth of cut is always lower than a percentage increase in water flow rate. Increase in water flow rate beyond a certain value may result in insignificant gain in particle velocity, higher pressure losses in supply lines, unacceptable environmental conditions (in mining or constructional applications), and in some cases reduced machined depth.

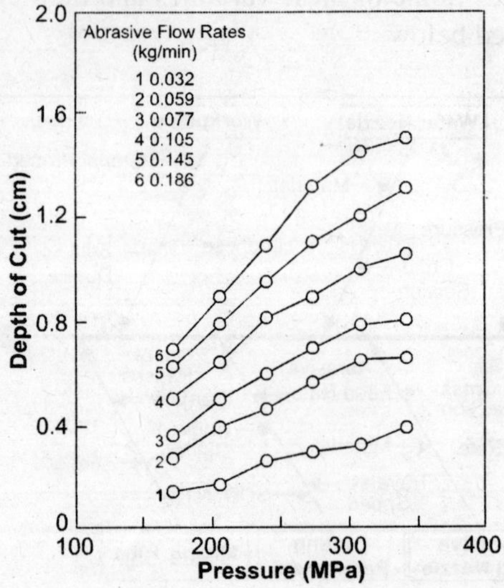

Fig. 6.5: Effect of water jet pressure on machined depth in mild steel workpiece for different abrasive flow rates (Garnet # 80, Nozzle Diameters (d_n) = 0.025 cm, traverse speed, u = 15 cm/min) [*Hashish*, 1986].

ABRASIVES

Abrasive Flow Rate

In abrasive jet machining where gas (usually air) is used as a propelling fluid, only small mass flow rates of abrasives (up to 100 g/min) can be achieved. In *AWJM,* water is used as a propelling fluid which enables high abrasive flow rates (0.1–5 kg/min) to be achieved, and makes it possible to accelerate abrasives to high velocities (over 300 m/s).

Machined depth (depth of cut) is proportional to the abrasive flow rate ($\dot{m}_a$) and square of particle velocity (V_p). However, an increase in abrasive flow rate beyond the critical value ($\dot{m}_{ac}$) would reduce machined depth (Figure 6.6). Increase in abrasive flow rate enhances wear rate of mixing nozzle and reduces mixing efficiency inside AWJ nozzle. Relationship between abrasive flow rate and depth of cut for various work materials is shown in Figure 6.6. Note that brittle and harder C.I. as workpiece material has got highest machined depth as compared to other two materials (Figure 6.6).

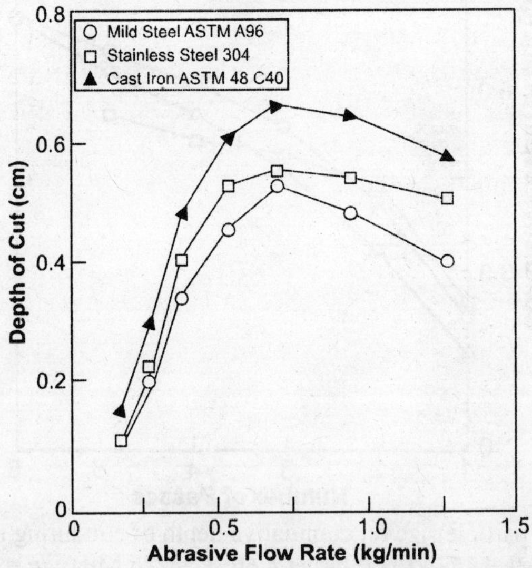

Fig. 6.6: Relation between abrasive flow rate and depth of cut for various workpiece materials, using garnet **# 60,** as abrasive material; traverse speed, $u = 15$ cm/min; nozzle diameter, $d_n = 0.025$ cm, $P = 207$ MPa. [*Hashish,* 1986]

Abrasive Particle Size

Commonly used abrasive particle size ranges from 100–150 grit. There is an optimum particle size for a particular feature to be created on the workpiece material and also for a particular nozzle mixing chamber configuration. Recently, efforts have been made to quantify the effect of abrasive particle size on depth of cut. Results have shown that an optimum particle size range exists for cutting different types of materials. Mesh size 60 is more effective as compared to other two sizes for relatively shallow depth of cut obtained with two passes (Figure 6.7) during machining of stainless steel [*Hashish,* 1986]. Approximate diameter (d_n in mm) of an abrasive particle of given mesh size (s) can be obtained from [*Malkin,* 1989] the following equation.

$$d_g(mm) = \frac{(0.6) \times 25.4}{s} \quad \text{... (6.10)}$$

Ignoring practical difficulties, the strategy recommended [*Hashish,* 1986] for deep cutting is to use different abrasive sizes with different number of passes to maximize the effect of abrasive particle size on the machined depth. However, the effect of abrasive particle size on nozzle wear should also be considered as one of the important selection criterion.

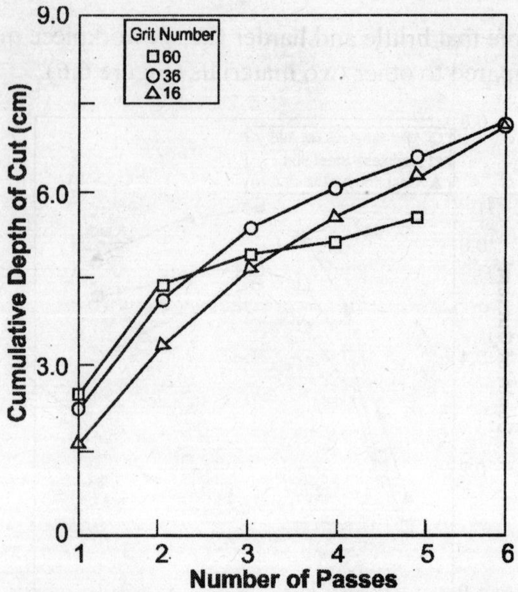

Fig. 6.7: Effect of particle size on cumulative depth of cut during multipass cutting of stainless steel 17–4 PH (Garnet # **60**; P = 276 MPa; dn = 0.071 cm; u = 20 cm/min; $\dot{m}_a$ = 1.8 kg/min) [*Hashish*, 1986].

Abrasive Material

Garnet, silica and silicon carbide are commonly used abrasives in AWJC. Type of abrasive to be used is determined after knowing *hardness of the workpiece* material. Higher the hardness of the workpiece material, harder should be the abrasive to be used. Complete recycling of the abrasives is not possible.

Machined depth is also affected by the type of the abrasives used (Figure 6.8). While selecting a type of abrasive for a particular application, one should consider cost, nozzle wear rate, environmental constraints, machining rate and strength of the particles.

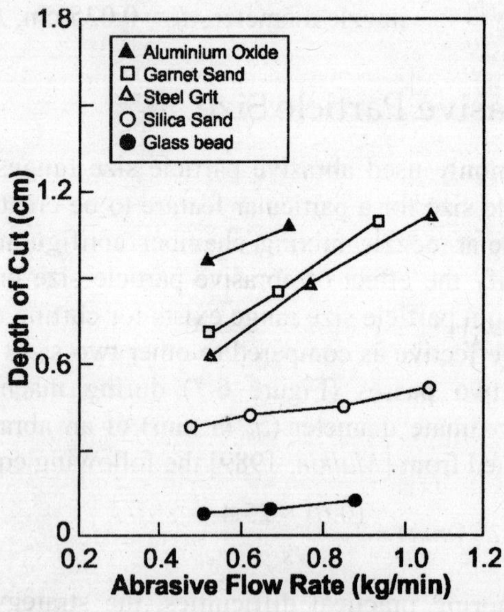

Fig. 6.8: Variation in depth of cut with a change in abrasive flow rate for various types of abrasive materials while using tool steel A2 as work material (P = 207 MPa; d_n = 0.051 cm; u = 25 cm/min) [*Hashish*, 1986].

CUTTING PARAMETERS

Traverse Speed

A decrease in traverse rate (U) increases the depth of cut as shown in Figure 6.9(a). However, if one goes below a minimum critical traverse rate then there would be hardly any increase in the depth of cut. Figure 6.9(b) shows an optimum traverse rate that exists for generation of a maximum kerf area (Traverse rate × depth of cut). However, in case of mild steel, the rate of kerf area generation decreases with an increase in traverse speed beyond the maximum value.

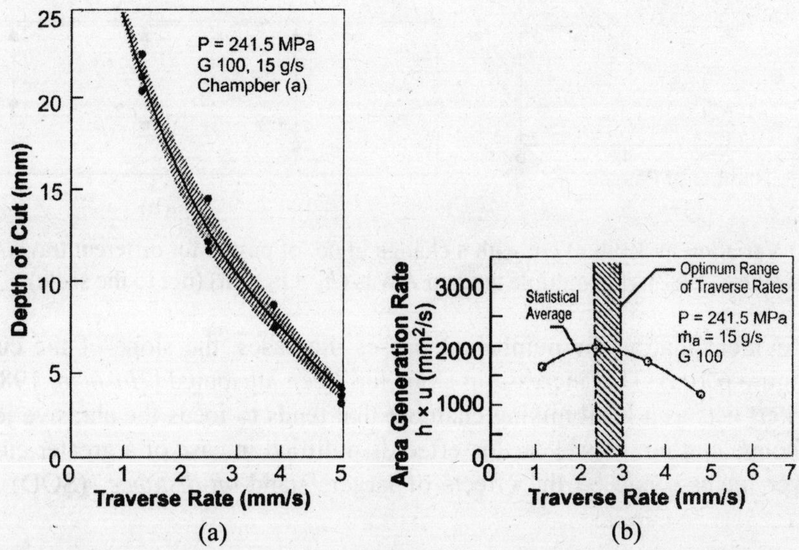

Fig. 6.9: Effect of traverse rate on (a) Depth of cut, (b) Area generation rate [*Hashish*, 1982]

Number of Passes

Multiple passes can be employed in the following two ways:

(i) A single abrasive water jet with multiple passes, and
(ii) Multiple tandem AWJs with a single traverse (pass).

In case of a single abrasive water jet, whole power and quantity of abrasive are used by a single jet while in the second case both the available power and the abrasives are divided among multiple jets. The distribution of power and abrasives among them are the considerations for overall optimization.

Multiple passes cutting requires special catching and containment precautions because the initial pass cutting results in the jet back flow and rebounding. Figure 6.10(a) shows a relationship between number of passes and cumulative depth of cut. Figure 6.10(b) shows a schematic arrangement of multiple tandem abrasive water jets. In this case, first jet is able to machine upto the depth of h_1, second jet increases its machined depth upto h_2 while the last that is third jet is able to machine upto the depth h_3 ($h_3 > h_2 > h_1$).

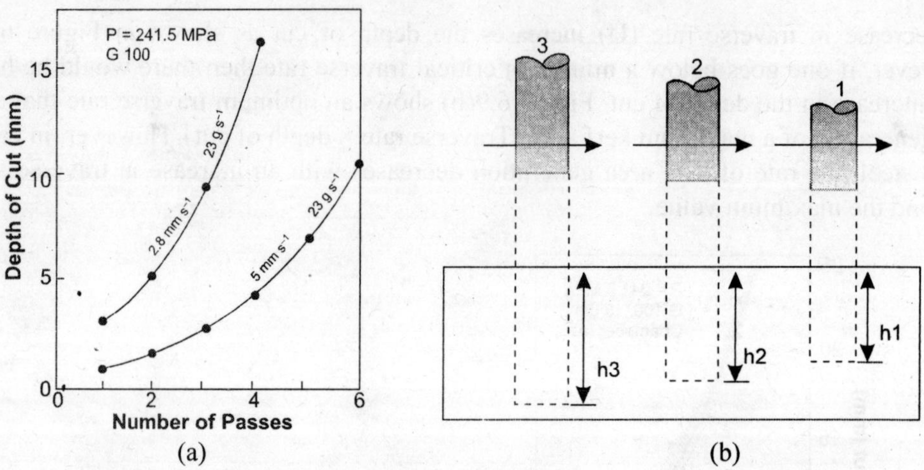

Fig. 6.10: (a) Variation in depth of cut with a change in no. of passes for different traverse speeds [*Hashish*, 1982], (b) Multiple tandem AWJs ($h_3 > h_2 > h_1$) (not to the scale).

It is also evident that as the number of passes increases, the slope of the curve also increases (Figure 6.10a). This increase in slope has been attributed [*Hashish,* 1982] to the fact that the kerf acts as a local mixing chamber that tends to focus the abrasive jet stream for more effective cutting. However, this effect is nullified in case of a greater cumulative depth at lower traverse rate by the effects of larger 'stand-*off-distance*' (SOD) and kerf friction drag.

Stand-Off-Distance (SOD)

An increase in stand-off-distance rapidly decreases machined depth (Figure 6.11). This has been explained [*Hashish*, 1982] by arguing that the liquid phase of the jet breaks up into droplets resulting in free abrasive particles. These free abrasive particles rebound upon impact that lead to a shallower penetration. There is an upper value of SOD beyond which the process will no longer do any cutting. However, in case of multi pass cutting the kerf made in the earlier passes helps the jet to remain intact. It is concluded that smaller is the stand-off-distance deeper is the cut.

Hashish [1989] has developed a model to predict the depth of machining. It has been reported that the correlation coefficient between the predicted and experimental values for different materials is over 0.9. Two properties *viz.*, flow stress of workpiece material and critical velocity of abrasive water jet are found to be more effective parameters (note that the flow stress has strong dependence on the modulus of elasticity) than others.

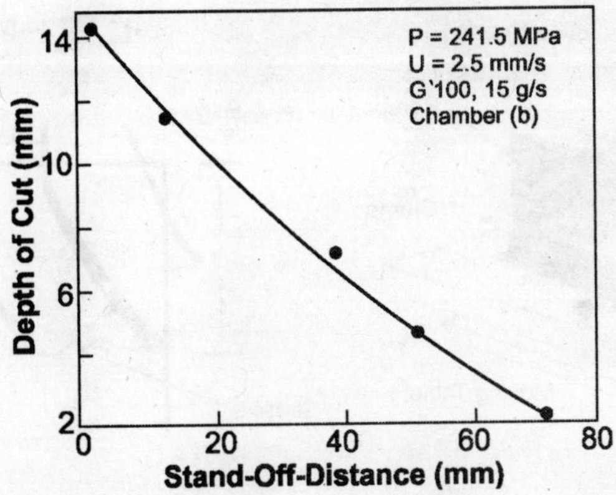

Fig. 6.11: Effect of stand-off-distance on depth of Cut [*Hashish*, 1982]

Visual Examination

Visual examination of the process was conducted on lexan, lucite and glass as work materials with the help of a movie camera. Based on the visual examination data, the mechanism of material removal has been attributed to two modes: cutting wear mode that occurs at shallow angles of impact, and deformation wear mode which occurs at lower part of the cut at larger angles of impact (Figures 6.12(a) and (b)). Nature of actual penetration of abrasive water jet in solids has been reported by *Hashish* [*June* 1988].

The quality of the machined feature also depends on whether the abrasive water jet is stationary or moving. For drilling operation, the jet is stationary and the erosion front is along the axis of the jet hence the cutting is 1-D only. The wall of the drilled hole is tapered and the diameter is non-uniform due to the nature of the distribution of energy and its interaction with workpiece material. To improve the quality of the generated feature, on-line control of dynamic parameters (say, jet pressure, abrasive flow rate etc.) and kinematic parameters (say, impact angle, stand-off-distance, etc.) is essential [*Hashish*, 1984].

For the operations such as cutting, milling etc., relative motion between the Abrasive Water Jet (AWJ) and workpiece is essential to cover the designed surface. In these operations, the erosion front moves along the axis of the AWJ and direction of movement of jet, Figure 6.12(b). The machined surface is characterized by the geometry of the Kerf and topography of the generated surface. As the AWJ keeps penetrating into the workpiece, the abrasive particle energy keeps on decreasing. As a result, the erosion rate perpendicular to the cutting front (direction-Y) decreases hence the leading edge of the AWJ bends in opposite direction causing a jet lag [Figure 6.12(b)]. A strategy of oscillating the AWJ has been proposed to overcome this problem.

Figure 6.12(c) shows a relationship between penetration rate and depth of cut as a function of machining time. It is found that a number of steps are formed below the steady state interface at the top. The step formation takes place as the jet traverses over the work-piece. Size of the step increases with increase in the machined depth.

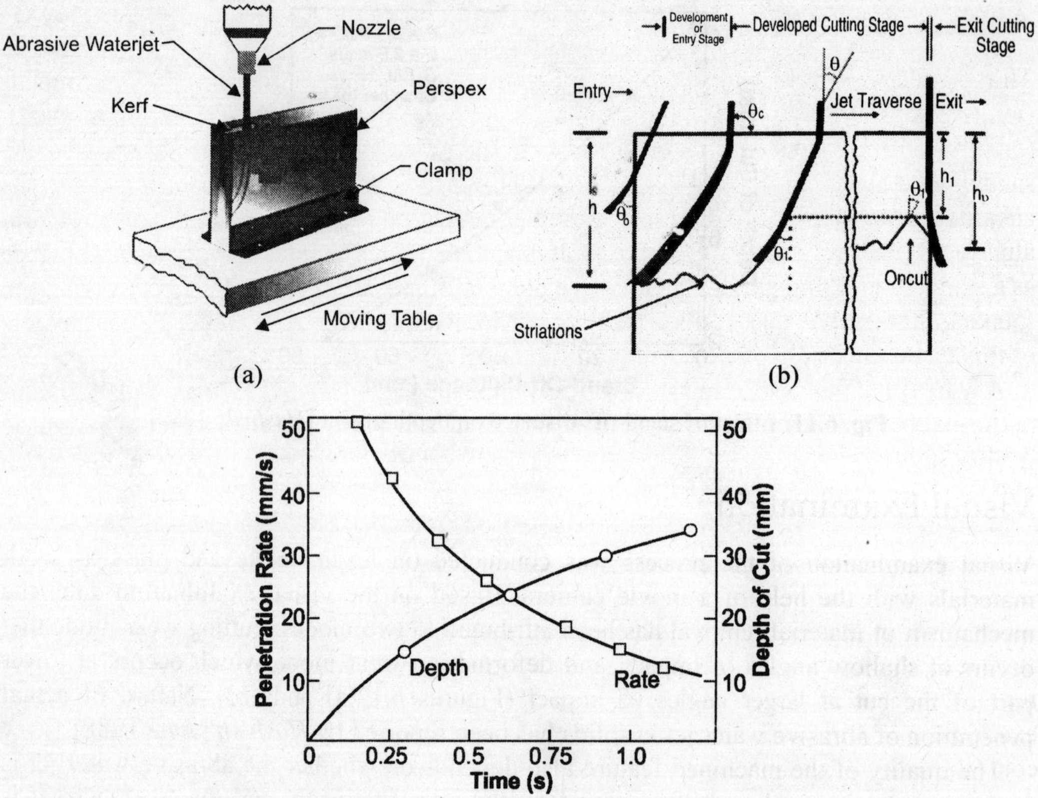

Fig. 6.12: (a) Penetration of abrasive water jet in solids, (b) Movement of erosion front along the axis of abrasive water jet and direction of movement of jet, (c) Variation in depth of cut and penetration rate with machining time [*Hashish*, 1989].

PROCESS CAPABILITIES

AWJM process has been used to cut even thick materials (200 mm) with a narrow kerf. Width of the kerf depends upon the workpiece hardness. For hard materials, the kerf narrows down towards the bottom while the reverse is true in case of a soft material. Machined surfaces have revealed [*Benedict*, 1987] no presence of any thermal or mechanical damage to them. In case of machining of glass, stray cutting may result in frosting.

Here, CNC system plays an important role in accurate cutting of different features on the workpiece. X, Y-axes motion of the abrasive water jet is computer controlled. Most systems have a motorized Z-axis that controls the movement of the cutting head to vary stand-off-distance. The water tank below the workpiece mounting table also serves as a catcher tank to collect the spent out abrasive, water and eroded workpiece material.

APPLICATIONS

This process has been employed to cut a wide range of materials including both metals (*viz*., copper and its alloys, lead, tungsten carbide, aluminum, etc.), and non-metals (graphite, silica, glass, acrylic, concrete, etc.). The process has been applied to machine the sandwiched honeycomb structural materials currently used in the aerospace industries. Its advantages are like omnidirectional cutting having no burrs. The edges of structural aluminum plate have also been successfully machined. This technique is getting acceptance as a standard tool for cutting materials in a number of *industries* like aerospace, nuclear, oil, foundry, automotive, construction and glass. The specific advantages claimed by promoters of this technique are economic and environmental.

AWJ cutting has been employed for decommissioning *nuclear facilities.* The optimum performance has been reported with 0.38 mm jet at 200 MPa with an abrasive flow rate of 0.54 kg/min. In some cases, this technique has proved to be economical as compared to the conventional bulk material removal methods.

Slotting is one of the common applications of AWJC. Various steels (stainless steel, mild steel, special alloy steel, etc.) have been cut into different shapes like plate, tube, corrugated structure etc. Kerf (slot) widths observed are in the range of 0.75–2.25 mm depending on the nozzle inner diameter and cutting parameters. Roughness of the cut surfaces varies with the machined depth and abrasive water jet cutting parameters. It is also reported [*Hashish*, Oct.–Sept. 1984] that no embedded abrasives were seen while examining the machined surfaces under Scanning Electron Microscope (SEM). Being a cold machining process, thermal stresses are not witnessed in the machined components. Aluminium also has been cut successfully.

The mechanism of material removal and the machined feature characteristics are governed by the relative motion between the abrasive water jet and workpiece material. In case of piercing and drilling operations, the abrasive water jet and workpiece both are stationary. Figure 6.13 shows a male and a female part (*4i LABORATORY*) cut from a metallic sheet.

Fig. 6.13: '*4i Laboratory*' (Male and Female parts) cut from a metallic sheet (*Courtesy:* 4i Lab. at IIT Kanpur)

ABRASIVE WATER JET MICROMACHINING (AWJMM)

Lately, it has been possible to apply Abrasive Water Jets (AWJs) for micromachining because the problems related to the feeding of abrasive particles to the cutting head have been largely solved to allow cutting down to 200 μm [*Miller*, 2014]. In AWJMM as in AWJM, one does not have Heat Affected Zone (HAZ), microcracks, and burrs on the machined surface. By this process, any kind of material can be micromachined. It has been possible to produce cutting jet diameter as small as 40 μm which means width/diam. of the cut will be slightly larger than 40 μm.

Millions of Abrasive Particles (APs) impact on the workpiece surface as isolated bodies because APs constitute less than 1% of AWJ volume. By volume, water droplets make up ≈10% and the remaining are air and water vapour [*Miller*, 2014]. Due to plastic deformation or brittle fracture, heat is generated during cutting process but the rate of heat removal is so high that it does not lead to any harmful effects (HAZ, etc.) on the machined workpiece.

Abrasive water jet can be produced in one of the two ways. Pressurized water is passed through a nozzle to produce high speed water jet and abrasive particles through a port entrain into jet. In the second method, abrasive particles are suspended in the pressurized water to form slurry which is passed through a cutting nozzle. One of these two ways is used in AWJMM as well as in AWJM.

Abrasive water jet micromachining center has been developed as reported by *Miller* [2014]. Linear motors for the X-axis and Y-axis, and a high precision ball screw for the Z-axis are used. A rigid machine frame is used in AWJMM center to which the motion system and workpiece support are attached. The catcher tank is isolated from the m/c structure to prevent disturbance from the jet energy dissipation which is otherwise transmitted to the m/c structure.

A wire EDM and wire ECM job shop may use a AWJMM center for drilling starter holes for the wire and then transfer the workpiece to a wire EDM or wire ECM m/c for high precision 3-D cuts. The CAD/CAM software for macro AWJM system can also be used for AWJMM center. The software allows machining instruction in G and M codes. Striations can be minimized by cutting at a comparatively slower speed.

BIBLIOGRAPHY

[1] Benedict, G.F. (1987). "Nontraditional Manufacturing Processes", *Marcel Dekker Inc.,* New York.

[2] Bitter, J.G.A. (1963). "A study of erosion phenomenon Part –I", Wear, Vol. 6, pp. 5–21.

[3] Bitter, J.G.A. (1963). "A study of erosion phenomenon Part –II", Wear, Vol. 6, pp. 169–190.

[4] Hashish, M. (1982). "The Application of Abrasive Jets to Steel Cutting", *Proc. 6th Int. Symp. on Jet Cutting Technol.,* Guildford, England, pp. 447–464.

[5] Hashish, M. (1982). "Steel Cutting with Abrasive Water Jets", *Proc. 6th Int. Symp. on Jet Cutting Technol.,* Guildford England, pp. 465–487.

[6] Hashish, M. (1983). "Experimental Studies of Cutting with Abrasive Water Jets", *Second U.S. Water Jet Symposium*, Rolla MO.

[7] Hashish, M. (1984). "Cutting with Abrasive Water Jets", *Carbide Tool J.*, pp. 16-23.

[8] Hashish, M. (Sept., 1986). "Aspects of Abrasive Water Jet Performance Optimization", *Proc. 8th Int. Symp. on Jet Cutting Technol.,* Durham (U.K.), pp. 297–308.

[9] Hashish, M. (June, 1988). "Visualization of the Abrasive Water Jet Cutting Process", *Exptl Mech.,* pp. 159–170.

[10] Hashish, M. (April, 1989). "A Model for Abrasive Water Jet (AWJ) Machining", *Trans. ASME, J. Engg. Mater and Technol.*, Vol. 111, pp. 154–162.

[11] Hashish, M. Halter, M. and McDonald, M. (May, 1985). "Abrasive Water Jet Deep Kerfing in Concrete for Nuclear Facility Decommissioning", *Proc. 3rd U.S. Water Jet Conf., Held at Univ. of Pittsburgh (Ed.Dr. N. Styler)*, pp. 123–143.

[12] Malkin, S. (1989). "Grinding Technology: Theory and Applications of Machining with Abrasives", Ellis Horwood, U.K.

[13] Miller, D.S. (2014). "Micromachinig with Abrasive Waterjets" (MAWs), In Introduction to Micromachining (Ed. V.K. Jain). Publised by Narosa Publishing House, New Delhi.

[14] Momber, A.W. and Kovacevic, R. (1998). Principles of Abrasive Water Jet Machining, Springer-Verlag London.

[15] Sitaram Chakravarthy, P. and Ramesh Babu, N. (1999). "A new approach for selection of process parameters in abrasive water jet cutting", *Materials and Manufacturing Processes,* Vol. 14(4), pp. 581–600.

SELF TEST QUESTIONS

Q1. Write True (T) or False (F).

(a) AWJM can't be applied to cut composite materials.

(b) Physical and mechanical properties of workpiece material do not affect performance of AWJM process.

(c) MRR obtained during AJM is higher than AWJM.

(d) Life of a nozzle made of sapphire is as high as 400 hr.

(e) It is difficult to fabricate multiple jets central feed nozzle as compared to single jet side feed nozzle.

Q2. Choose correct answer from the given choices.

(i) In AWJM process, an intensifier is used to:

(a) increase the pressure of water (b) increase the flow rate of abrasives

(c) to increase the velocity of water jet (d) none of these

(ii) It is a finishing process:

(a) AWJM (b) AJM (c) USM (d) AFM

Q3. Write the names of various elements of AWJC machine and explain them, in brief.

Q4. Why a need is being felt to develop a system which can directly use slurry instead of mixing abrasives and water in the mixing tube of the nozzle?

Q5. Answer the following questions, in brief.

(a) Some parts in an aerospace industry are made of asbestos. Because of high level of asbestos in the air (harmful to the health of workers) at cutting station, Osha shuts down Lockheed Georgia Company's mechanical cutting operation section.
You as incharge of the section are asked to suggest two of the suitable advanced machining techniques so that the work can be resumed in this section.

(b) Composite plates are to be cut into smaller pieces on large scale. Reasonably good finish is required. Which process(es) will you recommend?

(c) 5 mm diameter hole is to be drilled in a concrete block for a machine tool foundation. Which advanced machining process(es) would you recommend?

Q6. Write the differences (in table form) between WJM, AJM, USM, and AWJM processes (working principles, applications, limitations, and merits of the processes).

NOMENCLATURE

C_k	Characteristic velocity
d_g	Grit size
d_n	Nozzle diameter
E_d	Energy required to remove unit volume of WP material
m	Mass of particle
$\dot{m}_a$	Abrasive flow rate
$\dot{m}_w$	Water flow rate
$\dot{m}_f$	Air flow rate
P	Water jet pressure
P_c	Critical water jet pressure
$\dot{Q}_w$	Volumetric water flow rate
R_f	Particle roughness factor
s	Mesh size
u	Traverse speed
V_{ap}	Abrasive particle velocity
V_e	Maximum particle velocity
V_f	Flow velocity of air
V_{po}	Flow velocity of particles
V_o	Flow velocity of water
V_{oth}	Theoretical water jet velocity
V_{oexp}	Experimental water jet velocity
W_d	Material removed due to deformation wear

GREEK SYMBOLS

α	Impact angle
σ	Material flow stress
δ_v	Volume of workpiece material removed
μ	Efficiency coefficient
ρ_P	Density of abrasive particle
ρ_w	Workpiece density
η_T	Momentum transfer parameter

ACRONYMS

APs	Abrasive Particles
AJM	Abrasive Jet Machining
AWJC	Abrasive Water Jet Cutting
AWJM	Abrasive Water Jet Machining
AWJMM	Abrasive Water Jet Micromachining
HAZ	Heat Affect Zone
KE	Kinetic Energy
MRR	Material Removal Rate
SOD	Stand-off-Distance
USM	Ultrasonic Machining
WP	Workpiece

CHAPTER 6
AT-A-GLANCE
ABRASIVE WATER JET MACHINING (AWJM)

Introduction

- Jet of mixture of water & abrasive particles removes material.
- Pressure → 400 MPa. Upper limit of velocity → 900 m/s.
- Cutting in upper part of Kerf: Erosive action.
- Cutting in lower part of Kerf: Deformation.
- Used to machine:
  * Non-Metals : Ceramics, composites, rocks, etc.
  * Metals : Copper, aluminium, WC, lead, etc.
- Operations: Drilling, cutting, deburring, etc.
- Can cut any kind of material, good edge quality, adaptable for remote control, recycling of abrasives, etc.

Elements of AWJM System

- Pumping system.
- Abrasive feed system.
- Abrasive water jet nozzle.
- Catcher.

Pumping System

- Intensifier → 415 MPa, 75 HP motor
- High velocity jet.

Abrasive Feed System

- Delivers dry abrasives
- To control flow rate → Change control orifice diameter.
- Can't supply abrasives long distances → Use directly slurry → more power required.
- Water jet nozzle diam. → 75 to 635 μm.
- For high life of nozzle → Sapphire.

Abrasive Water Jet Nozzle

Functions
- Mixing of abrasives & water.
- Forming high velocity jet.

Materials
- WC, Boron Carbide, Sapphire.

Types

Single jet side feed nozzle:
- Simple to make.
- Rapid wear of exit part of nozzle.
- Non-optimal mixing efficiency.

Multiple jet central feed nozzle:
- Centrally located abrasive feed system.
- Surrounded by multiple water jets → converging annulus.
- Higher nozzle life & better mixing.
- Difficult & costly to fabricate.

Annulus water jet central feed nozzle:
- Abrasive feed system → in center.
- Abrasive jet surrounded by water jet.
- Cuts → wider & less accurate.

Catcher

*Stationary nozzle & moving workpiece.
- Long narrow tube placed under the point of cut.

*Moving nozzle & stationary WP.
- A water filled settling tank underneath the WP.
- High pressure water transmitted →
  - Flexible hose (press. < 24 MPa).
  - Rigid tubing (press. > 24 MPa).

Process Parameters

- Water → Flow rate and pressure.
- Abrasives → Type, size, and flow rate.
- Water nozzle and abrasive jet nozzle.
- Cutting parameters → Feed rate and stand-off-distance.
- WP material.
- Mixing tube → diameter & length.
- Angle of cutting.
- Traverse speed.
- Number of passes.

Water Jet Pressure and Water Flow Rate

- Critical pressure (P_c) → Below this pressure no cutting.
  → Different for different WP materials.
- Above a definite jet pressure → Machined depth tends to stabilize.
- Relationship between jet pressure and machined depth → Steeper with higher abrasive flow rate.
- Increased pressure → Higher nozzle wear rate
  → Higher cost of pump maintenance
  → Lower efficiency
- AWJs → Coherent hence more suitable for cutting.
- $\dot{Q}_w \, \alpha \sqrt{P}$
- $\dot{Q}_w \, \alpha d_n^2$

Abrasive Flow Rate

- Water → Propelling fluid enables high abrasive flow rate (up to 5 kg/min)
- Abrasive velocity → up to 300 m/s
- Machined depth $\alpha + \left(V_p^2,\ \dot{m}_a\right)$
- If abrasive flow rate above $\dot{m}_{ac}$ → Reduced depth
- Increase in $\dot{m}_a$ → ↑ Wear of mixing nozzle
  ↓ Mixing efficiency

Traverse Speed

- Over cut decreases with an increase in traverse speed.
- Traverse speed *vs* area generation rate has an optimum.

Abrasive Particle Size and Material

- Optimum particle size.
- Finer particles → For shallow depth of cut.
- Coarse particles → For large depth of cut.
- Different abrasibe sizes for different depths of cut.

Multiple Passes → Single water jet with multiple passes.
→ Multiple tandem jets with single pass.

* Increase in number of passes → ↑ Machined depth.

* Kerf acts as a local mixing chamber.

Stand-off-Distance

- ↑ Stand-off-distance → ↓ machined depth.

  ↓

  Rebound of free abrasive particles.
- Beyond upper value of SOD → No cutting.
- Smaller SOD → Deeper cut.

Visual Examination

- Visual examination using movie camera.

  ↓

  Two modes of material removal:
  - (i) Cutting mode → Shallow angle of impact.
  - (ii) Deformation mode → Larger angle of impacts.
- Penetration rate and depth of cut → Function of time.

Process Performance

- Can cut thick materials.
- Kerf width decreases as WP hardness increases.
- Machined surfaces → no thermal/mechanical damage.
- Machining of glass → stray cutting leads frosting of surface.

Applications

- Can machine both → Metals & non-metals.
- Omni-directional cutting with no burrs.
- Industries → aerospace, nuclear, foundries, construction, etc.
- Steel components cut → plates, tubes, corrugated structures, etc.

Micromachining

- In AWJMM → No HAZ, no μ-burrs, no μ-cracks.
- AWJ Diameter → ≈ 40 μm.
- Mechanism → Plastic deformation or brittle fracture as in AWJC.
- Abrasive water jet production → Two ways.
- CNC machines.
- CAD/CAM software for AWJMM.

PART–II

Thermoelectric Advanced Machining Processes

- Electric Discharge Machining (EDM)
- Wire Electric Discharge Machining (Wire EDM)
- Electric Discharge Grinding (EDG)
- Laser Beam Machining (LBM)
- Plasma Arc Machining (PAM)
- Electron Beam Machining (EBM)

Chapter Seven-A

ELECTRIC DISCHARGE MACHINING (EDM)

INTRODUCTION

Whenever sparking takes place between two electrical contacts usually a small amount of material is removed from each of the contacts. This fact was realized and the attempts were made to harness and control the *spark energy* to employ it for some useful purpose, *i.e.* for machining of metals. Lazeranko brothers (in USSR) found that female and male parts of an electric switch had craters formed on it (Figure 7.1). They asked the question to them self. Can we make constructive use of this destructive activity? Both brothers studied the phenomenon in detail. They attempted more than 40 times to develop the appropriate machine but failed. Finally, in around 43rd attempt, they developed a system and then patented a machine and named as 'Electro Spark Machining' machine (later on, it was named as "*Electric Discharge Machining*" (*EDM*) machine). It was found that the sparks of short duration and high frequency are needed for efficient machining. Further, it was also observed that if the discharge is submerged in liquid dielectric, the energy can be concentrated into a small area. *A relaxation circuit* was proposed in which electrodes (tool and workpiece) were immersed in the dielectric like kerosene, and they were connected to the capacitor.

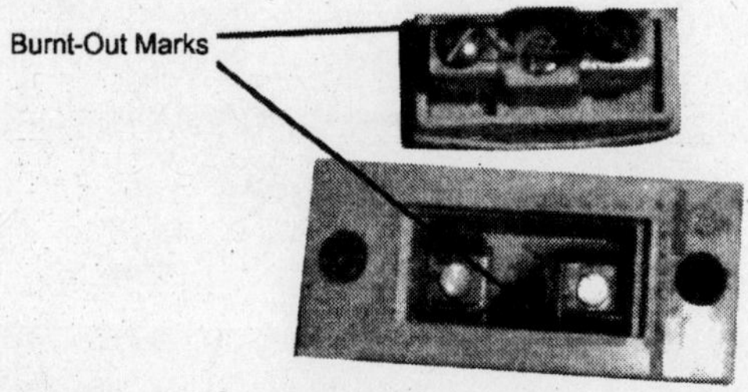

Fig. 7.1: Black Deposits/Burnt-Out Marks on Male and Female Parts of an Electric Switch

The capacitor is charged from a Direct Current (DC) source. Figure 7.2(a) shows RC circuit and Figure 7.2(b) shows voltage (V_{ct}) *vs* time, and current (i_{ct}) *vs* time relationships for the capacitor. As soon as the potential across the electrodes touches the breakdown voltage (V_b), the *sparking* takes place at a point of least electrical resistance. It usually occurs at the smallest Inter-Electrode Gap (IEG) (at surface roughness level of tool and workpiece). As shown in Figure 7.2(c), it would first occur at point *A* and then may be at point *B*. After each discharge, capacitor recharges and spark appears at the next narrowest gap. Occurance of each spark generates heat energy which is shared in different modes by workpiece, tool, dielectric, debris and other means of heat loss, as shown in Figure 7.3.

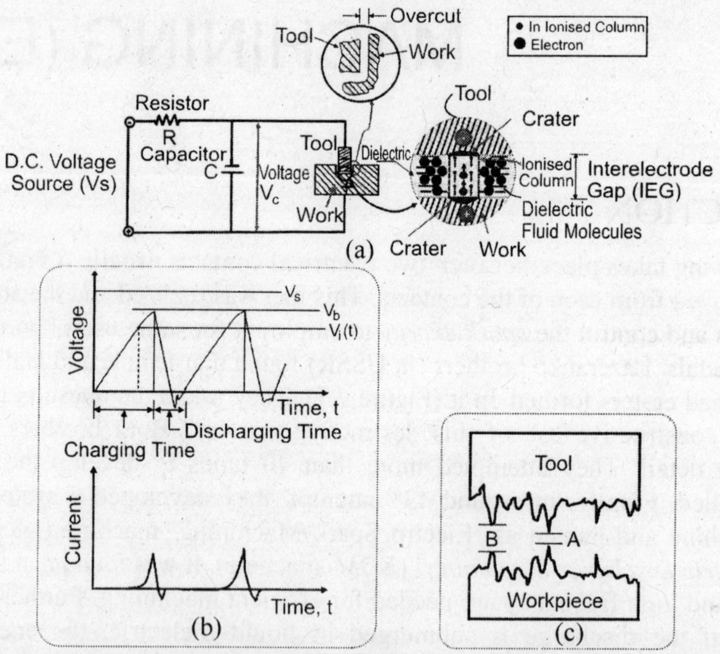

Fig. 7.2: (a) A schematic diagram of "Electric Discharge Machining" using relaxation circuit, (b) Voltage *vs.* time relationship and Current *vs.* time relationship in EDM using relaxation circuit, (c) A part of tool and workpiece surface roughness peaks to show where first spark (point A in this figure) may occur.

However, some researchers have theoretically proved that it is always not necessary that only one spark will be generated in each pulse. More than one spark also may be generated if they have certain minimum potential gradient.

The dielectric serves some important *functions, viz*., cools down the tool and workpiece, cleans (or flushes away) the IEG, and localizes the spark energy into a small cross sectional area. Energy content in each spark and frequency of sparking are also governed by the conditions in the IEG.

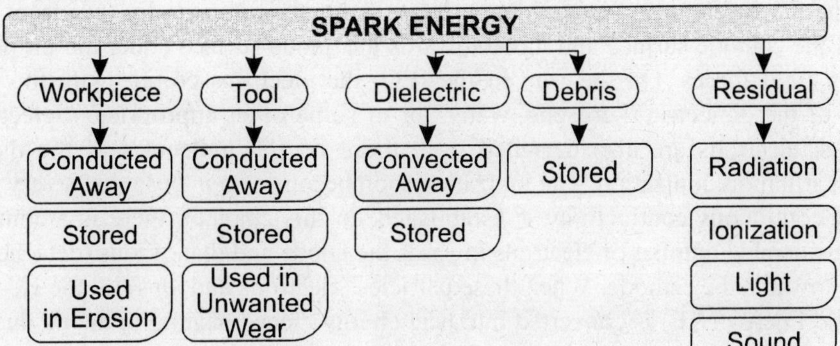

Fig. 7.3: Distribution of spark energy produced during EDM

WORKING PRINCIPLE OF EDM

EDM is a *thermoelectric process* in which heat energy of a spark is used to remove material from the workpiece. The workpiece and tool should be made of electrically conductive materials. A spark is produced between the two electrodes (tool and workpiece) and its location is determined by the narrowest gap between the two, Figure 7.2(c). Duration of each spark is very short. The entire cycle time is usually very small (100–200 μs). The frequency of sparking may be as high as many thousands of sparks per second. The area over which a spark is effective (or spark diameter) is also very small (usually, in the range of 100–200 μm). However, temperature of the area under the spark is very high. As a result, the spark energy is capable of partly melting and partly vaporizing material from the localized area on both the electrodes, *i.e.*, workpiece and tool. The material is removed in the form of *craters* which spread over the entire surface of the workpiece under the tool. Finally, the cavity produced in the workpiece is approximately the replica of the tool. To have the machined cavity as replica of the tool, theoretically the tool wear should be zero. To minimize wear of the tool, the operating parameters, tool material and polarity should be selected carefully.

Thermally eroded material from the electrodes is known as *debris*. Analysis of the debris has revealed that it is the mixture of irregular shaped particles (resulting from resolidification of the material removed in the molten state) as well as hollow spherical particles (resulting from the vapor condensation of the material removed in the form of vapor). Usually, the amount of material eroded from the tool surface is much smaller than that from the workpiece surface. A very small gap (even lesser than hundredth of a millimeter) between the two electrodes is to be maintained to have a spark to occur. For this purpose, a tool driven by the servo-system* is continuously moved towards or away from the work-piece.

During EDM, pulsed DC of 80–100 V and approximately 5 kHz is passed through the electrodes. It results in the intense electrical field at the location where surface irregularity provides the narrowest gap (Figure 7.2(c)). Negatively charged particles (electrons) break loose from the cathode surface and move towards the anode surface under the influence of the electric field forces. On the way in the IEG, the electrons collide with the neutral molecules of the dielectric (kerosene, water, air or some other appropriate dielectric). In this process, electrons are also detached from these neutral molecules of the dielectric resulting in still more ionization. The ionization soon becomes so intense that a very narrow channel of continuous conductivity is established. In this channel, there is a continuous flow of considerable number of electrons towards the anode and that of ions (detached from the anode) towards the cathode. When these particles (electrons and ions) hit the electrodes, their Kinetic Energy (KE) is converted into heat energy, hence heating of anode due to the bombardment of electrons and heating of cathode due to the bombardment of ions, take place. Thus, it ends up in a momentary current impulse resulting in the discharge which may be an arc or a spark. It is always desirable to have a spark rather than an arc during EDM, because the later damages both the electrodes (anode and cathode). The discharge/spark energy raises the localised temperature of the tool and workpiece to such a high value that it results either in melting, or melting as well as vaporization of a small amount of material from the surface of both electrodes at the point of spark contact. Figure 7.4(a) shows initiation of ionization process and Figure 7.4(b) shows fully developed ionized

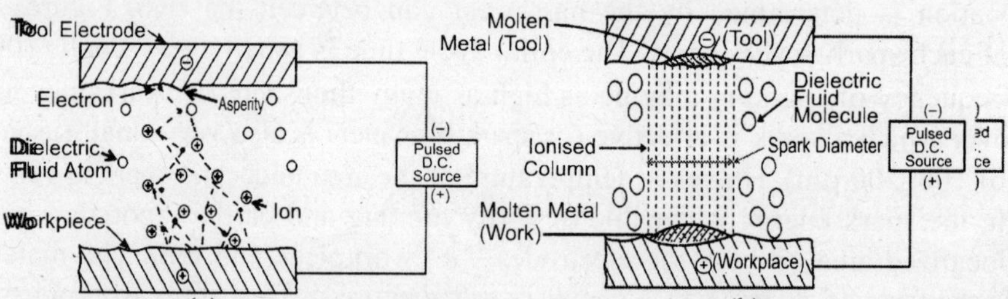

Fig. 7.4: (a) Initiation of ionization process showing movement of electrons towards the workpiece and ions towards the tool, (b) Shows fully developed ionized column in EDM with molten metal on the tool and workpiece both.

*It is an automatic device that uses error sensing negative feedback to correct the action of a mechanism (from Wikipedia).

column in EDM with molten metal on the tool and workpiece both. Once the molten metal is removed by flushing out dielectric, it will lead to the crater formation on the tool as well as workpiece. In fact, due to evaporation of dielectric the pressure in the plasma channel rises to a very high value (say, 200 atm) and it prevents the evaporation of superheated metal. As soon as the off-time of a pulse starts (when no current flows in the circuit), the pressure drops instantaneously allowing the superheated metal to evaporate.

The amount of material eroded from the workpiece and the tool depends upon the contributions (in the form of KE) of electrons and ions, respectively. The *polarity* normally used is straight (or normal polarity) in which the tool is -ve and workpiece is +ve (Figure 7.5a), while in reverse polarity the tool is +ve and workpiece is negative (Figure 7.5b). Movement of the tool towards the workpiece is controlled by a servo mechanism. The sparking takes place over the entire surface of the workpiece hence the (approximate) negative replica of the tool is produced in the workpiece. Usually, a component made by EDM process is machined in two stages, *viz.*, *rough machining* at high MRR with poor surface finish, and *finish machining* at low MRR with comparatively high surface finish (SF) that is low surface roughness value.

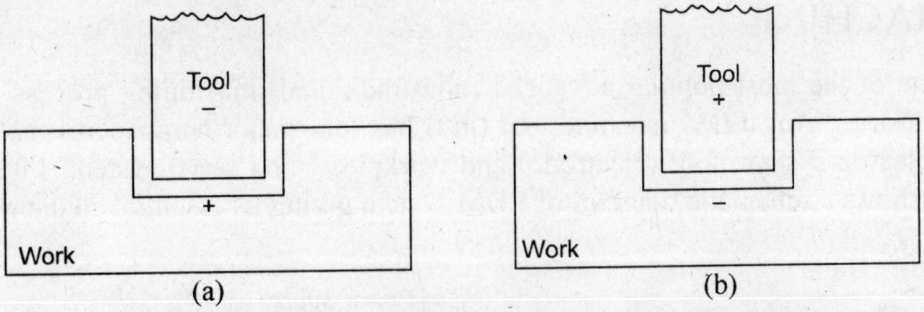

Fig. 7.5: (a) Normal polarity, (b) Reverse polarity in EDM

RC PULSE GENERATOR

RC pulse generator is shown in Figure 7.2. This generator has the following characteristics:

- Low Material Removal Rate (MRR) because of long idle time (charging time) and a short spark time.
- Improved surface finish achieved during finish machining, which is associated with further reduction in MRR.
- High Tool Wear Rate (TWR).

From Figure 7.6(a), it is evident that the peak current attained in RC circuit is very high. This high value of peak current results in a very high temperature of the electrodes which is not required and it may also result in thermal damage to both the electrodes, *i.e.*, workpiece and tool. *Controlled pulse generator* overcomes these problems [*Bhattacharyya*, 1973; and *Pandey and Shan*, 1980]. As shown in Figure 7.6(b), such generators can give low peak current, short idle time, and desired length of a pulse. It enables to select either rough machining conditions (high energy and low frequency of sparking), or finish machining conditions (low energy and high frequency of sparking).

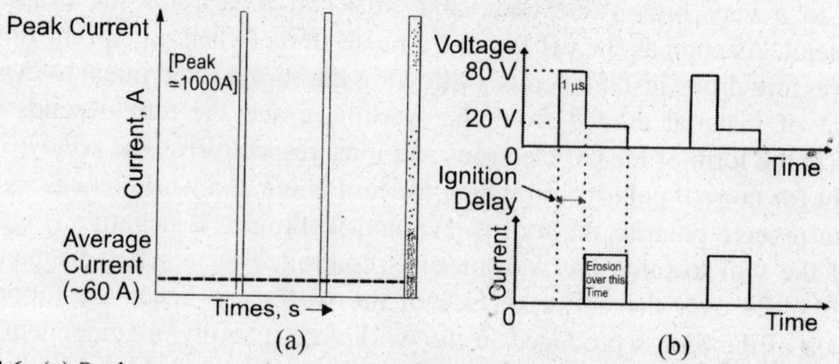

Fig. 7.6: (a) Peak current attained and its rapid decline on spark initiation in RC circuit (b) Typical voltage and current characteristic for a controlled pulse generator.

EDM MACHINE

EDM is one of the most popular advanced (non-traditional) machining process used in various industries. Any EDM machine tool (m/t) has four major components, *viz.* power supply, dielectric system, tool (electrode) and workpiece, and servo system, Figure 7.7. Figure 7.8 shows a schematic diagram of EDM system giving its essential elements.

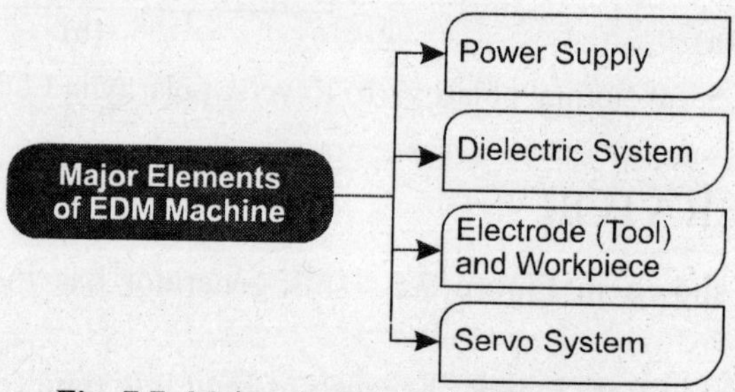

Fig. 7.7: Major elements of EDM machine

Power Supply

A Solid state *rectifier* converts Alternating Current into pulsed Direct Current (DC) used to produce sparks between tool and workpiece. A fraction of DC power is used to generate a square wave signal with the help of a digital multi-vibrator/oscillator. This signal triggers a bank of power transistors that act as high speed switches to control the flow of remaining DC power. It creates high power pulsed output responsible for generating sparks between the electrodes.

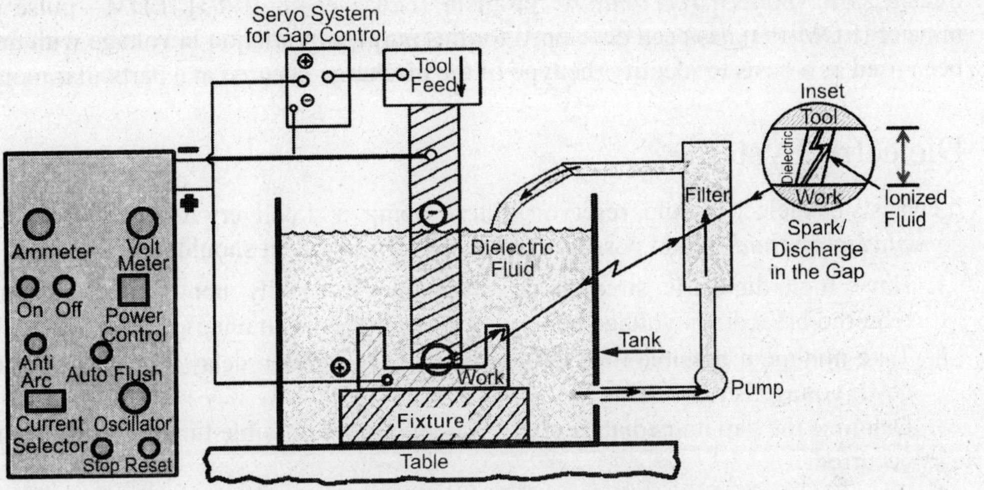

Fig. 7.8: Schematic diagram of EDM system

For a given input voltage, a spark can be generated only at or below a certain gap between the tool and workpiece (that is inter electrodes gap-IEG). EDM power supply senses the voltage between the electrodes and then sends the relevant signal to the *servo system*, which adjusts/maintains the desired gap value between the electrodes.

Power supply should also be able to control the *parameters* such as pulse voltage, current, pulse duration, duty cycle (the ratio of on time to pulse time), pulse frequency and electrode polarity. Duty cycle (or duty factor) controls the amount of time the energy is on during each pulse (Figure 7.9). EDM power supplies are also equipped with cut-off protection circuit. As soon as an arc, due to a short circuit, between the electrodes occurs immediately the cut-off circuit terminates the power. The power is also cut-off due to over voltage and over current.

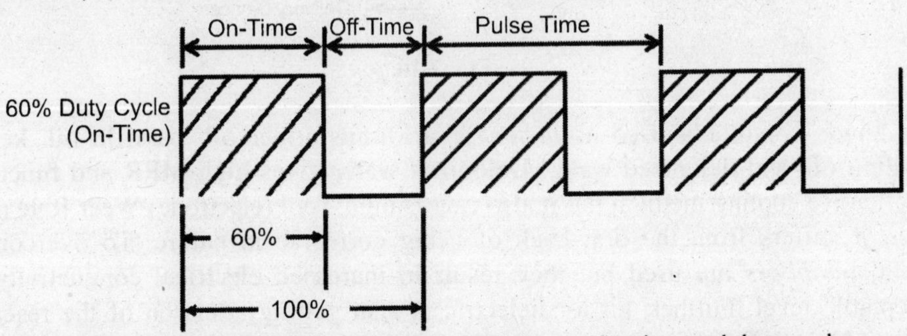

Fig. 7.9: An Illustration of on-time, off-time, pulse time and duty cycle

During EDM, one of the four states can occur, *i.e.*, normal discharge, free discharge (arc), short circuit and open circuit. The pulse detection and discrimination have been treated as a *"pattern recognition"* problem [*Davw et al.,* 1983]. EDM—pulse discriminator (EDM-PD) has been developed for this purpose. Variation in voltage with time has been used as a basis to identify the type of the discharge occured at a particular moment.

Dielectric System

It consists of dielectric fluid, reservoir, filters, pump, and delivery devices, Figure 7.10. *A good dielectric fluid* should possess certain *properties, viz.*, it should:

1. Have high dielectric strength (*i.e.*, remain electrically non-conductive until the required breakdown voltage between the electrodes is attained).
2. Take minimum possible time to breakdown (i.e., ignition delay time) once the breakdown voltage is reached.
3. Deionize the gap immediately (i.e., takes minimum possible time) after the spark has occurred.
4. Serve as an effective cooling medium.
5. Have high degree of fluidity.

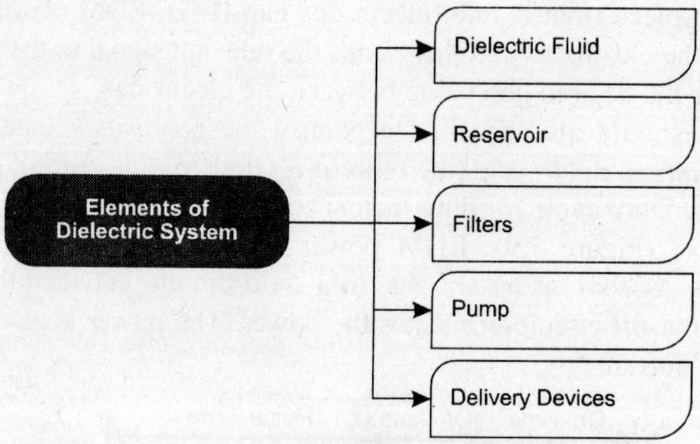

Fig. 7.10

The fluids commonly used as *dielectric* are transformer oil, paraffin oil, kerosene, lubricating oil, and deionized water. Deionized water gives high MRR and functions as more effective cooling medium but it also causes high Tool (electrode) Wear Rate (TWR). Further, it suffers from the drawback of being corrosive in nature. To overcome this problem, *inhibitors* are used but they result in increased electrical conductivity to an unacceptable level. Further, air as dielectric is also getting attention of the researchers because it is considered as more environment friendly compared to other commonly used dielectric fluids.

Deionized water is commonly used as dielectric in wire-EDM and drilling of small diameter holes (micro drilling). *Filtration* of dielectric fluid before re-circulation is highly essential so that a change in its insulation property during the process is minimal. Dielectric is filtered to remove debris as fine as 2 μm [*Albert*, 1981].

The concentration of the *debris* in the gap increases rapidly as the machining progresses. These wear particles should be removed from the gap so that the fresh dielectric enters in the IEG for spark discharges to take place. Increase in the contamination of dielectric by debris, results in decrease in the breakdown intensity of the electric field [*Bruyn*, 1960]. Hence, IEG at the entrance of the clean fluid is narrower than at the exit of the flow. This affects the capability of accurate *reproduction* of the EDM process. It is possible to correct such errors only if it could be feasible to predict quantitatively the error of shape. However, complex shaped electrodes with sharp edges or protruding points further add for poor reproduction accuracy. Hence, the flushing is decisive for both, process efficiency and product quality.

Effective flushing of dielectric removes by-products from the IEG. Ineffective flushing results in low MRR and poor surface finish. The effective flushing may increase MRR as much as by a factor of 10 or so [*Koshy et al.*, 1993]. Poor flushing ends up with stagnation of dielectric and build-up of machining residues which apart from low MRR also lead to short circuits and arcs. A good flushing system is the one that forces the dielectric to reach the locations with the narrowest gap where the sparking is desirable to occur. It is felt that the adequate flushing in case of blind cavities is difficult.

Various *methods* have been proposed by different researchers [*Masuzawa et al.*, 1983; *Mitskevich* 1965; *Erden*, 1982; *Kremer et al.*, 1985; *Bruyn*, 1960 and *de Bryun*, 1968] to ensure proper and adequate *flushing* of the gap and thereby obtain a better process output. Suction through electrodes, pressure through electrodes, jet flushing, alternating forced flushing, ultrasonic vibration of electrodes, and rotating electrode flushing [*Koshy et al.*, 1993] are some of the dielectric flushing techniques (Figure 7.11) reported in the literature. If it is a deep hole/cavity machining then jet flushing becomes less efficient Figure 7.11(g). It is also reported [*Albert*, 1981] that more than one modes of flushing may be used for effective cleaning of the gap. In some cases, hypodermic needles have been used as nozzles for flushing in tight corners.

In case of *machining of blind cavities*, flushing through a hole in the tool is most effective but it gives rise to a protruding bump (or spike) Figure 7.12(a). This is the situation similar to the one seen during ECM [*Sreejith et al.*, 1994]. To avoid this problem a rotating tool with eccentric holes (off-centered holes) for dielectric supply can be used Figure 7.12(b). Jet flushing is less effective hence it should be used only when none of the other methods can be used due to tool or workpiece configuration. In case of inflammable dielectric fluid, the workpiece should always be immersed in the dielectric fluid to minimize any chance of accidental fire. As machining continues, the dielectric gets contaminated with more and more amount of debris. Presence of such debris in the IEG, results in bridging the gap. It yields more number of arcs. Arcs are undesirable in EDM because they damage both tool and workpiece. Occurrence of such arcs can be eliminated by proper filtration of the dielectric as well as appropriate flushing of the IEG.

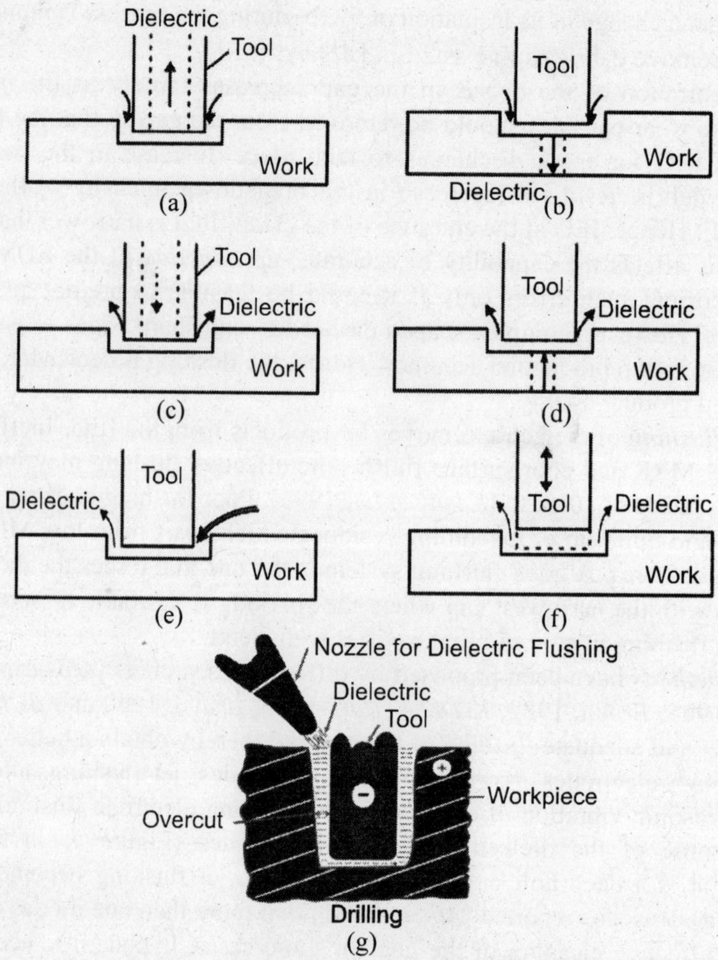

Fig. 7.11 Various methods for dielectric flushing (a) Suction through electrode, (b) Suction through workpiece, (c) Pressure through electrode, (d) Pressure through workpiece, (e) Jet flushing, (f) Periodic cycling of electrode [*HMT, Bangalore, Catalogue*], (g) Jet flushing (same as in (e)).

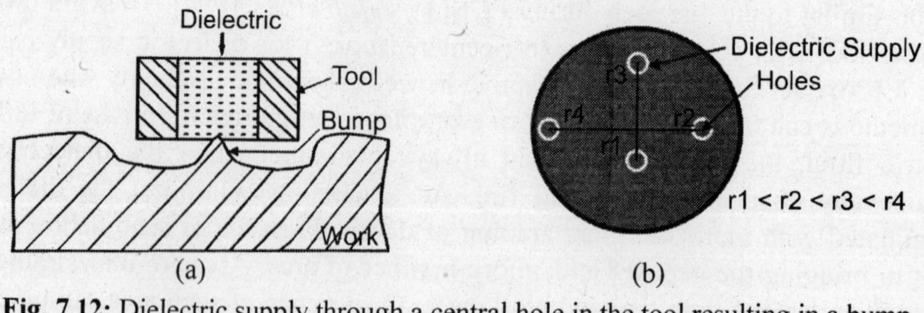

Fig. 7.12: Dielectric supply through a central hole in the tool resulting in a bump, (b) Accentric dielectric supply holes in the rotating tool which avoids the presence of bumps.

Electrodes

Both tool and workpiece are electrodes in EDM. However, in general, the word 'electrode' is used only for tool. The *material* to be used as tool electrode should possess the desirable properties like easily machinable, low wear rate, good conductor of electricity and heat, cheap and readily available. Graphite (easily machinable, low wear rate, and high conductivity), copper, brass (highly stable and relatively low wear rate), copper tungsten (low wear rate, expensive, and cannot be easily shaped), cast aluminum, copper boron, and silver tungsten are some of the materials that are used for making tools for EDM. However, copper and graphite are more commonly used because they can be easily machined and they have higher thermal and electrical conductivity. A graphite electrode with finer grains gives lower TWR, better surface finish, and higher MRR. However, its brittleness is an undesirable characteristic because of which it is prone to breakage.

During electro-spark sinking operation, overcut is unavoidable. To account for electrode *shape degeneration* due to uncontrolled wear of tool is a serious problem encountered during tool design for EDM. The problem is difficult due to unequal degeneration of geometrical form of the tool. This change in electrode shape and size cannot be compensated by over feeding the (roughing) electrode as it is possible in case of electrode with shorter length. As a result, higher *'finishing allowance'* (volume) will be provided leading to higher time required for finishing cut(s) during EDM.

As discussed above, *erosion of tool electrode* in EDM is *inevitable*. Recent developments in pulse shape and tool material, and selection of proper machining conditions have led to a very low electrode wear rate (or TWR). Hence, it is a normal practice to have the tool electrode for finish electro-discharge machining, conferring closer to the required shape and size of the workpiece. The tool gets tapered and its corners rounded off due to shape degeneration during EDM. Figure 7.13(a) shows *overcut* obtained during Electro-Discharge Drilling (EDD) operation. Its value is governed by machining parameters, work material properties, and also tool material properties. Overcut should be taken care of at the stage of tool design. A properly designed tool can produce a straight sided hole with the desired dimension. The shape and size of the designed/redesigned tool can be obtained with the help of empirical models developed using experimental data for the given tool material,

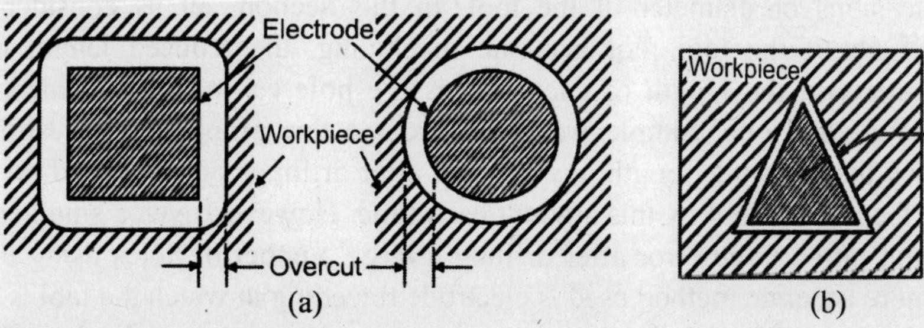

Fig. 7.13: (a) Schematic diagram for overcut obtained during electric discharge drilling (EDD), (b) Electric discharge trepanning operation.

workpiece material and machining conditions. Similar procedure can be developed for tool design in EDM as done for tool design for ECM [*Reddy et al.*, 1988].

The model developed [*Jain,* 1989] for the prediction of tool electrode wear lacks in generality mainly because of involvement of so many variables. As the electric field density is highest at the corners, they wear (or round off) faster than the remaining part.

The tool wear is usually quantified by *'wear ratio'* (ratio of the volume of tool material removed to the volume of workpiece material removed during the same period). Electrode wear is more at the corners/edges than rest of the tool. Depending upon the tool and workpiece materials, machining conditions and type of operation, the wear ratio may vary in the range of 0.05:1 to 100:1. Under certain situations, a through hole may be produced by employing *trepanning* operation (Figure 7.13(b)) rather than drilling because it is very fast compared to the drilling operation. In electro spark trepanning operation, only a small thickness of material around periphery of the cavity is removed by machining in place of machining the whole material equal to the volume of the cavity.

Servo System

A servo system is used to maintain a predetermined gap between the tool and workpiece. There is a gap voltage sensor in power supply, which sends signals to the servo system. As soon as it senses that the gap between the electrodes has been bridged by some electrically conductive material (say, debris), a signal is sent to the servo system to reverse the direction of movement of the tool/electrode. The servo system keeps the tool reciprocating towards and away from the workpiece until the dielectric fluid cleanse the gap, and clears it off the foreign material. If the dielectric flushing system is inefficient in cleaning the interelectrode gap, the machining cycle time will be long and vice versa. Two of the requirements of the servo system are sensitivity for small movements, and enough power to overcome the weight of ram, and electrode and flushing forces.

Electrode Refeeding

Tool wear during Electric Discharge Drilling (EDD) reduces length of the tool as well as other two dimensions or diameter of the tool. In this section, let us consider mainly reduction in length of the tool. For accurate machining, the reduced length must be compensated otherwise the length of each successive hole cavity to be drilled will be shorter than the desired one. Simple way to overcome this problem is to estimate the amount by which the electrode length has reduced after drilling each hole and, in turn, to increase the in feed of the tool by this amount each time. However, even a small error will finally compound into a large error after drilling a large number of holes using the same tool. Another more accurate method used is electrode refeeding in which the tool is brought back every time, after drilling a hole, to a reference point where from the depth of drilling is measured. The electrode refeeding can be done manually also but perferably automatically.

CNC-EDM

Computer Numerical Control (CNC) of EDM machine tool (m/t) has made it very efficient and versatile. Tolerance upto 25 μm are easily achieved [*Albert,* 1982]. It offers the provision of fine finish settings which flatten the microscopic peaks on the EDM'd surface. It eliminates bench-work by proper selection of EDM parameters. Helical gears have been made [*Albert*, 1982] with relative ease by using the canned cycle. With CNC-EDM, it is possible to run jobs unattended overnight or even days together. It is also possible to minimize the setting up time. Operator need not to babysit the EDM m/c and hence can use his/her time elsewhere in a more productive way, *viz*., planning of another job. This is possible only if an *Adaptive Control* (AC) is used in CNC-EDM. It monitors power supply parameters, as well as does sampling of electrical conditions versus time for each pulse so that the m/t can be protected against arcing. By the use of CNC of EDM m/c, mechanical orbiters have been replaced by software orbiters by which orbiting pattern of any shape can be made. Orbiting electrodes agitate the dielectric fluid which results in a better flushing. Exploiting CNC-EDM with AC facility, the m/t can be programmed to retract electrode from the workpiece periodically so that the flow of clean dielectric takes place in the gap. Thus, improved flushing works hand-in-hand with adaptive control. By the use of CNC-adaptive control, it is possible to monitor and control the discharge energy which permits to machine at *'no-wear setting'*, which improves the repeatability of the process.

The *Automatic Tool Changer* (ATC) employed on EDM machines makes it possible to change electrodes automatically in any programmed sequence, and with minimum loss of production time. Hence, one can load the tool on tool magazine** for rough, semifinish, or finish cuts. EDM with CNC can employ a DC servo drive for quill (or tool) rotation, *i.e.*, fourth axis (c-axis) rotation. CNC-EDM [*Quinlan*, 1983; *Albert,* 1983] has a 16 stations automatic tool changer. CNC feature is utilized to run the machines unattended for days together. Whenever a specific tool is required, automatic tool changer (or a lever arm) exchanges the used tool with the new tool to be used. This helps in reducing the tool change time. Nowadays, CNC-EDM machines are more popular, in advanced industries, than conventional type of EDM machines without CNC feature. Figure 7.14 shows a Z-axis NC machine with limited features. In this case, only Z-axis along with machining conditions (voltage, current, pulse on-time and pulse off-time) are controlled using NC part program.

Recent systems have automatic *electrode-zeroing* feature [*Anon*, 1981]. Sensors automatically determine the correct starting point, *i.e.*, zero point. Every time, the cutting starts from the predefined location/coordinates. This facility permits for compensation for the wear of electrode so that the depth is controlled accurately on the successive batches of production. Further, with today's CNC-EDM systems, programming is easier and faster because they are *menu driven.* These systems utilize English language, prompts and

**Tool magazine is a device of CNC-EDM m/c which can house as many as 150 differently types of the tools (electrodes) and ATC exchanges the used tool with the new tool to be used.

commands. Some of the systems have graphic display facility as well. Some other program controlled features associated with CNC-EDM are *centering, edge finding, flushing, and slope compensation.*

Fig. 7.14: Z-NC EDM machine [*Courtesy:* Electronica, Pune (India)]

ANALYSIS

Analysis of R-C Circuits

Charging voltage and charging current:

The relaxation type circuit (Figure 7.15) consists charging, discharging and sparking sections. The charging current (i_{ct}) flowing in the charging circuit at time 't' is given by:

$$i_{ct} = \frac{V_o - V_{ct}}{R_c}. \qquad \text{... (7.1)}$$

Also,

$$i_{ct} = C\frac{d\,V_{ct}}{dt}. \qquad \text{... (7.2)}$$

where, V_o is the supply voltage, and V_{ct} is the charged voltage of the condenser at time 't'. R_c is charging resistance and C is capacitance of the condenser.

From the above,

$$\frac{d\,V_{ct}}{(V_o - V_{ct})} = \frac{1}{R_c C}dt. \qquad \text{... (7.3)}$$

Integrate both sides,

$$\ln\,(V_o - V_{ct}) = -\frac{t}{R_c C} + K_1. \qquad \text{... (7.4)}$$

where, K_l is a constant of integration. Use the boundary condition ($V_{ct} = 0$ at $t = 0$) to evaluate K_1. It gives,

$$K_1 = \ln V_o \quad \text{... (7.5)}$$

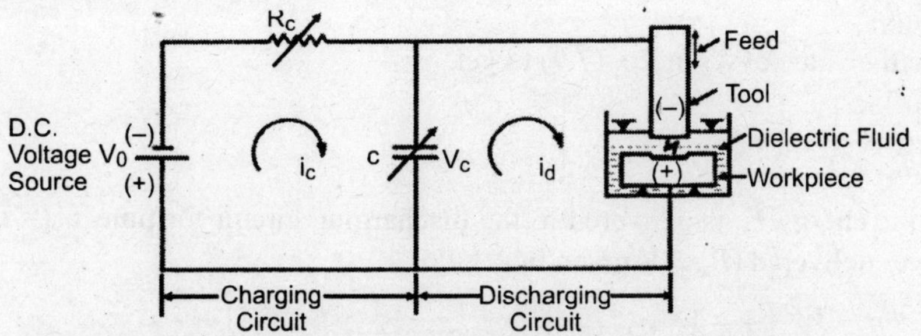

Fig. 7.15: Various elements of a relaxation circuit with normal polarity

Substitute the value of K_1 in Eq. (7.4) and simplify to get,

$$V_{ct} = V_o\left(1 - e^{-\frac{t}{R_C C}}\right) \quad \text{... (7.6)}$$

The time required by the condenser to attain 0.638 times of its charging voltage (= 0.638 V_o) is called "time constant" ($t = R_c C$) [*Bhattacharyya*, 1973].

Substitute the value of V_{ct} from Eq. (7.6) in Eq. (7.1) and simplify to get Eq. (7.7).

$$i_{ct} = \frac{V_o}{R_c}\left(e^{-\frac{t}{R_C C}}\right) \quad \text{... (7.7)}$$

Power Delivered to the Discharging Circuit

The energy delivered to the discharging circuit at any time t is given by,

$$dE_n = i_{ct}\, V_{ct}\, dt$$

Substituting the values of V_{ct} and i_{ct} from Eqs. (7.6) and (7.7), respectively in the above equation, we get

$$dE_n = \frac{V_o}{R_c}\left(e^{\frac{-t}{R_C C}}\right).V_o\left(1 - \left(e^{\frac{-t}{R_C C}}\right)\right)dt. \quad \text{... (7.8)}$$

Integrate Eq. (7.8) to get,

$$E_n = \frac{V_o^2}{R_c}\left[-\tau\ e^{-t/\tau} + \frac{\tau}{2}\ e^{-2t/\tau}\right] + K_2, \quad \text{... (7.9)}$$

where, $\tau = R_c\,C$. K_2 is a constant, and it can be evaluated by using boundary condition ($E_n = 0$ at $t = 0$). We get,

$$K_2 = \frac{V_0^2 \tau}{2R_c}$$

Substitute the value of K_2 in Eq. (7.9) to get,

$$E_n = \frac{V_o^2 \tau}{R_c}\left[\frac{1}{2} + \frac{1}{2}e^{-2t/\tau} - e^{-t/\tau}\right]. \qquad \text{... (7.10)}$$

Suppose the energy E_n is delivered to the discharging circuit for time τ_c (= t) then the average power delivered (P_{avg}) is given by,

$$P_{avg} = \frac{En}{\tau_c} = \frac{V_o^2}{R_c x}\left[\frac{1}{2} + \frac{1}{2}e^{-2x} - e^{-x}\right] \qquad \text{... (7.11)}$$

where, $x = \tau_c/\tau$. The condition for the maximum power to be delivered to the discharging circuit is given by,

$$\frac{dP_{avg}}{dx} = 0. \qquad \text{... (7.12)}$$

After solving Eq. (7.12), we get, $x = 1.26$ [*Bhattacharyya*, 1973]. Substitute this value of x in Eq. (7.6) to get,

$$V_{ct} = V_o\left(1 - e^{-1.26}\right) \approx 0.72\ V_o. \qquad \text{... (7.13)}$$

Thus, the discharging voltage for the maximum power delivery is about 72% of the supply voltage (Figure 7.16). In other words, for the maximum power delivery through the gap, the breakdown voltage and the supply voltage should follow the following relationship,

$$\boxed{V_b \approx 0.72\ V_0}$$

Current in the Discharging Circuit

Let R_s represent the sparking resistance (other resistances are assumed to be negligible hence neglected). Then, the current i_d at any time 't' is given as follows,

$$i_d = -\frac{dq}{dt} = -C\frac{dV_{ct}}{dt}. \qquad \text{... (7.14)}$$

Also, $i_d = \dfrac{V_{ct}}{R_s}$. ... (7.15)

Therefore,

$$\frac{V_{ct}}{R_s} = -\ C\frac{dV_{ct}}{dt}.$$

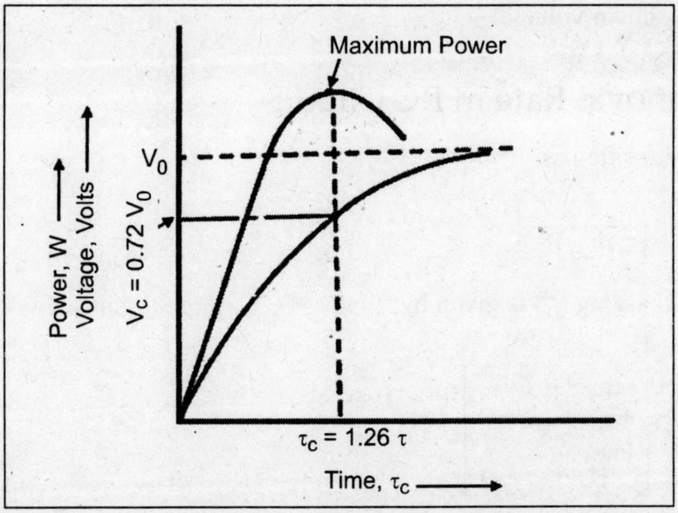

Fig. 7.16: Schematic diagram showing the condition for maximum power delivery for the RC circuit

or, $\dfrac{dV_{ct}}{V_{ct}} = -\dfrac{dt}{R_s C}$.

After integration of the above equation,

$$\ell n\left(V_{ct}\right) = -\frac{t}{R_s C} + K_3.$$

Constant K_3 is evaluated using boundary condition, $V_{ct} = V_{co}$ at $t = 0$.

$$K_3 = \ell n\ V_{co}.$$

Therefore,

$$V_{ct} = V_{co} e^{-t/R_s C}$$

Hence,

$$\boxed{i_d = \frac{V_{ct}}{R_s} = \frac{V_{co}}{R_s} e^{-t/R_s C}} \qquad \text{... (7.16)}$$

Energy dissipated across the IEG is given by,

$$\boxed{W_d = \frac{1}{2} C V_b^2.} \qquad \text{... (7.17)}$$

where, V_b is breakdown voltage.

Material Removal Rate in RC Circuit

Eq. (7.6) can be rewritten as,

$$t = R_c C \ \ln\left(\frac{1}{1 - V_{ct}/V_o}\right).$$

Frequency of charging (f_c) is given by,

$$f_c = \frac{1}{t} = \frac{1}{R_c C}\left[\frac{1}{\ln \dfrac{1}{(1 - V_b/V_o)}}\right].$$

Material removal rate should be proportional to the total energy delivered in the sparking per second. It is based on the assumption that every power pulse generates a spark and each spark removes equal amount of material.

$$MRR \propto \frac{1}{2} C \ V_b^2 f_c$$

Substitute the value of f_c, and let K_4 be a constant of proportionality. Then,

$$MRR = K_4 C \ V_b^2 . \frac{1}{R_c C}\left[\frac{1}{\ln \dfrac{1}{(1 - V_b/V_o)}}\right] \qquad \text{... (7.18)}$$

Thus, MRR $\propto 1/R_c$, *i.e.* R_c should be decreased to increase MRR, however, at very low value of R_c, it will result in arcing. The minimum value of the resistance (R_c) that will prevent arcing is known as *critical resistance*.

Value of the critical resistance can be proved [*Bhattacharyya*, 1973] to be equal to,

$$R_{\min} = \sqrt{\frac{L}{C}}. \qquad \text{... (7.19)}$$

Above equation is not obeyed during experiments. Based on the experimental results, it is found that,

$$R_{\min} \geq 30\sqrt{\frac{L}{C}}. \qquad \text{... (7.20)}$$

Surface Finish

In EDM, each spark results in approximately spherical crater formation on the surface of the workpiece. Hence, the center line average (H) value of surface finish will be a function of crater depth (h) (Figure 7.17) and frequency of sparking (f_c) as follows.

$$H \propto \frac{h}{f_c}.$$

The volume of the material removed ($\propto h^3$) per discharge will be proportional to the energy delivered during sparking ($=\frac{1}{2}CV_b^2$). Hence,

$$h^3 \propto \frac{1}{2}CV_b^2.$$

or, $h \propto C^{1/3} \ V_b^{2/3}$.

Therefore, from the above equations,

$$H \propto \frac{V_b^{2/3} C^{1/3}}{f_c},$$

or, $$H = \frac{K_5 V_b^{2/3} C^{1/3}}{f_c}, \qquad \text{... (7.21)}$$

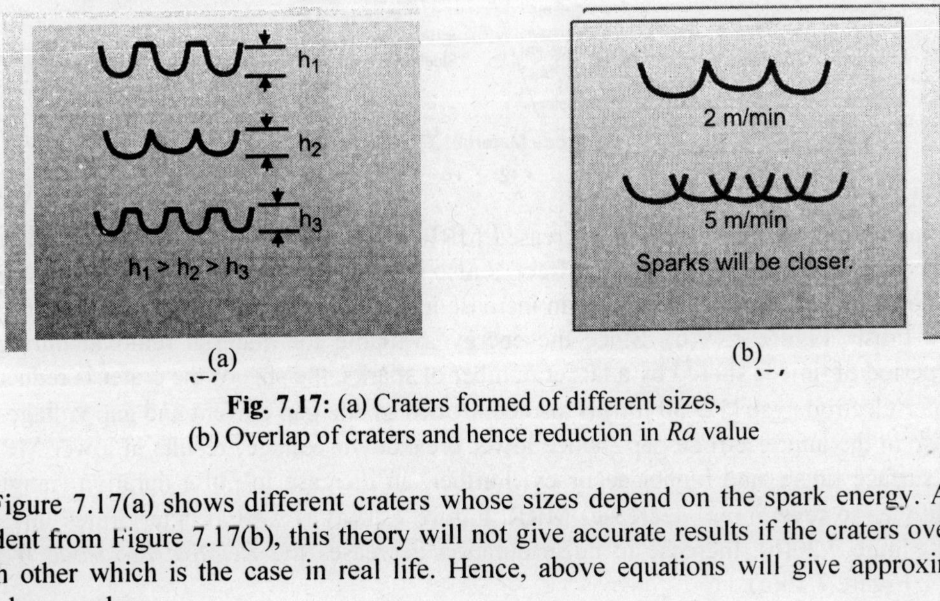

Fig. 7.17: (a) Craters formed of different sizes,
(b) Overlap of craters and hence reduction in *Ra* value

Figure 7.17(a) shows different craters whose sizes depend on the spark energy. As is evident from Figure 7.17(b), this theory will not give accurate results if the craters overlap each other which is the case in real life. Hence, above equations will give approximate roughness value.

But, $f_c \propto R^{m} C^{n}$ (for *RC* type circuit).

Therefore,

$$\boxed{H = K_6 V_b^{2/3} C^{n+1/3} R^{m}.} \qquad \text{... (7.22)}$$

However, from different experimental results it is concluded that,

$$\boxed{H = K_7 V_b^{0.5} C^{0.31\sim0.36}} \qquad \text{... (7.22)}$$

(*Note:* Take *C* in Farads to get *H* in μm.)

PROCESS VARIABLES

Figure 7.18 shows a diagram which indicates different dependent (responses) and independent variables (input parameters) of the EDM process.

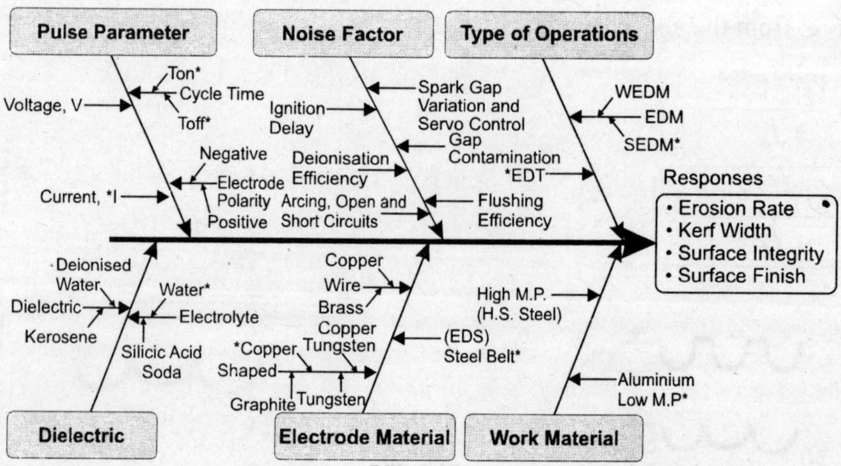

Fig. 7.18

An increase in current results in increased MRR as well as increased surface roughness Figures 7.19(a) and 7.19(b). Similar effect is also observed when spark voltage (or break down voltage) is increased. However, an increase in spark frequency results in an improved surface finish, Figure 7.19(c). Since the energy available for material removal during a given period of time is shared by a larger number of sparks, the size of the crater is reduced. The inter-electrode gap (12–50 μm) is also controlled by the gap current and gap voltage. A decrease in the interelectrode gap (hence lower breakdown voltage) results in lower MRR, better surface finish, and higher accuracy. Further, an increase in pulse duration (ranging from few μs to several ms) decreases MRR, Figure 7.20(a) as well as deteriorates surface finish, Figure 7.20(b). Increase in pulse duration decreases the *Relative Electrode Wear* (REW), Figure 7.19(c).

$$\left(\text{REW} = \frac{\text{Vol. of tool material eroded over the test duration}}{\text{Vol. of workpiece material eroded over the test duration}} \times 100 \right)$$

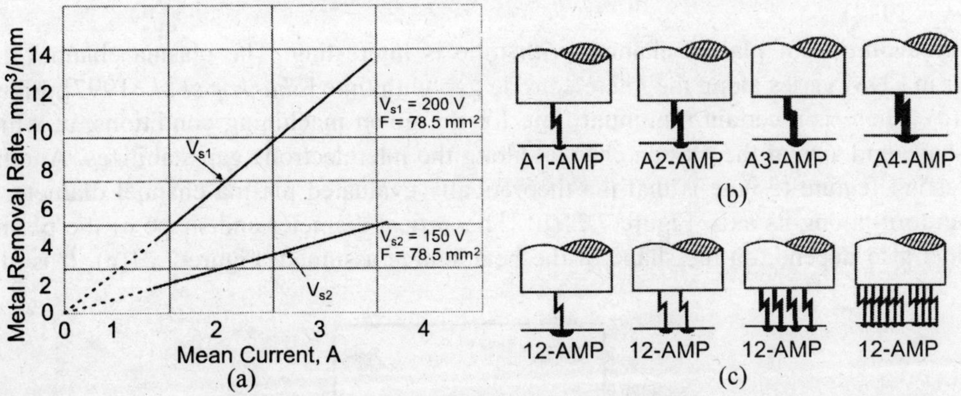

Fig. 7.19: (a) Effect of mean current on MRR. Dielectric = Kerosene, Tool = Brass, Work = Low carbon steel, F = Tool electrode area [*Pandey and Shan,* 1980], (b) Effect of current on surface finish (or crater size), (c) Effect of frequency of sparking on surface finish (or crater size)

Figure 7.20(a) shows the relationship between MRR and pulse duration for different current values and electrode materials with different polarities. A decrease in MRR with increase in pulse duration is attributed to unsteady machining conditions observed for pulses greater than 20 μs.

Figure 7.20(b) shows that copper tools, in general, yield better surface finish as compared to brass tools. Experiments have also been conducted using tap water and the mixture of tap and distilled water as dielectric during EDM. Figure 7.20(c) shows the relationship between REW and pulse duration. In case of copper electrode with negative

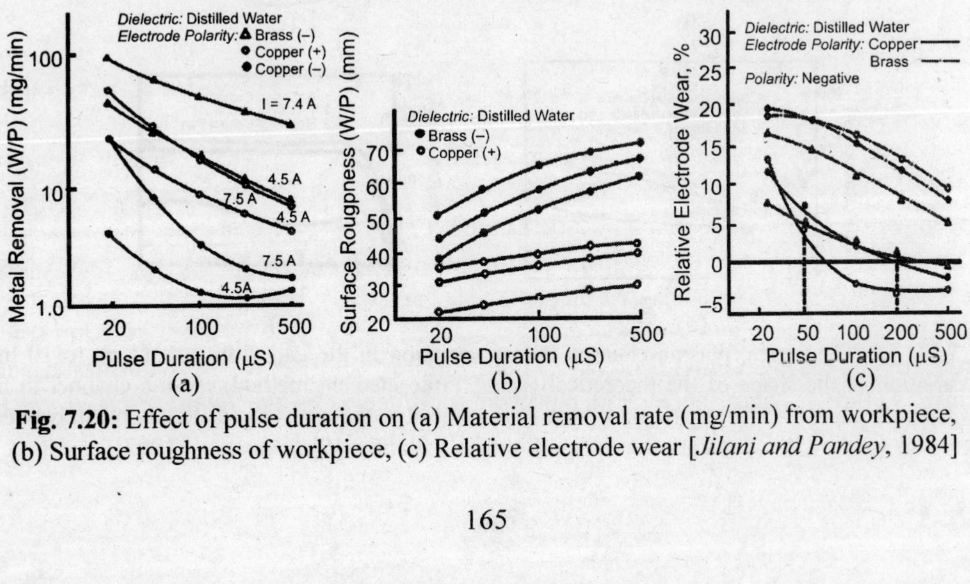

Fig. 7.20: Effect of pulse duration on (a) Material removal rate (mg/min) from workpiece, (b) Surface roughness of workpiece, (c) Relative electrode wear [*Jilani and Pandey*, 1984]

polarity for the specified conditions, there seems to be a pulse duration for which REW is negative (copper as cathode material) which can be attributed to the phenomenon of metal deposition on cathode. However, no appropriate explanation exists for this [*Jilani and Pandey*, 1984].

The evolution of plasma channel with time is interesting. The plasma channel diameter in EDM varies along the interelectrode gap with time [*Shankar et al.,* 1997], Figure 7.21(a). There is a certain minimum time for the given machining conditions, at which the shape and size of the plasma channel along the interelectrode gap stabilizes. Another interesting feature to note is that the theoretically evaluated plasma channel diameter is not uniform along its axis, Figure 7.21(b). Theoretical diameter and shape of the plasma channel also depend, on the shape of the heat source assumed, Figure 7.21(c). It is also

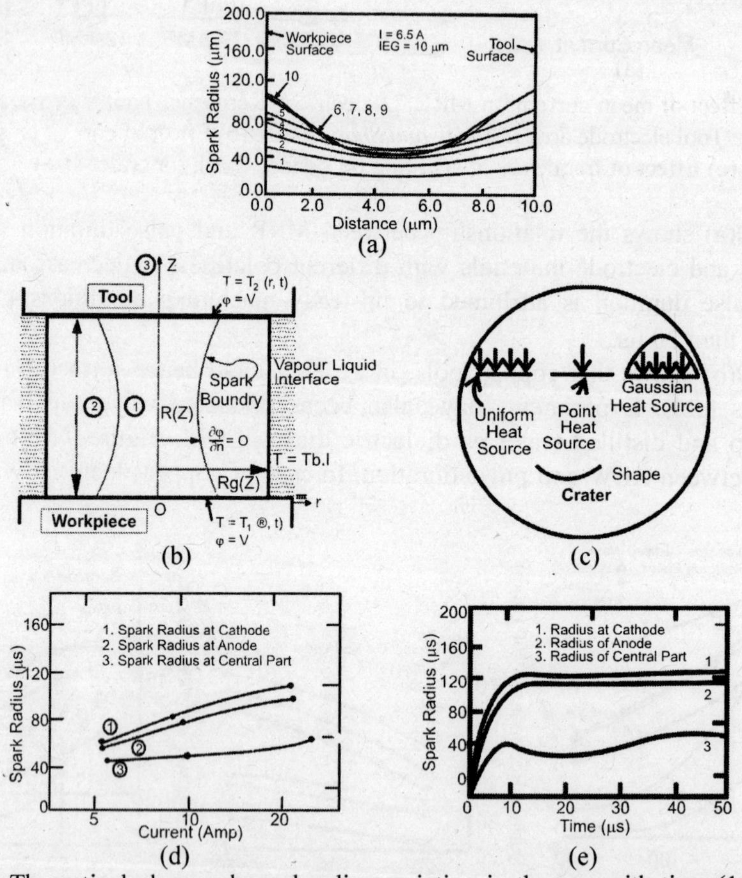

Fig. 7.21: (a) Theoretical plasma channel radius variation in the gap with time (1 μs to 10 μs), (b) Variation in the shape of the theoretical (using finite element method) plasma channel in the interelectrode gap, (c) Different types of heat sources used in the analysis, (d) Relationship between current and spark radius, (e) Variation of spark radius with time [*Shankar et al.,* 1997].

different at anode and cathode which is governed by the thermal properties of these electrodes. It has been reported [*de Bruyn*, 1968] that major part of the tool erosion occurs during first few microseconds of the pulse. Hence, use of a pulse with low initial current could reduce tool wear. It can be noted that the spark radius at anode, cathode and centre of the channel varies with current (Figure 7.21(d)) and with time Figure 7.21(e).

Pulsed power for EDM can be obtained either by using relaxation circuit or high power transistor circuit. In the first case, the spark energy is directly related to the inter-electrode gap characteristics during charging period (pause/off period). Dielectric (say, tap water) which would give low breakdown potential will give low energy stored in the condenser hence low energy per spark.

In case of *EDM generators* with high power transistor circuits, the use of tap water and deionized water as dielectric has been recommended [*Godinho and Noble*]. Use of water as dielectric makes the EDM process more safe (*i.e.*, without any risk of fire or explosion). Specially, in case of deep electric discharge machining, the use of water as dielectric in place of mineral oil has shown improved results (Figure 7.22). Therefore, the value of electrodes wear ratio (tool wear rate – TWR/workpiece wear rate – MRR) is lower for the case of tap water as compared to mineral oil. However, the comparison shown in Figure 7.22 is not accurate enough because the machining conditions (pressure and polarity) in two cases are different. Both these parameters affect the measured responses.

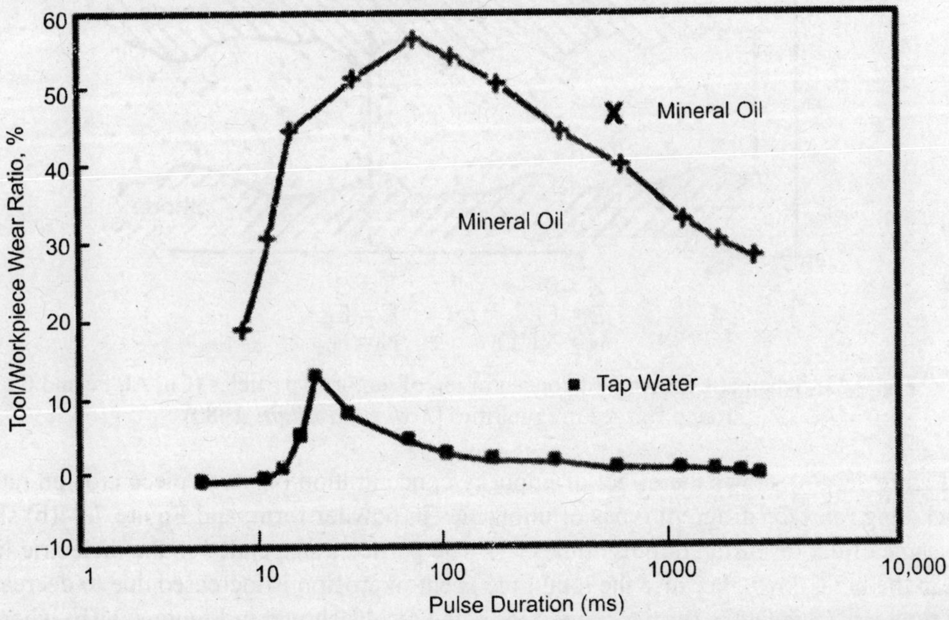

Fig. 7.22: Comparison of EDM performance while using tap water and mineral oil as dielectrics. dielectric pressure = 60 psi (Tap water); = 50 psi (Mineral oil); Tool polarity –ve (Tap water) and +ve (Mineral oil) [*Godinho and Noble*].

Dielectric Contamination and its Effects

Dielectric during EDM gets contaminated by the debris produced due to erosion of tool and workpiece, and decomposited products of the dielectric itself. *Dielectric contamination* has been reported [*Erden and Bilgin*, 1980] to influence breakdown, time lag (or, ignition delay), short circuit, machined surface characteristics, spark gap and machining rate (or machining performance, in general).

The impurity particles tend to gather around surface irregularities due to *electric field concentration* around such points (Figure 7.23). Theoretically, it has been shown [*Erdin and Bilgin*, 1980] that the larger impurity particle size results in shorter time lag and vice versa. Experiments have also been conducted to investigate the effects of dielectric contamination on the performance of the EDM process. For this purpose, commercially pure powders (market grade) of Cu, Al, Fe and C were used. Non-metal powders have different effect compared to metal powders due to their different electrical properties. Electrodes are ground and polished prior machining to minimize possible *surface effects*. Presence of impurities/contaminations decreases the time-lag; some of the open circuit pulses are replaced by the effective discharges thus improving the EDM performance. However, such effects are not observed when copper is used as electrode material.

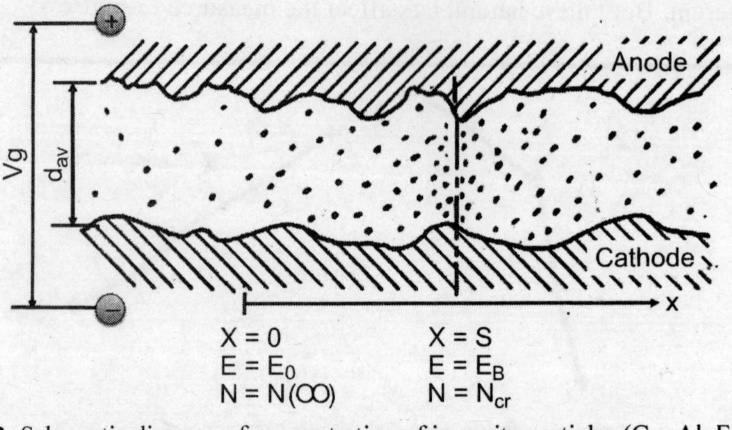

Fig. 7.23: Schematic diagram of concentration of impurity particles (Cu, Al, Fe and C) around surface irregularities [*Erdin and Bilgin*, 1980]

Figure 7.24(a) shows the effect of impurity concentration on work piece erosion rate (or machining rate) for different types of impurities in powder form, and Figure 7.24(b) shows the same effect for different pulse times (t_S). The particles suspended in the dielectric liquid aid in the breakdown, because the total time spent in erosion is increased due to decrease in the time-lag. This fact is further supported by the results shown in Figure 7.24(b) where the effect of impurity concentration on machining rate is more pronounced at longer pulse times. In case of short pulses, some of the voltage pulses applied across the IEG end before the breakdown conditions are satisfied. It is reverse in case of long pulses that is why the number of pulses with effective discharges is significantly increased leading to higher MRR. Further, IEG is reported to increase by an increase in concentration of the impurity in dielectric liquid. Machining rate is increased at low concentration of impure particles but remained almost constant for higher values of impurity concentration. As explained above, tool wear rate is also influenced by the kind and concentration of impurity in the dielectric, Figure 7.24(c). Note that all the impurity contents discussed above (Figure 7.24) are electrically conductive in nature. Electrically non-conductive in nature impurity content would give different effects as compared to electrically conductive impurity content.

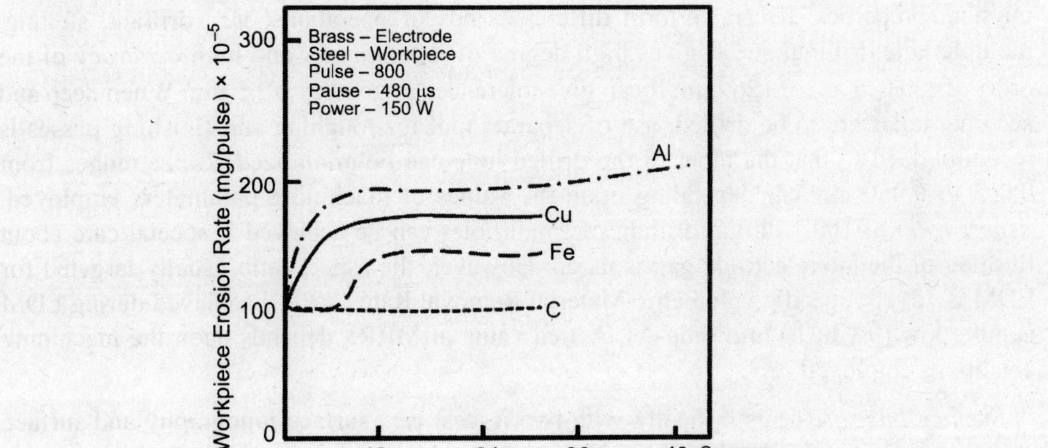

Fig. 7.24: Effect of impurity concentration on (a) Workpiece erosion rate for various kind of impurities, (b) Change in workpiece erosion rate for different pulse rates, (c) Tool erosion rate for various kinds of impurities [*Erdin and Bilgin,* 1980]

Microcrack length has been found to depend on the discharge power in general, and on the pulse duration, in particular. Longer pulse durations (≥ 800 μs) result in abnormally long cracks which may extend into the HAZ.

PROCESS CHARACTERISTICS

Physical, mechanical and metallurgical properties of the workpiece do not significantly influence the performance of the process. EDM can be employed for machining any electrically conductive material irrespective of its hardness and other mechanical and physical properties. It can perform different kinds of operations, *viz.*, drilling, slotting, multiple hole drilling, etc. It gives high degree of *repeatability* and high *accuracy* of the order of ± 0.025 to ± 0.126 mm. It can give tolerances as good as ± 2.5 μm. When deep and accurate holes are to be drilled, use of separate tool for roughing and finishing passes is recommended so that the taper of the drilled hole can be minimized. *Taper* ranges from 0.005 to 0.050 mm/cm depending upon the values of machining parameters employed. *Aspect ratio* of 100:1 during drilling of small holes can be achieved if special care about flushing of the interelectrode gap is taken. However, the aspect ratio usually targeted for EDM is 15–20. Specific volumetric Material Removal Rate *(MRR_{vs})* achieved during EDM is quite low (0.1 to 10 mm^3/min-A). Actual value of MRR_{vs} depends upon the machining conditions employed.

Surface integrity deals basically with two issues, *viz.*, surface topography and surface metallurgy (*i.e.,* possible alterations in the surface layers after machining which include surface and sub-surface defects). Surface integrity greatly affects the performance, life and reliability of the component.

Microscopic study of the machined component reveals three kinds of layers, *viz.*, recast layer, Heat Affected Zone (HAZ), and converted layer, Figure 7.25(a). The remaining part is parent material. If the molten material from the workpiece is not flushed out quickly, it gets resolidified and hardened due to cooling effect of the dielectric, and gets adhered to the machined surface. This thin layer (about 2.5 to 50 μm or so) is known as *"recast layer"*. It is extremely hard (approximately 65 HRC) and brittle. The surface is porous and may contain microcracks. Such surfaces should be removed before using these parts or putting them in the sub-assembly/assembly. The layer next to the recast layer is called *"Heat Affected Zone"* (HAZ which has approximately 25 μm thickness). Heating, cooling, and diffused material are responsible for the presence of this zone. Thermal residual stresses, grain boundary weaknesses and grain boundary cracks are some of the characteristics of this zone. *Conversion zone* (or converted layer) is identified below the HAZ and it is characterised by a change in grain structure from the original structure. Figure 7.25(b) shows traces of the redeposited material and Figure 7.25(c) shows a hole drilled by EDM process [*Indurkhya*, 1985].

Excessive local thermal expansion and subsequent contraction may result in *residual tensile stresses* in the machined part. Such experiments were conducted [*Barash*, 1959] on mild steel and hard rolled brass. It was found that the thickness of the affected layer was lesser than 25 μm. Thermal residual stresses in the electrical discharge machined parts have been evaluated using finite element method [Shankar, 1997; Yadav *et al.*, 2004].

The *"surface finish"* achieved during EDM is influenced by the chosen machining conditions. New generation EDM machines are capable to produce surface finish as good as 0.18 μm – 0.25 μm. Surface finish is mainly governed by pulse frequency and energy per spark. The texture of eroded surface has been analysed by *Kahng and Rajurkar* [1977], and a schematic illustration of roughing and subsequent finishing is given in Figure 7.26. It is observed that the application of higher discharge energy results in deeper HAZ and subsequently deeper cracks. Spark eroded surfaces have been examined [*Barash*, 1959] to study after-machining characteristics.

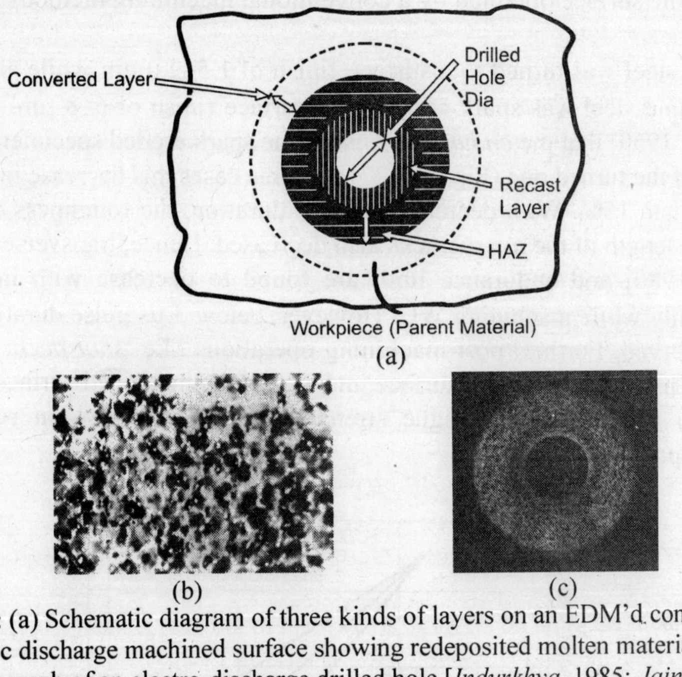

Fig. 7.25: (a) Schematic diagram of three kinds of layers on an EDM'd component, (b) Electric discharge machined surface showing redeposited molten material (grey), (c) Photograph of an electro-discharge drilled hole [*Indurkhya*, 1985; *Jain*, 1989].

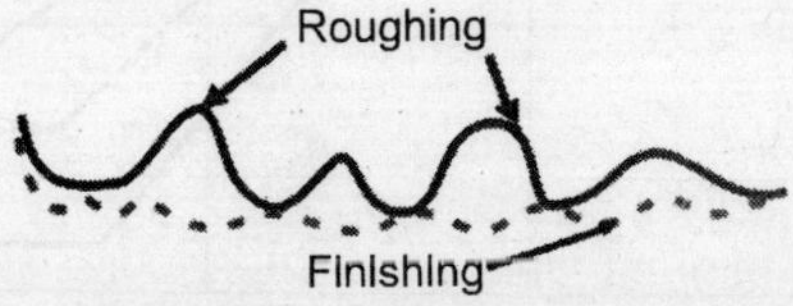

Fig. 7.26: Schematic diagram of roughing and subsequent finishing operation [*Kahng and Rajurkar*, 1977]

Interelectrode Gap Cleaning

Interelectrode Gap (IEG) or gap cleaning is one of the crucial factors for good *EDM*. At the end of a spark, cooling starts and the spark, plasma channel and vapour bubbles collapse. As a result, relatively cool dielectric fluid rushes into this zone of a spark. This leads to the explosive expulsion of the molten material from the electrodes. This process of removal of molten material is greatly assisted by *forced flushing* of the gap by dielectric.

EDM usually gives machined workpiece surface of *matte appearance*. Such surface is achieved because of removal of material from the workpiece in the form of craters. Craters produced during EDM are approximately a part of a sphere in shape having the depth of the crater as 0.10 to 0.02 times of its diameter. The typical values of diameter and depth of craters are 12.50 μm and 1.25–0.25 μm, respectively. The ED machined surface is fully covered with such craters resulting in non-directional surface (Figure 7.25(c)), which is different from the surface obtained by a conventional machining methods say, turning and milling.

Normalized steel was turned to a surface finish of 1.5–2.0 μm while another specimen made of the same steel was spark eroded to a surface finish of 6–8 μm. It has also been found [*Barash*, 1959] that the *endurance limit* of the spark eroded specimen was about 15% less than that of the turned one (Figure 7.27). In some cases this decrease in endurance limit may be more than 15%. With decrease in pulse duration, the roughness of the machined surface and the length of the microcracks also decreased. Hence, transverse rupture strength [*Heuvelman*, 1980] and endurance limit are found to decrease with increase in pulse duration specially while machining WC. However, below 3 μs pulse duration, microcracks are hardly observed. Further, post-machining operations like '*shot peening*' [*Neema and Pandey*, 1980] helps in improving surface integrity of the machined surface. This operation (shot peening) also enhances fatigue strength and stress corrosion resistance of the machined component.

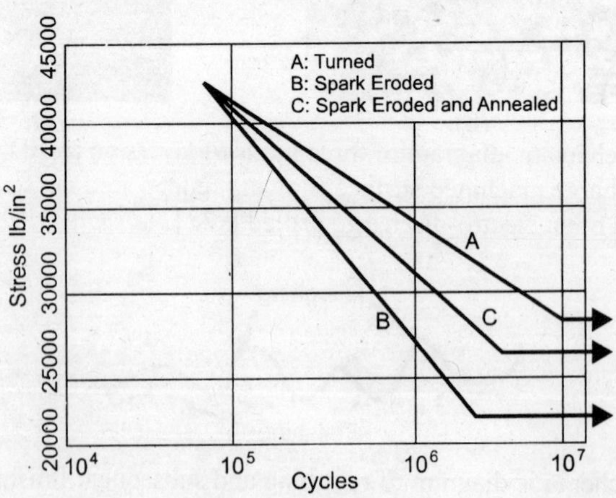

Fig. 7.27: Effect of type of machining operation on the endurance limit [*Barash*, 1959]

APPLICATIONS

EDM can be employed to machine any material (hard, tough, brittle, etc.) provided it has some minimum electrical conductivity (Figure 7.28). The manufacture of hardened steel *dies* is also the field of application of EDM, other than aerospace, automobile, tools, and machine tool components. It is used for making through cavities, and miniature holes.

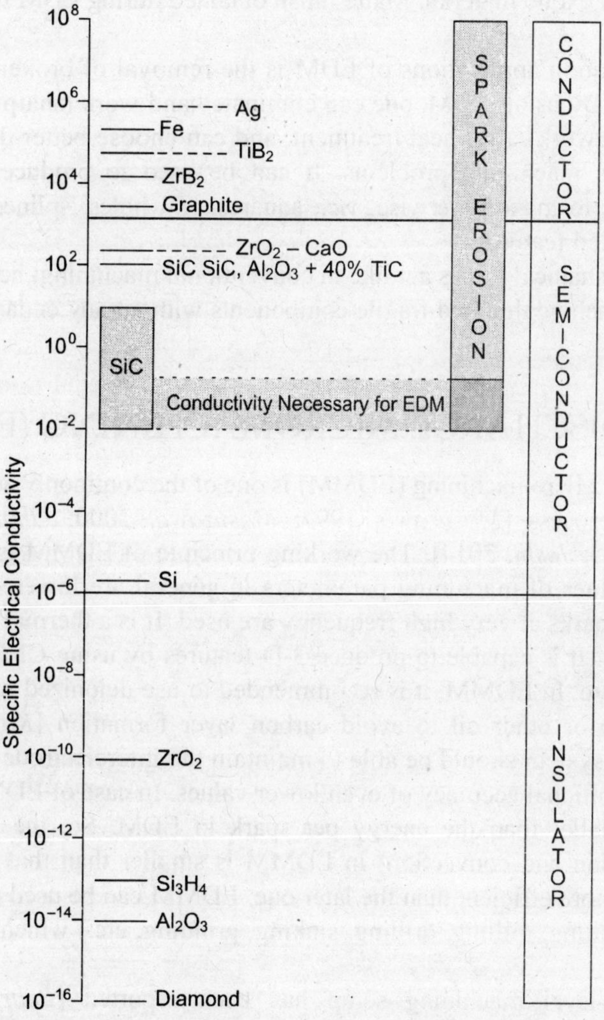

Fig. 7.28: Specific electrical conductivity of various materials [*Konig and Dauw*, 1988]

Application of EDM in die and mold engineering is exemplary. It is used in making components for plastic injection molding m/c [*Albert*, 1981]. EDM can be used for making dies for molding, casting, forging, stamping, coining, forming, etc. It is also used for making dies for extrusion, wire drawing, etc., which require through holes. It is employed for making tiny holes, orifices (say, 50 μm) and fragile features (micro sized slots). In some cases of making of intricate shaped dies, the time has come down to 50% or so. Some of the industries claim that because of EDM technology die changes are faster. ED machined dies are free of burrs and have higher life as compared to the dies made by using conventional methods. It permits the use of more durable die materials, *viz*., carbide, hardened steel, and exotic material. Matte finish obtained during EDM minimizes polishing time required.

One of the common applications of EDM is the removal of broken taps, drills, studs, reamers, pins, etc. By using EDM, one can eliminate hand work (sharp details); we can do EDM (including rework) after heat treatment, and can choose better die material without worrying about its machining problems. It can be used to produce shapes which are extremely difficult to make otherwise, *viz*., squares, '*D*' holes, splines, narrow slots and grooves, and blended features.

In EDM, no mechanical forces act like in conventional machining; hence the process can be employed to machine thin and fragile components without any endanger of damage due to the forces.

ELECTRIC DISCHARGE MICROMACHINING (EDMM)

Electric Discharge Micromachining (EDMM) is one of the commonly advocated tool based micromachining process [*Masuzawa*, 1997; *Masuzawa,* 2000; *Uhlmann et al.,* 2005; *Dhanik et al*., 2005; *Joshi*, 2014]. The working principle of EDMM is the same as EDM except that the values of machining parameters in general are lowered down. Normally, very low energy sparks at very high frequency are used. It is a thermal non-contact micro-machining process. It is capable to produce 3-D features by using CNC electric discharge micromachining m/c. In EDMM, it is recommended to use deionized water as dielectric in place of Kerosene or other oil to avoid carbon layer formation [*Kagaya et al.,* 1990]. Further, the servo system should be able to maintain the interelectrode gap of the order of 1 μm or so and positional accuracy of even lower values. In case of EDMM, the energy per spark is much smaller than the energy per spark in EDM. So, the loss of energy (by conduction, radiation and convection) in EDMM is smaller than that in EDM hence the earlier process is more efficient than the later one. EDMM can be used to perform all those operations (say, cutting, milling, drilling. sinking, grinding, etc.) which can be done on an EDM setup.

A micro-/nano-level machining setup has been reported [*Beltrami et al.,* 2004] (8 mm × 8 mm × 8 mm) having a resolution of 5 nm using electromagnetic motors for actuation in X, Y, Z axes. The errors that could occur are 20 nm. One of the important issues in EDMM is the fabrication of high aspect ratio tool electrode (cathode). Four different methods (wire-electric discharge grinding, rotating sacrificial disk, stationary sacrificial block and self drilled micro hole) have been discussed in the ref. [*Joshi*, 2014].

Making of high aspect ratio (say, 15–16) holes has been reported [*Puranik and Joshi*, 2008] using normal as well as reverse polarity with different accuracies of the drilled holes. To improve the accuracy by using the easy removal of the debris, a horizontal machine in place of a vertical machine has been advocated [*Masuzawa et al.*, 1989]. Machined depth, tool wear, and machined feature accuracy are affected by the machining conditions, and circuit parameters (capacitance, and resistance). It is concluded [*Han et al.*, 2004] that the available smallest spark energy has to be further reduced and the positional accuracy has to be further improved in the EDMM system to go for a nanometer size feature development.

In EDMM and EDM, the machined feature's accuracy depends upon tool wear as well as machining conditions and tool-work material properties. Attempts have been made to quantify the tool wear resistance as follows [*Tsai and Masuzawa*, 2004].

$$\text{Electrode wear resistance Index} = \rho c \lambda T_m T_b. \qquad \ldots (7.23)$$

Where, λ = thermal conductivity, ρ = electrode density, T_m = melting temperature, T_b = boiling temperature, c = specific heat.

The main factor which affects the accuracy of the deep hole is the time for which EDM/EDMM takes place at different locations along the hole-axis. If it is a constant feed rate (assumption) then the time for which point A and point B are machined, varies almost linearly, Figure 7.29(b). Because of this machining time variation, hole diameter is larger at the top and smaller at the bottom. Sparking (or removal of material) at point A will stop when the IEG becomes greater than the maximum gap at which the sparking can take place under the given machining conditions. To stop sparking deliberately, design the tool upside down replica of the hole as shown in Figure 7.29(c). This should be done scientifically by first developing the tool erosion model for the process and then use it to design the tool or compute the taper or difference in radius of the tool along its axis. Follow the process as given in the ref. [*Reddy et al.*, 1988].

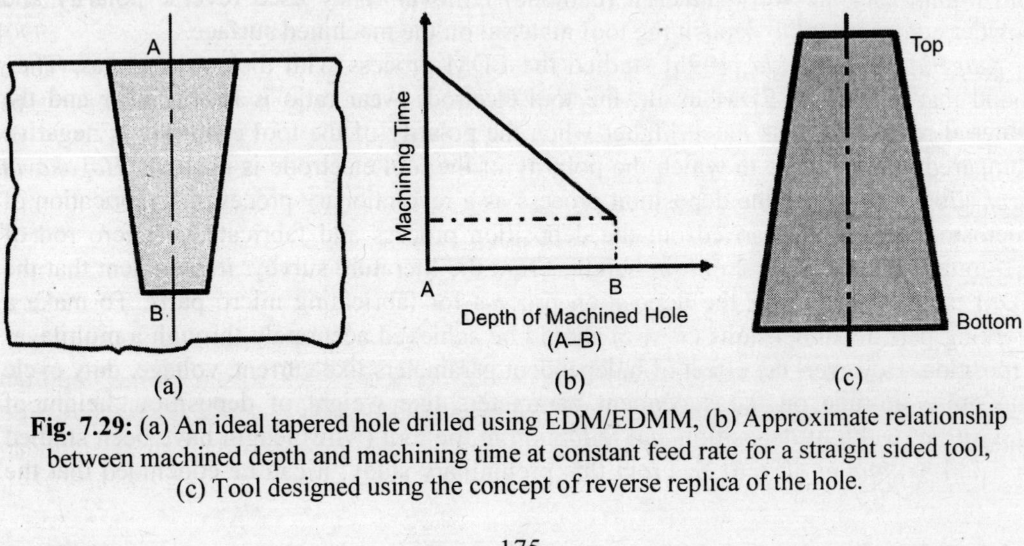

Fig. 7.29: (a) An ideal tapered hole drilled using EDM/EDMM, (b) Approximate relationship between machined depth and machining time at constant feed rate for a straight sided tool, (c) Tool designed using the concept of reverse replica of the hole.

ELECTRIC DISCHARGE MICRODEPOSITION (EDMD)

There are basically two approaches for the fabrication of any part: 'top to bottom' and 'bottom to top'. In the first one, the material is being removed by attrition process(es) from the original blank to give the desired shape and size while in the second case, the material is deposited layer by layer to achieve the desired shape and size of the part as per design specifications. In the first case (top to bottom), mechanical and physical properties of the macro/meso part are the same as that of the original blank. But, in the second case mechanical and physical properties of the macro/meso parts are not very well predictable and they are different from that of the bulk material. It requires a lot of research work yet to be done in this direction. In both cases of fabrication (macro and micro), micro/nano parts have different mechanical and physical properties as compared to those of the bulk material.

The beauty of the EDM process is that it can be used to fabricate a part by both techniques, by removal of material as well as by deposition of material, irrespective of the size of the part, that is meso/macro part as well as micro part. So far, we have seen that during EDM material removal takes place mainly from the workpiece (or anode) and we try to minimize material removal from the tool to maintain the accuracy of the machined parts.

In case of ED deposition, we deliberately remove the material from the tool and deposit it on the substrate to form the micro feature/micro structure.

According to *Meshcheriakov* [1970], in electric discharge machining there is positively charged state which is responsible for driving molten metal towards the cathode. The anodic material transfer towards the cathode is a significant phenomenon, giving rise to the concept of surface modification. *Gangadhar et al.* [1991] conducted studies to prove that by employing suitable parameters, surface modification by metal transfer from tool electrode to the machined surface is possible by electric discharge machining process using EDM setup. The major criterion they applied is maximizing the tool (anode) erosion and minimizing the work material (cathode) removal. They used reverse polarity and powder compact tool to deposit the tool material on the machined surface.

Kunieda and Yoshida [1996] studied the EDM process with dielectric as gas. They found that in case of EDM in air, the tool electrode wear ratio is much lower and the material removal rate is much higher when the polarity of the tool electrode is negative compared with the case in which the polarity of the tool electrode is positive. *Hayakawa et al.* [2003] described the deposition process as a revolutionary process for fabrication of microstructures. They carried out the deposition process and fabricated a micro rod of 0.14 mm diameter and 2.2 mm in height. From the literature survey, it is evident that the EDM is also suitable for the deposition process for fabricating micro parts. To make a working part, its dimensions (*x, y, z*) should be achieved accurately through a multilayer deposition. However, the effect of independent parameters like current, voltage, duty cycle and pulse on-time on the dependent parameters like weight of deposition, height of deposition, width of deposition and reduction in the tool (wire) length have been studied in detail by *Jain et al.* [2012]. From this preliminary study, it can be concluded that the

Electric Discharge Deposition process can be used as a future alternative to various macro/micro fabrication techniques. Figure 7.30(a) shows the effect of voltage on the height of the developed feature. Figure 7.30(b) shows the effect of current on the width of the deposition. Figure 7.31 shows four different microfeatures developed using electrodischarge deposition process.

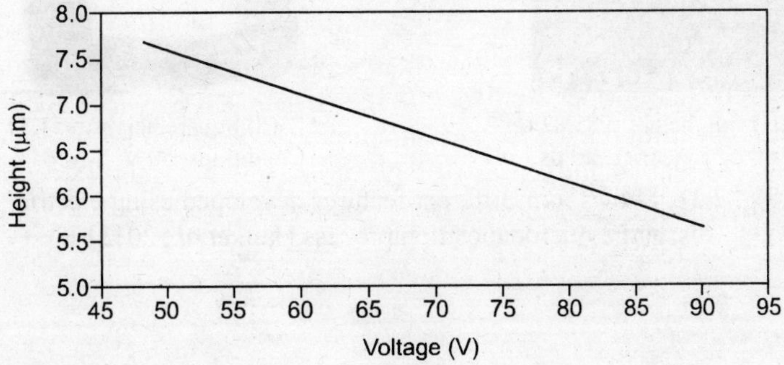

(a)

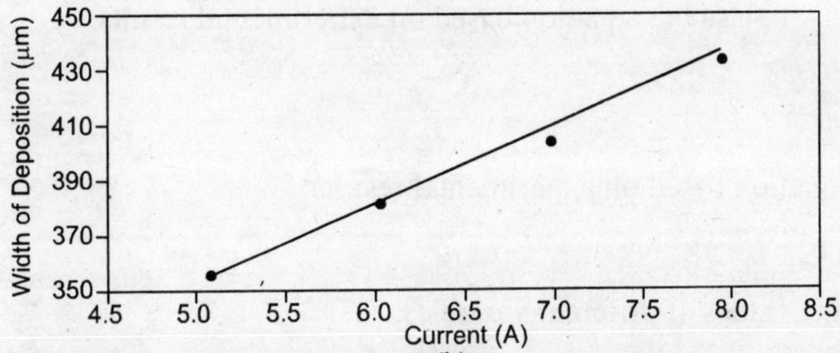

(b)

Fig. 7.30: (a) Effect of voltage on deposition height; current = 7 A, duty cycle = 56%, pulse on-time = 20 μs, (b) Effect of current on width of deposition; current = 7 A, duty cycle = 56%, pulse on-time = 20 μs [*Jain et al.*, 2012].

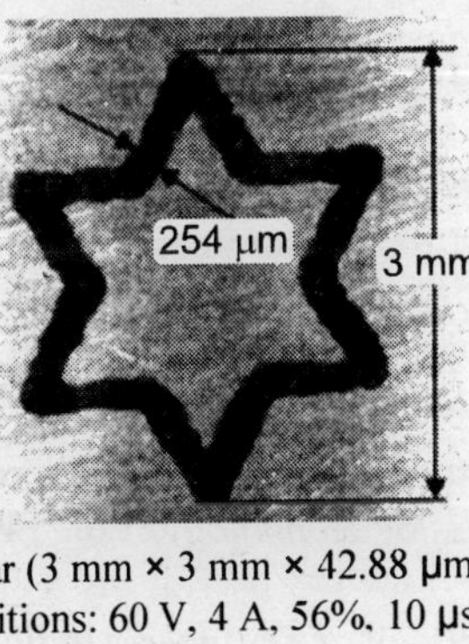

(a) Star (3 mm × 3 mm × 42.88 μm)
Conditions: 60 V, 4 A, 56%, 10 μs

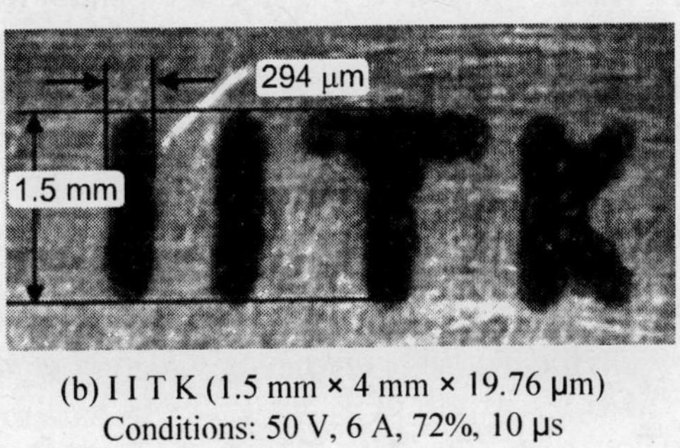

(b) I I T K (1.5 mm × 4 mm × 19.76 μm)
Conditions: 50 V, 6 A, 72%, 10 μs

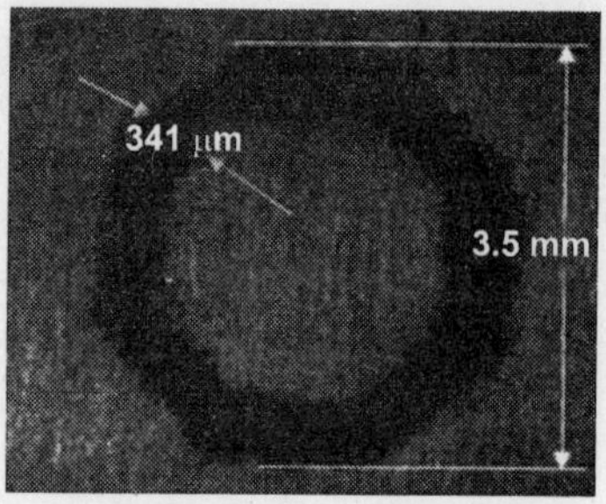

(c) Octagon Side 1 mm height = 55.82 µm
Conditions: 60 V, 5 A, 56%, 20 µs

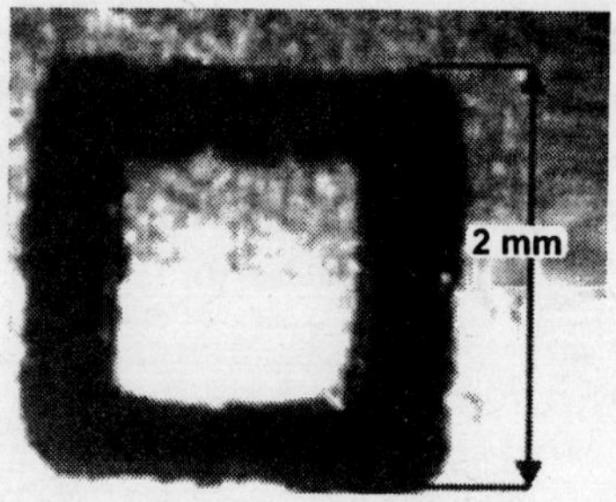

(d) Square height = 51.51 µm
Conditions: 50 V, 5 A, 64%, 20 µs

Fig. 7.31: Shows four different features developed using electric discharge microdeposition process [Jain *et al.*, 2012]

PROBLEM

EDM is used to machine a metallic sheet. Calculate surface finish value if C = 15 µF, V_b = 130 V, K_6 = 4.0. Use the equation based on experimental results.

Solution

Surface finish equation based on experimental results:

$$H_{CLA} = K_6\ V_b^{0.5} C^{0.33}$$

Substituting the values of different variables,

$$H_{CLA} = 4.0 \times (130)^{0.5} \times \left(15 \times 10^{-6}\right)^{0.33}$$

$$H_{CLA} = 1.16\ \mu\text{m}.$$

BIBLIOGRAPHY

[1] Albert, M. (1981). "EDM: Close Control for a Closer Shave", *Modern Machine Shop,* pp. 69–86.

[2] Albert, M. (1982). "How CNC has Changed EDM", *Modern Machine Shop*, pp. 1–11.

[3] Barash, M. (Feb. 1959). "Some Properties of Spark Eroded Surfaces", *Microtechnique*, pp. 51–54.

[4] Benedict, G.F. (1986). "Non-traditional Manufacturing Processes", *Marcel Dekker, Inc.* New York.

[5] Beltrami, I., Joseph, C., Clavel, R., Bacher, J. and Bottinelli, S. (2004). "Micro and nano electric discharge machining", *Journal of material processing technology*, Vol. 149, pp. 263–265.

[6] Bhattacharyya, A. (1973). "New Technology", *The Institution of Engineers* (India), Culcutta.

[7] CIRP Scientific Technical Committee E. (1969). "Summary Specifications of Pulse Analysers for Spark Erosion Machining", Vol. 2, No. 2.

[8] Davw, D.F., Snoeys, R. and Dekeyser, W. (1983). "Advanced Pulse Discriminating System for EDM Process Analysis and Control", *Annals CIRP*, Vol. 32, pp. 541–549.

[9] de Bruyn, H.E. (1960). "Some Aspects of the Influence of Gap Flushing on the Accuracy in Finishing by Spark Erosion", *Annals CIRP*, Vol. 18, pp. 146–151.

[10] de Bryun, H.E. (1968). "Effect of a Magnetic Field on the Gap Cleaning in EDM", *Annals CIRP*, Vol. 26, pp. 93–95.

[11] Dhanik, S. and Joshi, S.S. (2005). "Modeling of a Single RC-pulse Discharge in Micro-EDM", Transactions of ASME, *Journal of Manufacturing Science and Engineering,* Vol. 127, pp. 759–767.

[12] Erden, A. and Bilgin, S. (1980). "Role of Impurities in EDM", *Proc. 21st Int. Machine Tool Design and Research Conference,* Swansea, pp. 345–350.

[13] Erden, A. (1982). "Role of Dielectric Flushing on EDM Performance", *Proceedings of the 23rd Int. Machine Tool Design and Res. Conference,* pp. 283–289.

[14] Gangadhar, A., Shunmugam, M.S. and Philip, P.K. (1991). "Surface modification in electro discharge processing with powder compact tool electrode", *Wear,* Vol. 143, pp. 45–55.

[15] Godinho, L.H. and Noble, C.F. "The Use of Water as Interelectrode Medium in Pulsed EDM", pp. 63–66 (personal communication).

[16] Han, F., Wachi, S. and Kunieda, M. (2004). "Improvement of machining characteristic of micro-EDM using transistor type isopulse generator and servo feed control", *Precision Engineering,* Vol. 28, pp. 378–385.

[17] Hayakawa, Ori, Itoigawa, Nakkamura and Matsubara (2003). "Fabrication of microstructures using EDM deposition", *ISEM XIV Proceeding,* pp. 783–792.

[18] Hewuvelman, C.J. (1980). "CIRP Technical Reports", *Annals CIRP,* Vol. 29, No. 2, pp. 541–544.

[19] Indurkhya Gopal (1985). "Some Investigations into electro discharge drilling process" *M.Tech Thesis, Deptt. of Mech. Engg.,* I.I.T. Kanpur.

[20] Jain, V.K. (1989). "Analysis of Electro Discharge Drilling of a Blind Hole in HSS Using a Bit Type of Tools", *Microtecnic,* No. 2 , p. 34.

[21] Jain, V.K., Shashank; Sidpara, Ajay and Jain, Himanshu (2012). "Some aspect of micro-fabriaction using electro-discharge deposition process", Int-sym. on Flexble Automation (ISFA 2012). June 18–20, 2012, St. Louis, Missouri, USA, Paper # ISFA2012 – 7108, pp. 419–424, do::10.115/ISFA2012-7108.

[22] Jilani, S.T. and Pandey, P.C. (1984). "An Analysis of Surface Errors in Electrical Discharge Machining", *Wear*, Vol. 84, pp. 265–284.

[23] Jilani, S.T. and Pandey, P.C. (1984). "Experimental Investigations into the Performance of Water as Dielectric in EDM", *Int. Journal of Machine Tool Design and Research,* Vol. 24, No. 1, pp. 31–43.

[24] Joshi, S.S. (2014). "Micro-electric discharge micromachining", in the book, Introduction to Micromachining (Ed: V.K. Jain). Narosa Publishers, New Delhi.

[25] Kahng, C.H. and Rajurkar, K.P. (1977). "Surface Characteristic Behaviour Due to Rough and Fine Cutting by EDM", *Annals of the CIRP*, Vol. 25, No. 1, pp. 66–82.

[26] Kagaya, K., Oishi, Y. and Yada, K. (1990). "Micro electro-discharge machining using water as a working fluid 2: Narrow slit. fabrication", *Precision Engineering, Butterworth & Co (Publishers) Ltd.*, pp. 213–217.
[27] Koing, W. and Dauw, D.F. (1988). "EDM- Future Step towards the Machining of Ceramics", *Annals of the CIRP,* Vol. 36, No. 2, pp. 623–631.
[28] Kremer, D., Lebrun, J.L., Hosori, B. and Moisan, A. (1989). "Effects of Ultrasonic Vibrations on the Performance in EDM", *Annals CIRP*, Vol. 38 pp. 199–202.
[29] Kunieda, M., Tanihuchi, N. and Yoshida, M. (1997). "Electric discharge machining in gas", *Annals CIRP,* Vol. 46, pp. 143–146.
[30] Masuzawa, T. and Heuvelman, C.J. (1983). "A Self Flushing Method with Spark Erosion Machining", *Annals CIRP*, Vol. 32, pp. 109–111.
[31] Masuzawa, T., Tsukamoto, J. and Fujino, M. (1989). "Drilling of deep microholes by EDM", *Annals CIRP*, Vol. 38(1), pp. 195–198.
[32] Masuzawa, T. and Tonshoff, H.K. (1997). "Three-dimensional micromachining by machine tools", *Annals CIRP,* Vol. 46(2), pp. 621–628.
[33] Masuzawa, T. (2000). "State of the art micro-machining", *Annals CIRP*, Vol. 49(2), pp. 621–628.
[34] McGeough, J.A. (1988). "Advanced Machining Methods", Chapman and Hall, London.
[35] Neema, M.L. and Pandey, P.C. (1980). "Investigation of the Performance Characteristics of Cold-worked Machined Surfaces", *Wear*, Vol. 60, pp. 156–166.
[36] Pandey, P.C. and Shan, H.S. (1980). "Modern Machining Processes", Tata McGrawHill, New Delhi (India).
[37] Puranik, M.S. and Joshi, S.S. (2008). "Analysis of accuracy of high-speed aspect ratio holes generated using micro-EDM drilling", *Proceedings of the Institution of Mechanical Engineering,* Part *B, Journal of Engineering Manufacturing*, Vol. 222, pp. 1453–1464.
[38] Quinlan, J.C. (1983). "Harnessing the Spark the Evolution in EDM", Tooling and Production, pp. 36–42.
[39] Reddy, M.S., Jain, V.K. and Lal, G.K. (1988). "Tool design for ECM: Correction Factor Approach", *Trans. ASME, J. Engg. for Ind.*, pp. 111–118.
[40] Shankar, Prem, Jain, V.K. and Sundararajan, T. (1997). "Analysis of Spark Profiles During EDM Process", *Machining Science and Technology*, Vol, 1, No. 2, pp. 195–216.
[41] Snoeys, R. and Van Dijck, F. (1962). "Plasma Channel Diameter Growth Affects Stock Removal in EDM", *Annals of the CIRP,* Vol. 21, pp. 39–40.
[42] Sreejith, P.S., Jain, V.K. and Lal, G.K. (1994). "Investigations into the Characteristics of Spike Obtained During EC Drilling of Blind Holes in HSS", *Processing of Advanced Materials,* Vol. 4, pp. 66–69.
[43] Tsai, Y.Y. and Masuzawa, T. (2004). "An index to evaluate wear resistance of the electrode in micro-EDM", *Journal of material processing technology,* Vol. 149, pp. 304–309.
[44] Uhlmann, E., Piltz, S. and Doll, U. (2005). "Machining of micro/miniature dies and moulds by electrical discharge machining–Recent developments", *Journal of material processing technology*, Vol. 167, pp. 488–499.
[45] Vitlin, V.B. (1966). "Electro-Contact Abrasion as a Finishing Metal Cutting Process", *Machines and Tooling*, Vol. 48, No. 4, pp. 32–34.
[46] Yadav, Vinod, Jain, V.K. and Dixit, P.M. (2004). "Theoretical analysis of thermal stresses in Electro-discharge Diamond Grinding", *Machining Science and Technology*, Vol. 8(1), pp. 119–140.

SELF TEST QUESTIONS

Q1. Write True (T) or False (F).

1. Usually the life of a die made using EDM process is higher than the die made using ECM process.
2. To provide the controlled electrical resistance in the inter-electrode gap during EDM, an appropriate insulating fluid is flooded between the tool and workpiece.
3. Tool life in EDM is infinite.
4. Dielectric fluid is reused during EDM.
5. EDM can not be used to drill a slanted hole in the aluminum alloy workpiece.
6. Flash point of the dielectric used in EDM should be high enough.

Q2. Make the most appropriate choice.

1. Surface damage to the machined component is insignificant in case of:
   (a) EDM (b) ECM (c) Grinding (d) PAM
2. During calculation of MRR in EDM, supply voltage was used as 60 V in place of the actual supply voltage as 40 V. What is the ratio of actual and calculated MRR? Assume that the condition for maximum power delivery to the discharging circuit is satisfied.
   (a) 1.5 (b) 0.44 (c) 2.25 (d) none
3. The current used during EDM is:
   (a) AC (b) DC (c) Pulsed AC (d) Pulsed DC
4. Location of spark generated during EDM is:
   (a) random (b) governed by dielectric strength
   (c) governed by surface finish of tool and workpiece both (d) none of these
5. For maximum power delivery through the discharge circuit during EDM, ratio of the breakdown voltage to supply voltage is:
   (a) 0.62 (b) 1.0 (c) 0.65 (d) 0.72
6. Feed to the electrode during EDM is given:
   (a) to maintain constant velocity of dielectric flow in the IEG
   (b) to maintain constant IEG (c) to maintain constant current density
   (d) none of these
7. Relative Electrode Wear (REW) during EDM is:
   (a) <100% (b) >100% (c) 100% (d) any one of these
8. Value of surface roughness with the increase in breakdown voltage:
   (a) increases (b) decreases (c) remains unchanged
9. Discharging time (in RC circuits) as compared to charging time is:
   (a) more (b) less (c) equal (d) no definite relationship

10. Servomechanism operates when there is:
    (a) short circuit (b) high IEG (c) low IEG (d) all of these
11. Poor flushing during EDM results in:
    (a) low MRR (b) short circuit (c) both (d) none
12. In EDM, a graphite electrode composed of very fine grains results in surface finish of the machined component as:
    (a) improved (b) deteriorated (c) no effect
13. If the flushing system of EDM m/c is inefficient the machining cycle time will be:
    (a) longer (b) shorter (c) no effect (d) no definite trend
14. Wear of tool during EDD process would result in:
    (a) deeper hole (b) lesser diameter hole (c) both (d) none
15. Electrode refeed system is used to compensate for wear in:
    (a) length (b) diameter (c) both (d) none
16. During EDM, the charged voltage of the condenser is increased from V1 to V2 (V2 = 4V1). What will be surface roughness at V2 if its value at V1 is equal to Ra1:
    (a) 2 Ra1 (b) 2.5 Ra1 (c) 4 Ra1 (d) 0.25 Ra1 (e) none
17. In the above question in place of changing the charged voltage the frequency is changed from f1 to f2 ($f_2 = 0.5$ f1). The new surface roughness is:
    (a) 2 Ra1 (b) 0.5 Ra1 (c) 4 Ra1 (d) 0.25 Ra1 (e) none
18. Which process would you recommend to make cooling holes (diam. = 0.5 mm, and depth to diameter ratio may be as high as 65) in a turbine blade.
    (a) EDM (b) AJM (c) USM (d) PAM (e) none
19. For high MRR, dielectric to be used in EDM should have dielectric strength:
    (a) high (b) low (c) moderate (d) any of these

Q3. Review Questions.

1. Explain the following terms:
   (a) duty factor, (b) ignition delay, (c) flash point, (d) relative electrode wear or wear ratio, (e) Heat Affected Zone (HAZ), (f) dielectric strength, (g) time constant, (h) critical resistance in RC circuit.
2. What are the functions of an 'adaptive control system' used for EDM?
3. With the help of a neat sketch, explain the mechanism of material removal in EDM.
4. Find the condition for maximum power delivery to the discharging circuit in EDM.
5. Sketch different feasible dielectric flushing techniques applicable in case of ED drilling.
6. Derive the relationship for surface roughness in EDM.
7. Write short note on the following: Resolidification of molten metal in EDM.

8. When using water as dielectric during machining of HSS by EDM process, which specific problem do you expect?
9. Sketch the effects of following parameters on MRR during EDM: (a) resistance, (b) current density, (c) pulse energy, (d) capacitance.
10. Illustrate the effect of (a) magnitude of current and (b) frequency of sparking on the shape and size of the craters formed during EDM.
11. A through hole (diam. = 10 mm) in a plate (5 mm thick) is to be drilled using EDM. Estimate the time required for the process. Assume $R = 50\ \Omega$, $C = 10$ mF, $V_s = 200$ V, $V_b = 150$ V, $K_1 = 0.18$ to obtain MRR in mm^3/min.
12. For RC circuit, adjusted for maximum power delivery condition, the following data are available: $R = 250\ \Omega$, $C = 25$ mF and supply voltage = 65 V. Calculate charging current and frequency of discharge when the circuit is closed.
13. A 10 mm diameter hole is to be drilled in a 5 mm HSS plate by EDM using RC circuit. The required surface finish is 20 μm. Determine specifications of a capacitor to be used when supply and discharge voltages are 220 V and 150 V, respectively. Use resistance as 50 Ω and value of the constant as 0.4. Also, estimate time required to complete the job. (Use surface finish in μm to get the value of C in Farads for $k = 0.4$). Use constant of proportionality as k.

NOMENCLATURE

C	Capacitor
c	Specific heat
E	Energy
f	Frequency
h	Crater depth
H	Center line average value of surface finish
i	Current
$K, K_1, \ldots, K_6$	Constants
m, n	Exponents
L	Inductance
P	Power
R	Resistance
Tb	Boiling temperature
Tm	Melting temperature
t	Time
V	Voltage

GREEK SYMBOLS

ρ	Tool material (electrode) density
λ	Thermal conductivity

ACRONYMS

AC	Adaptive Control
ATC	Automatic Tool Changer
CNC	Computer Numerical Control
DC	Direct Current
EDM	Electric Discharge Machining
EDD	Electric Discharge Drilling
Hz	Hertz
IEG	Inter Electrode Gap
KE	Kinetic Energy
MRR	Material Removal Rate
m/c	Machine
m/t	Machine Tool
NC	Numerically Controlled (Numerical Control)

REW	Relative Electrode Wear
SF	Surface Finish
T_{bt}	Boiling Temperature of Liquid
TWR	Tool Wear Rate
V_b	Breakdown Voltage
W/P	Workpiece

SUBSCRIPTS

Avg	Average
b	Break down condition
c	Charging (circuit)
ct	Charging at time 't'
d	Discharging (circuit)
Min	Minimum
o	Supply condition
s	Sparking (zone)

CHAPTER 7
AT-A-GLANCE
ELECTRIC DISCHARGE MACHINING (EDM)

- Sparking between Two Electrical Contacts → Loss of Material
- Harness → Control the Spark Energy to Employ for Machining
- If Discharge Submerged in Dielectric → Energy Concentrates into a Small Area.
- At Breakdown Voltage → Sparking Occurs at a Point of Least Electrical Resistance → (At Nerrowest Gap)
- Recharging/Sparking Frequency → Many Thousands Per Second
- Functions of the Dielectric → Cooling of Electrodes
  → Concentration of Spark Energy
  → Medium for Sparks to Occur

Working Principle

- Localised Spark Energy → Very High Temperature → Melting and Vaporization of Tool and Workpiece material
- Material Removal → By Crater Formation Over Entire W/P Surface
- Servosystem to Maintain Gap → Moves Tool Towards/Away From Workpięce

How Sparking Takes Place?

- Intense Electrical Field at the Narrowest Gap
- Negatively Charged Particles (Electrons) Detach from Cathode → Move Towards the Anode
- Electrons Collide with Neutral Molecules of:

  Dielectric → Ionisation — Electrons Move Towards Workpiece / Ions Move Towards Tool

  Narrow Channel of Continuous Conductivity (Plasma Channel)

  ↓

  Momentary Current Impulse (Discharge/Spark)

  ↓

  Spark Energy Raises Temperature → Melting and Vaporisation

  ↓

  Metal is Removed from Tool and Workpiece → Craters Formed

  - Frequency of Sparking → Many Thousands/Second
  - To Minimise Wear of Tool → Machining Conditions and Polarity are Selected Carefully

RC Pulse Generator

- Low MRR ← Long Idle Time, Short Spark Time
- Improved Surface Finish
- High TWR
- Attains High Peak Current and Then High Rate of Decline

  ↓

  Very High Temperature ← Not Needed

  ↓

  Thermal Damage to WP and Tool
- Controlled Pulse Generator overcomes this Problem
- Enables to Select Rough Machining (High Energy and Low Frequency) or Finish Machining (Low Energy and High Frequency)

EDM Machine Elements
- ⇒ Power Supply
- ⇒ Dielectric System
- ⇒ Electrode, Workpiece
- ⇒ Servo System

Power Supply

- Solid State Rectifier → AC to DC
- High Power Pulsed Output → Sparking
- For a Given Voltage → Sparking at or Below a Minimum Gap
- EDM Power Supply and Servosystem Control → Desired Gap Value
- Also Controls Parameters → V, I, Pulse Duration, Duty Factor, Frequency and Polarity
- Equipped with Cut-Off-Protection Circuit → in Case of Over-Voltage or Over-Current → Power Cuts Off

Dielectric System
- ⇒ Dielectric Fluid
- ⇒ Reservoir
- ⇒ Filters
- ⇒ Pump
- ⇒ Delivery Devices

- Good Dielectric Fluid
  - High Dielectric Strength
  - Minimum Ignition Delay Time
  - Minimum Deionisation Time
  - Effective Coolant
  - High Degree of Fluidity

- Dielectric Fluids
  - Transformer Oil
  - Paraffin Oil
  - Lubricating Oil
  - Kerosene Oil
  - Deionised Water → High MRR, High TWR, Corrosive, Good Coolant → Used in Wire EDM
- Effective Flushing
  - Removes "By – Products" From Gap
  - High MRR and Good SF
  - Avoids Short Circuits ← "Absence of Residues"
- Techniques for Proper Flushing
  - Suction through Tool/Workpiece
  - Pressure through Tool/Workpiece
  - Jet Flushing
  - Rotating Disk Electrode
  - Alternating Forced Flushing
- Machining of Blind Cavities
  - Flushing through a Hole in the Tool → Spike Formation

    ↓

    Can be Avoided by Rotating a Tool with Eccentric Hole(s)
  - Jet Flushing Less Effective
  - Inflammable Dielectric → Workpiece Immersed in Dielectric → Minimizes Chances of Fire
  - Arcs are Undesirable → Caused Due to Concentration of Debris in IEG

    ↑

    Requires Proper Filtration/Flushing
  - Electrodes → Tool & Workpiece → Both Electrically Conductive

    Tool Material Requirements
    - Easily Machinable
    - Low Wear Rate
    - Good Conductor of Electricity and Heat
    - Cheap and Readily Available
- Tool Materials → Graphite, Copper, Brass, Copper Tungsten, etc.
  - Graphite
    - Lower TWR
    - Higher MRR
    - Better SF
    - Brittleness → Prone to Breakage

- ▸ Overcut

  Overcut during EDD is a Function of
  - Machining Conditions
  - Workpiece Material
  - Tool Material
- ▸ Take Care at Design Stage
- ▸ Through Hole
  - Only for Comparatively Small Thickness of Material
- ▸ Servo System
  - To Maintain Pre-Determined Gap between Electrodes
  - Gap Bridged by Electrically Conductive Material → Sends Signal to Servo System → Reverses Direction of Tool/WP → Movement
  - Reciprocates towards Workpiece Until Dielectric Flushes Gap
  - If Inefficient Flushing → Very Long Cycle Time due to Tool Oscillation
- ▸ Tool Wear
  - Compensation for Reduced Tool Length → Tool Wear

    ↓
  - Increase Infeed after Each Hole
  - Electrode Refeeding → Brings Tool to the Reference Point

- Automatic Tool Changer (ATC)
  - ▸ Recent Innovation
  - ▸ Many Tools Mounted on a Magazine → Called by CNC Program
  - ▸ Reduces Tool Change Time
  - ▸ CNC-EDM More Popular

Process Variables

- Increase in Current or Spark Voltage → Increased MRR → Higher Surface Roughness
- Increase in Spark Frequency → Improved SF ← Energy Shared by More # of Sparks → Decreased Crater Size
- Low IEG → Low MRR, High SF, Better Accuracy
- Decreased Pulse Duration → Low Value of MRR and SF. High Electrode Wear

Process Characteristics

- Workpiece Material → Electrically Conductive
- Operations → Drilling, Slotting, Die Sinking, etc.
- Accuracy → ± 0.025 to ± 0.126 mm

- Deep Accurate Holes → Roughing/Finishing Passes ← To Minimize Taper (0.005 to 0.050 mm/cm)
- Upper Limit of Aspect Ratio → 100 : 1 ← Special Care of Flushing
- MRR_v → Low
- Matte Surface
- Kinds of Layers
  - Recast Layer → 2 to 50 μm → Extremely Hard and Brittle ← Remove before Use
  - HAZ → ~25 μm. Residual Stresses, Grain Boundary Cracks, etc.
  - Converted Layer → Change in Grain Structure

Applications

- Any Material → Electrically Conductive
- Aerospace, Automobiles, Tools and Die Making Industries
- Thin Fragile Components ← No Danger of Damage

Electric Discharge Micromachining (EDMM)

- Working Principle of EDMM & EDM → Same
- Low Energy/Spark and High Frequency
- Thermal Non-Contact Micromachining
- Gap 1 μm or so
- Fabrication of High Aspect Ratio Tool: Four Methods
- Machined Feature Characteristics = (Machining Conditions, Tool Wear & EDM m/c)
- Electrode Wear Resistance Index
- Deep Microhole Drilled Accuracy → (Machining Time along the Hole Axis)

Electric Discharge Microdeposition (EDMD)

Fabrication Approaches

- "Top to Bottom" → Attritious Processes
- "Bottom to Top" → Layer by Layer Deposition Processes/Additive Manufacturing
- Bottom to Top Approach → Mechanical & Physical Properties are Different from Bulk Material
- Use Reverse Polarities
- Maximize Tool (Anode) Erosion & Minimize Workpiece (Cathode) Erosion.

Chapter Seven-B

WIRE ELECTRIC DISCHARGE MACHINING (WIRE EDM)

INTRODUCTION

Wire Electric Discharge Machining (Wire EDM) was introduced in late 1960s. Wire EDM is one of the most exciting allied process developed in the class of Advanced Machining Processes (AMPs). There is no physical contact of wire with workpiece hence only hydraulic forces act on the workpiece. In this process, part's quality does not depend on the skill of the operator because the operation is controlled by CNC part program and in that sense it is automatic. Most machines run in a "light out" environment, and it increases productivity since the machining is unattended. It is a single step process, and majority of the workpieces come out of the machine without the need for secondary operation(s).

In wire EDM, a wire (about 0.05–0.30 mm diameter) is used as an electrode (cathode), workpiece as an anode and deionised water as dielectric. The workpiece and tool (wire) material should be electrically conductive. The material is removed by a discrete series of discharges/sparks that take place between the workpiece and the moving wire. The working principle or the mechanism of material removal in wire EDM remains the same as that in

EDM *i.e.*, material is removed by melting, or melting and vaporization of the electrodes' material. Here, material is removed simultaneously from both electrodes (*i.e.*, anode and cathode). A part of the dielectric also gets evaporated due to the heat of the sparks. A nozzle is employed to inject dielectric in the machining area and electrodes (wire and workpiece) are connected to a pulsed DC power supply. Heat generated due to sparking results in the melting of both workpiece and wire material, and sometimes, a part of the material may even vaporize like in EDM. A constant gap between the tool (wire) and workpiece is maintained with the help of a computer controlled positioning system. This system is also used to cut through complicated contours specially in difficult-to-machine materials. This process gives high degree of accuracy and good surface finish. Figure 7.32 shows details of the wire EDM system.

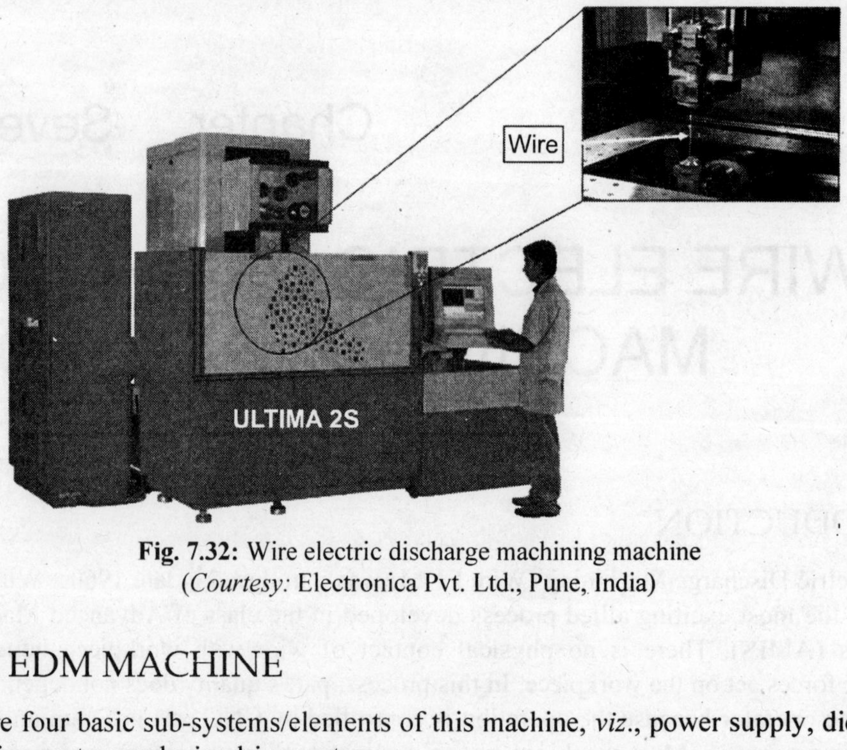

Fig. 7.32: Wire electric discharge machining machine
(*Courtesy:* Electronica Pvt. Ltd., Pune, India)

WIRE EDM MACHINE

There are four basic sub-systems/elements of this machine, *viz.*, power supply, dielectric, positioning system and wire drive system.

1. Power Supply System

Wire EDM power supply system differs from the conventional EDM power supply system basically in pulse frequency mode which is about 1 MHz in case of wire EDM. It results in reduced crater size *i.e.*, better surface finish. However, because of very small wire size, it usually cannot carry heavy current.

2. Dielectric System

An efficient way of dielectric delivery is to provide a stream of deionized water along the axis of the wire in the machining zone. Cost of the dielectric per unit production can be lowered down by reusing it after proper filtration. About 5 μm size disposable paper filters are used. To minimize rust formation on the machined part, special additives are mixed with dielectric. The resistivity and temperature of the dielectric should be maintained as per machine specifications. After filteration of the used water, the cleaned water is measured for its conductivity, and if required it is passed through the vessel containing a mixed bed of anion and cation beads. The debris and reaction products added to the dielectric during the process change its conductivity/resistivity.

3. Positioning System

Usually, positioning system is a Computerized Numerical Control *(CNC) two-axes table*. However, it operates in an Adaptive Control (AC) mode. In case, the wire approaches very near to the workpiece or the gap is bridged by debris, and it may result in a short circuit, the positioning system is capable to sense it before any short circuit takes place. In a very short period, it takes corrective action to re-establish proper cutting conditions in the gap. Machine movement is accomplished by precision lead screws with recirculating ball bearings on the axes.

4. Wire Drive System

This system serves two purposes; it continuously delivers fresh wire, and always keeps the wire under appropriate tension so that it moves in the machining zone as a straight wire. Later is important from the point of view of quality of the machined surface. For example, it helps in minimizing taper, streaks as well as vibration marks on the machined surface. It also minimizes wire breaking due to short circuit during machining. On the way while moving to the machining zone, wire is guided by sapphire or diamond "wire guides". As the wire moves towards the take up spool, it passes through a series of tensioning rollers. To enhance the productivity as well as to make the machine *to run unattended*, modern wire EDM machines are equipped with the devices which permit *automatic reloading* of wire after it has broken up during its use [*Anon*, 1981].

Advances in Wirecut EDM

In today's wire EDM m/c, it is possible to program relative motion between the wire and workpiece such that the wire follows a complex path in two axes. Hence, it is possible to use this m/c for making dies for stamping, fine blanking and extrusion. It is also possible to tilt the wire in the positions other than perpendicular to the X and Y axes. Such machines also have the facility for speedy automatic realignment of wire. Hence, it is possible to perform 3-D cutting using wire EDM machine in which two additional axes (U and V, Figure 7.33(a)) have been introduced. The drive motors which tilt the wire in front or back (V, axis) and left or right (U, axis) directions are controlled by the programmed commands in CNC wire EDM machine.

The extent to which the wire can be tilted (say ± 20° or so) tells about the limitations of the m/c. In a versatile CNC wire EDM m/c, all the four axes (X, Y, U and V) can be controlled simultaneously. Such machines usually have closed loop control arrangement for work table and wire guide positions. This facility has been used to make in one stroke, a draft or relief in the mold, land on the die, and constant radius corners along the tapered walls.

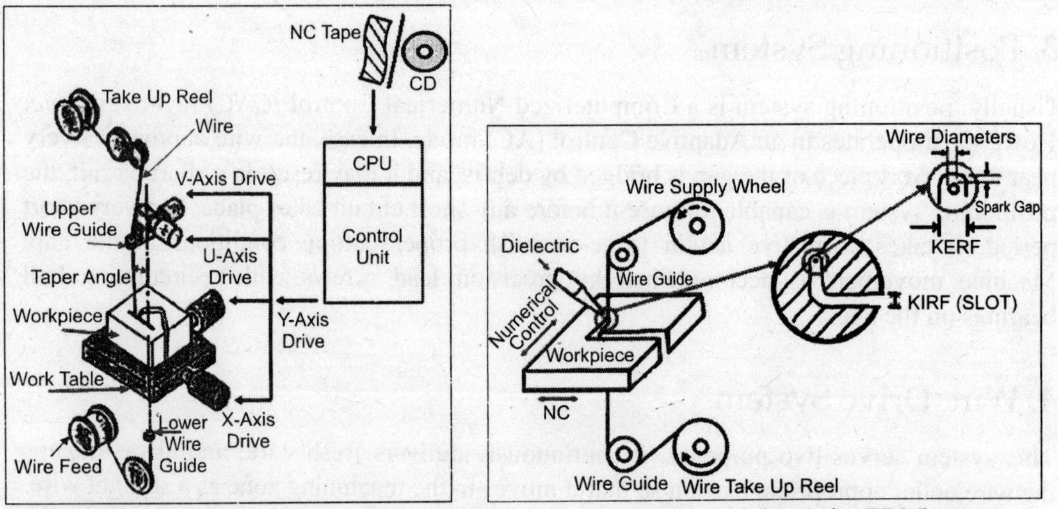

Fig. 7.33: (a) Schematic illustration of wire electric discharge machining (wire EDM), set-up, (b) A Simplified schematic diagram of wire EDM showing its some working details (Electrical connection to the tool and workpiece are not shown for simplicity).

CNC features have been incorporated with wire cut EDM machine tool. Some features of wire cut EDM system are automatic wire threading, restarting and improved tensioning devices. In CNC wire cut EDM m/c, an operator can online input for wire compensation. It is also possible to interrupt the process and on-line change the machining parameters (V, I, etc.) and settings, if required.

Machining and wire threading are controlled through the CNC part program. If a wire breaks during machining, the m/c returns to the start point, re-threads the wire and moves through the program path to the position where it broke, power up and starts cutting as if the wire had never broken. Further, automatic wire threading allows multiple openings to be cut in die blocks, progressive dies, prototype workpiece, etc. automatically and unattended.

PROCESS VARIABLES

Most of the variables that control the process are common in case of EDM (or ED die sinking) as well as wire EDM. The linear cutting rate in wire EDM is dependent on the thickness of the workpiece but not on the complexity of the cut. Wire speed may be as high as 40 mm/s. Process input parameters include pulse time (including pulse on-time and pulse off-time), gap voltage, current, dielectric flow rate, wire feed rate, wire tension, and table speed. The output responses include MRR, surface roughness, kerf width, and wire wear rate. Figure 7.34 shows different process input parameters which control the responses/ output.

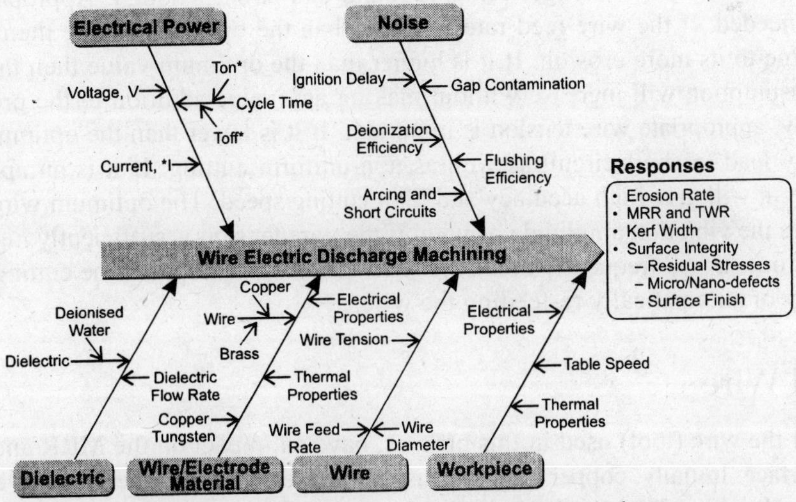

Fig. 7.34: Wire EDM process input parameters and responses

Dielectric

The dielectric acts as a coolant as well as flushes out the debris and reaction products from the machining gap. Water is a likely substitute for kerosene as dielectric in EDM. It is an attractive proposition because of its easy availability, low cost, desirable thermal properties, low viscosity and pollution-free working. It gives higher MRR and better surface finish under the identical machining conditions.

Deionized (DI) water has low viscosity, no fire hazard, high cooling rate and high MRR. That is why *(DI)* water is used as dielectric in most of the wire EDM systems. Low viscosity of the dielectric helps in efficient flow specially in complex shaped components while high cooling rate yields very thin recast layer. DI water gives high MRR and high TWR also. Dielectric flow rate is important to flush out the debris from the gap. High flow rate of dielectric also helps in avoiding short circuiting due to the presence of debris in the inter electrode gap (IEG).

Tool Electrode (or wire): Wire erosion also takes place because a spark takes place between the tool (Wire) and workpiece. It should be kept to a minimum to avoid the wire breakage. Wire Erosion Rate (WER) is influenced by the energy of each spark, % of total energy of each spark going to the wire, wire feed rate, dielectric flow rate and sparking frequency.

However, wire is not reused hence high TWR does not adversely affect the process performance unless it is extremely high which will lead to the wire breakages. Large diameter (0.15–0.30 mm) wires, used in wire EDM, are made of copper or brass while small diameter wires are usually made of molybdenum steel. Wire may be as smali as 0.05 mm. Wire is discarded after it has been used once because sparking takes place at its leading surface, hence, it no longer remains round and strong enough. Appropriate wire feed rate is needed. If the wire feed rate is lower than the optimum value, then the wire may break due to its more erosion. If it is higher than the optimum value then the cost of the wire consumption will increase without making any value addition to the product. In the same way, appropriate wire tension is important. If it is lower than the optimum value, the wire may lead to short circuit as well as non-uniform cutting. If it is an appropriate one then only it will give high accuracy and high cutting speed. The optimum wire tension will minimize the vibration amplitude of wire. If the wire tension is sufficiently higher than the optimum one, it will frequently break the wire, resulting in stopping the cutting process and manually or automatically re-feeding the wire.

Stratified Wire

Properties of the wire (tool) used in this process, have an impact on the MRR and quality of the cut surface. Initially, copper wire was used because of its high thermal and electrical conductivity. However, low tensile strength of copper wire was made responsible for wire breakage when enough tension was applied. Also, due to high thermal conductivity of copper wire, a good part of the spark heat was transferred to the wire in place of workpiece which could be utilized to melt and vaporize the workpiece. That is why wires of different materials are available for selection. However, nowadays, "*stratified wires*" are used as tool electrode in wire EDM. These wires are made of copper core with a thin layer of zinc over it (Figure 7.35). Such a wire can carry more current compared to bare copper wire hence the stratified wire gives higher MRR. A stratified wire for wire EDM was developed by Charmiles Corporation of Switzerland. This wire has a layer of zinc alloy on the outside and a core of drawn copper wire on the inside just like electrical wire is having cladding of vinyl plastic for insulation. The wire is used only once and then scrapped because it is not

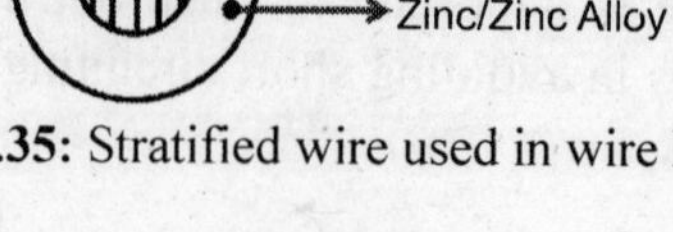

Fig. 7.35: Stratified wire used in wire EDM

very expensive. Further, it also gets eroded during the machining process hence it becomes weak and non-uniform in diameter.

A wire can carry *heavier load* if it can absorb more amount of heat without breaking. The heavier load means a spark with higher energy (higher value of voltage or current, or both) hence higher MRR resulting in higher cutting speed. In the stratified wire, zinc melts and some part of it even vaporizes at a temperature lower than the melting temperature of copper. In other words, core of the stratified wire remains intact, and it gets no hotter than the temperature at which the outer layer of the zinc vaporizes. Hence, more *powerful sparks* can be created when using stratified wire as compared to the case when copper wire is used without cladding. MRR is related to current density which depends on the surface area of the wire engaged in cutting.

In the tiny space (0.025 mm or so), workpiece is eroded by a violent, extremely localised and momentary action of the spark. Gas bubbles are rapidly formed during this process; and on the way to escape out, these bubbles press against the wire in the gap. This pressure results in *bowing of* the *wire* which tends to drag on bends and sharp corners as it advances. As a remedy, wire guiding and feeding mechanism should be designed such that the wire can withstand higher tension. Therefore, the adverse effect of gap condition on the quality of the product can be minimized. Large sized spools are used which can supply wire for machining for long period of time, say, 60 hr or so [*Albert*, 1982].

PROCESS CHARACTERISTICS

Wire EDM is extremely accurate but it needs skin cuts to be made. Depending upon the wire diameter, very small radius corners can be produced. Accuracies of the order of ± 6 μm or even lower, in some cases, can be achieved. However, uniformity of wire diameter, temperature as well as resistivity of dielectric should be closely controlled.

CNC feature of a wire EDM m/c offers the capabilities of scaling, mirror imaging, rotation, axis exchange, etc. This saves the time of a part programmer and the programming becomes comparatively easy. Adaptive control has helped in improving spark gap sensitivity. The shortest possible reaction time of the servo motor can avoid short circuiting. It also helps in reducing wire breakage.

Surface Roughness

It produces very high level of surface finish because the crater size is very small and the craters are overlapping on each other. Surface roughness is controlled by energy per spark, frequency of sparking, wire feed rate and workpiece table speed. Wire EDM process produces *matte finish.* Thousands of tiny craters on the machined surface help in retaining lubricating oil and result in increased *die life*. *Surface finish* of the order of 0.1 μm can be achieved in finish pass. To some extent, wire tension and dielectric flow rate also affect the machined surface roughness [*Scott*, 1991]. Various researchers [*Spedding and Wang*,

1997; *Kumar and Agarwal*, 2012; *Mahapatra and Patnaik*, 2007] have been trying to optimize the process input parameters to minimize the surface roughness value of the machined surface. As a principle, higher peak current is applied during roughing operation as compared to finishing operation. The volume of material removed in each spark depends on the energy per spark and flushing efficiency. In case of multiple passes, the crater size achieved on the final machined surface in the final pass decides the surface finish.

Material Removal Rate: As in EDM, MRR in Wire EDM is also governed by the energy of the spark, which in turn depends on the voltage and current including on-time. Another variable which affects MRR is feed rate of work table. With today's wire EDM systems, machining rate for definite materials has gone up from 12.50 cm^2/hr to about 40 cm^2/hr.

Kerf Width: Kerf width and its uniformity across the cut, decide part's accuracy. It is governed by MRR at different locations along the cut. In EDM and wire EDM both, MRR, TWR (wire erosion ratio or WER), and surface roughness depend on the crater size formed by each spark and % of the molten material flushed out from the interelectrode gap.

APPLICATIONS

Wire EDM has been employed for making dies for various types of components. It is possible to control tolerances very effectively. The process is also used for fabrication of press tools and tool electrodes for EDM. It is widely used for making 3-D components as shown in Figure 7.36.

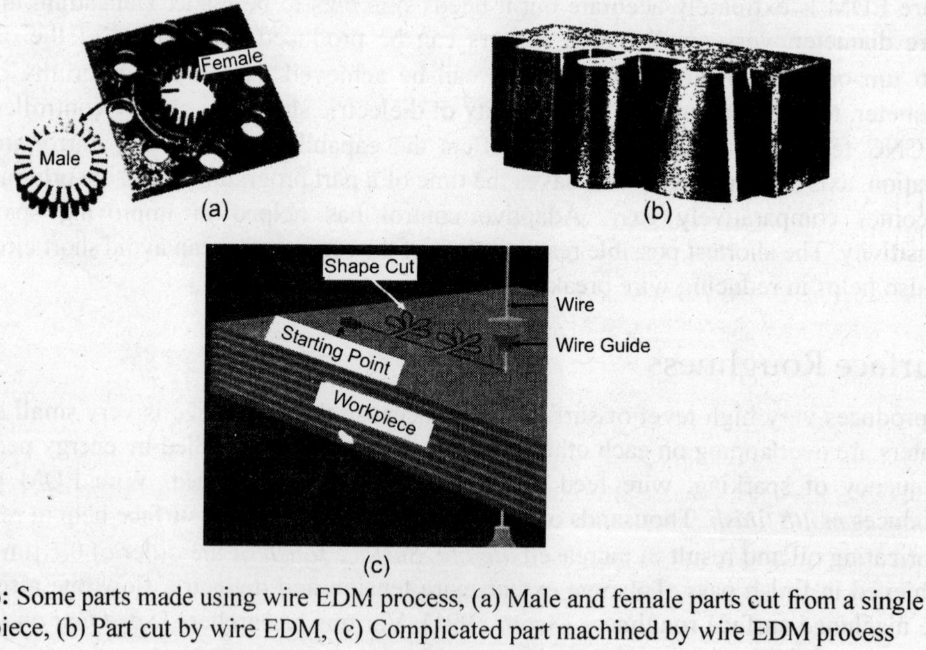

Fig. 7.36: Some parts made using wire EDM process, (a) Male and female parts cut from a single workpiece, (b) Part cut by wire EDM, (c) Complicated part machined by wire EDM process

PROBLEM

Male and female parts (external gear and internal gear) are made from a block of Aluminium using wire EDM process (see figure below). The wire diameter used is 60 μm and the interelectrode gap maintained is 10 μm. What is the expected difference in the values of radius r_1 and radius r_2 of the two gears? Please see Figure given below, and write the assumptions if any.

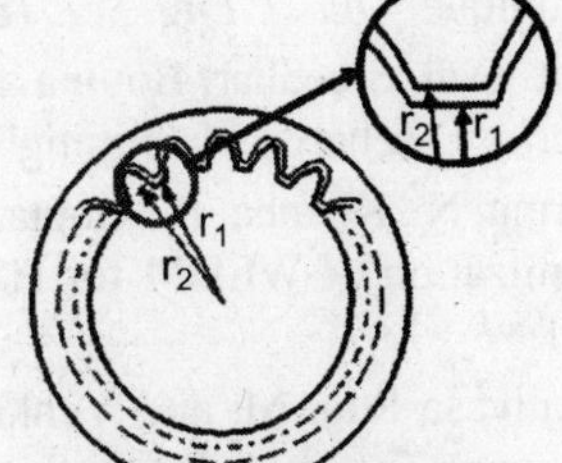

Solution

Let us make wire cutting diagram with the given details.

From the diagram,

$r_2 = r_1 + 10 + 60 + 10 = (r_1 + 80)$ μm

Here, the difference between r_2 and r_1 is,

$= (r_1 + 80 - r_1)$ μm

$\therefore \boxed{r_2 - r_1 = 80 \text{ μm.}}$

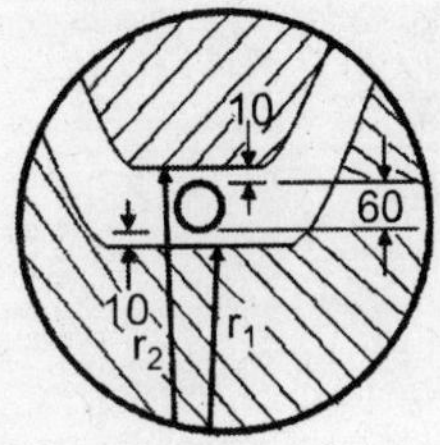

Assumption

IEG and wire tension remain constant everywhere including corners.

BIBLIOGRAPHY

[1] Albert, M. (1982). "Stratified Wire for TWEDM", *Modern Machine Shop,* pp. 96–102.

[2] Albert, M. (1983). "Taper Cutting: Value Added TWEDM", *Modern Machine Shop*, pp. 65–69.

[3] Anon (Feb. 1981). "Wire Cut EDM Can Cut It", *Production*, pp. 62–65.

[4] Banerjee, S., Prasad, B.V.S.S.S. and Mishra, P.K. (1993). "A simple model to estimate the thermal loads on an EDM wire electrode", *J. Mater. Process Tech.,* Vol. 39, pp. 305–317.

[5] Guo, Z.N., Lee, T.C., Yue, T.M. and Lau, W.S. (1997). "Study on the Machining Mechanism of WEDM with Ultrasonic Vibration of the Wire", *J. Mater. Process Tech.*, Vol. 69, pp. 212–221.

[6] Kuriakose, S. and Shunmugam, M.S. (2005). "Multi-objective Optimization of Wire-Electro-Discharge Machining process by Non-dominated sorting genetic Algorithm", *J. Mater. Process Tech.,* Vol. 170, pp. 133–141.

[7] Moultan and Donald, B., "Wire EDM, the Fundamental", *EDM Network Sugar Grove,* IL USA.

[8] Puri, A.B. and Bhattacharyya, B. (2005). "Modeling and Analysis of white layer depth in wire-cut EDM process through response surface methodology", *Int. J. Adv. Manuf. Tech.*, Vol. 25, pp. 301–307.

[9] Rajurkar, K.P. and Wang, W.M. (1993). "Thermal Modeling and Online Monitoring of Wire-EDM.", *J. Mater. Process Tech.*, Vol. 38, pp. 417–430.

[10] Saha, P., Singh, A., Pal, S.K. and Saha, P. (2009). "Soft Computing model based prediction of cutting speed and surface roughness in wire electro-discharge machining of tungsten carbide cobalt composite", *Int. J. Adv. Manuf. Tech.,* Vol. 39(2), pp. 74–84.

[11] Saurav, D. and Siba, M.S. (2010). "Modeling, Simulation and Parametric Optimization of Wire EDM Process Using Response Surface Methodology Coupled with Grey-Taguchi Technique", *Int. J. Eng. Sci. Tech.,* Vol. 2(5), pp. 162–183.

[12] Scott, Dan, Sreedhar, Boyina and Rajurkar, K.P. (1991). "Analysis and Optimization of Wire Electric Discharge Machining", *Int. J. Prod. Res.*, 29:11, pp. 2189–2207.

[13] Sharma, N., Khanna, R., Gupta, R.D. and Sharma, Renu (2013). "Modeling and multi-response optimization of WEDM for HSLA by RSM", *Int. J. Adv. Manuf. Tech.,* Vol. 67, pp. 2269–2281.

[14] Sreenivasa Rao, M. and Venkaiah, N. (2013). "Review on Wire-cut EDM Process", *Int. J. Advanced Trends in Comp-Sc. and Engg.,* Vol. 2(6), pp. 12–17.

SELF TEST QUESTIONS

Q1. Make the most appropriate choice.

1. Stratified wires are used so that they can,
   (a) Withstand more forces (b) Carry more heat energy
   (c) Both of these (d) None of these
2. The diameter of the wire used usually in wire EDM is,
   (a) 50 to 300 μm (b) 1–10 μm (c) 1–2 mm (d) None of these
3. Power supply used in wire EDM process is,
   (a) A.C. (b) D.C. (c) Pulsed D.C. (d) None of these
4. Frequency of sparking in wire EDM is,
   (a) 300–400 kHz (b) 1 MHz or above (c) 500–900 Hz
5. Which is/are not used as dielectric in wire EDM,
   (a) Kerosene (b) Tap water (c) Deionized water (d) Paraffin oil
6. The cutting rate in case of wire EDM is given as around,
   (a) 12–40 cm^2/hr (b) 12–40 cm^3/hr (c) 12–40 cm/hr (d) None of these
7. What is repeatedly used in wire EDM,
   (a) Dielectric (b) Tool (c) Wire (d) Cathode

Q2. (a) Explain the working principle of wire EDM.
(b) Explain how the stratified wire works?
(c) What is bowing of wire in wire EDM?
(d) Sketch a wire EDM system showing its different elements and explain its working principle.
(e) Make a neat sketch showing kerf width, wire diameter and spark gap, in wire EDM process.
(f) Can you make 3-D components using wire EDM? Explain how?
(g) Draw different elements of wire drive system and explain how the bowing of the wire can be avoided or minimized?
(h) Explain how a stratified copper wire can carry higher load compared to a simple copper wire?
(i) What will happen in a stratified wire if the inner core is made of zinc and outer cladding is done of copper? Will it be able to carry more current than a simple copper wire of the same dimension? Explain it.
(j) Draw a neat *fish bone* diagram indicating process input variables and responses.

ACRONYMS

AMPs	Advanced Machining Processes
AC	Adaptive Control
CNC	Computer Numerical Control
DC	Direct Current
DI	Deionized
EDM	Electric Discharge Machining
MHz	Mega Hertz
IEG	Inter Electrode Gap
MRR	Material Removal Rate
m/c	Machine
NC	Numerically Controlled (Numerical Control)
WER	Wire Erosion Rate
TWR	Tool Wear Rate

CHAPTER 7B
AT-A-GLANCE
WIRE ELECTRIC DISCHARGE MACHINING (WIRE EDM)

Introduction

- Workpiece and Tool Materials → Electrically Conductive
- Pulsed DC Power Supply
- Heat Generated → Spark between Workpiece and Wire
- Wire as Tool → 0.05–0.30 mm dia
- Stratified Wires
- Dielectric → Deionised Water
- Material Removal → Melting and/or Vaporisation
- Constant IEG → Computer-Controlled Positioning System
  ↓
  Can Cut Complicated Contours/3-D Contours
- High Degree of Accuracy and Good SF (Surface Finish)

Wire EDM Machine

- Power Supply System
- Dielectric System
- Positioning System
- Wire Drive System

Power Supply System

- Pulse Frequency → 1 MHz or More → Reduced Crater Size → Better SF
- Small Wire Size → Low Current Carrying Capacity

Dielectric System

- Water as Dielectric
  - Low Viscosity → Efficient Flow
  - Capable to Reach Difficult to Reach Areas
  - No Fire Hazard
  - High Cooling Rate
  - High MRR and High TWR also (TWR Does Not Affect Performance unless it breaks)

- Reused after Filtration → 5 µm Size Disposable Filter
- Additives to Minimize Rusting

Positioning System

- CNC → 2 or More – Axes Table ← Adaptive Controlled Mode
- In Case of Short Circuiting (IEG → Too Small → Wire and Workpiece Too Close) → Senses and Moves Back to Re-Establish Proper Cutting Gap Conditions

Wire Drive System

Functions
- Delivers Fresh Wire
- Keeps Wire Always Under Tension

- For High Quality Machining → Avoid Bending of Wire → Minimizes Wire Breakage Frequency, Taper, etc.
- Wire Guided by Wire Guides → Sapphire or Diamond
- Wire Movement Towards Take Up Spool → Series of Tensioning Rollers
- Automatic Reloading of Broken Wire
  - To Enhance Productivity
  - To Run the m/c Unattended
- Wire Material
  - Small Diam. (< 0.15 mm) Molybdnum
  - Large Diam. (0.15–0.3 mm) → Copper or Brass
- Wire Discarded After Used Once
- Due to Wear, it No Longer Remains Straight

Advances in Wirecut EDM

- With CNC Wire EDM, Possible to have → Automatic Wire Threading, Restarting, Control of Input Variables Online, Changing Machining Parameters Online
- Follow Complex Path → Making Dies, 3-D Cutting, U and V Additional Axes, Possible to Tilt the Wire
- Closed Loop Control → Working Table and Wire Guide Position, Draft on Mold, etc.

Stratified Wire

- Stratified Wire → Copper Core Claded with Thin Layer of Zinc
- Carries Higher Current → Higher MRR → Also Higher TWR
- Used Only Once → Worn Out → Scrapped

- Escape of Gas Bubbles → Bowing of Wire → Inaccurate Machining & Short Circuiting
- Remedy → Wire under Higher Tension
- Large Size Spools → Wire Available for 60 hr or More

Process Variables

- Power Supply, Dielectric, Wire, Workpiece, Noise Factors
- Linear Cutting Rate → $\oint$ (Thickness of Workpiece But Not the Complexity of Cut)
- Highest Wire Speed → 40 mm/s

Process Characteristics

- Matte Surface
- Craters Help in Retaining Lubricating Oil → Increased Die Life
- SF in Finish Pass → 0.1 μm
- Tolerances → ± 6 μm

Applications

- Dies
- Press Tools
- Electrodes, etc.

Chapter Seven-C

ELECTRIC DISCHARGE GRINDING (EDG)

INTRODUCTION

Hybrid machining processes, which are combination of two or more machining processes, are gaining popularity in machining of advanced engineering materials like titanium and its alloys, cemented carbides, polycrystalline diamonds, metal matrix composites, etc. These hybrid processes are developed to exploit the potential advantages and to restrict the disadvantages associated with an individual constituent process. Usually, the performance of a hybrid process is better than the sum of their individual performance with the same parameters setting [*Konig et al.*, 1990; *Yadav et al.*, 2002].

Grinding can be classified basically into two categories, *viz.*, conventional (or mechanical) grinding and hybrid grinding. *Hybrid grinding* includes those processes in which mechanical grinding is assisted by one or more (advanced) machining techniques. Some of the hybrid grinding processes are Electrochemical Grinding (ECG), Ultrasonic Grinding (USG), Electric Discharge Grinding (EDG), Electric Discharge Diamond Grinding (EDDG), Electrochemical Spark Grinding (ECSG), etc.

Electric Discharge Grinding (EDG) can be classified in three categories based on the type of the grinding wheel used. First one is called EDG without having any abrasive particles embedded in the wheel. It is better to call it as "*EDM with a rotating disk electrode*", rather than calling it EDG. Second category is known as Electric Discharge Diamond Grinding (EDDG) where diamond particles are used as abrasive particles embedded in the grinding wheel made of electrically conducting material (metal). Third category is Electric Discharge Abrasive Grinding (EDAG). This name is also used by some researchers where other than diamond abrasive particles are embedded in the grinding wheel. However, in this book, a small introduction is given to the first one, and second and third categories are merged under the heading "Electric Discharge Grinding (EDG)" where abrasive grinding wheel with any kind of abrasive particles (diamond or other) can be used. Here, working principle remains the same in both the cases *i.e.*, grinding wheel with diamond or any other abrasive particles. It is also to be noted that the EDG without abrasive particles in the grinding wheel is more close to EDM than EDG.

ELECTRIC DISCHARGE MACHINING WITH ROTARY DISK ELECTRODE

EDM with a rotary disk electrode uses EDM *machine*. In the EDM m/c, an attachment is fixed to the ram, which has a rotating disk electrode and a provision to rotate it at different speeds. In this case, grinding wheel (more correctly, a disk) having no abrasive particles is made of electrically conductive material like graphite which is less expensive and easy to machine. The *servo-system* of this m/c maintains a constant inter-electrode gap (IEG). If debris somehow block the gap, the electrode is retracted back and forth by the servo system to allow the debris to be flushed out of the (IEG) by the dielectric.

Electric discharge machining with a rotating disk electrode is the process which works on the same *principle* as EDM. A part of the rotating wheel (cathode) and workpiece (anode) both are immersed in the dielectric, and they are connected to the pulsed DC power supply, Figure 7.37. The rotating motion of the wheel ensures the effective flow of dielectric in the IEG, and hence flushing of the gap with fresh dielectric can be eliminated.

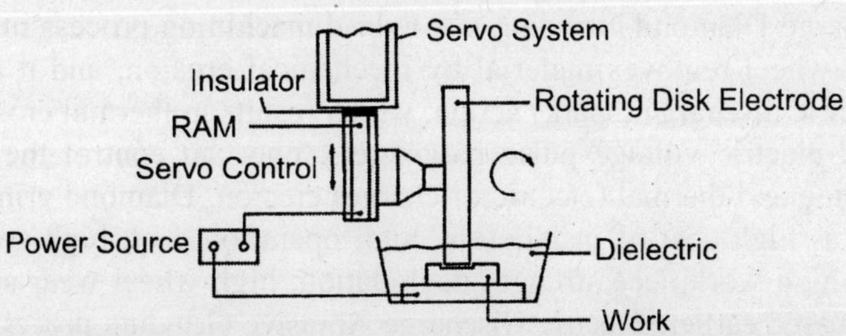

Fig. 7.37: Schematic diagram of electric discharge grinding with rotating disk electrode [*Khosy et al.*, 1996].

Mechanism of material removal is exactly the same as that of EDM except that the rotary motion of the tool (*i.e.*, disk) helps in effective ejection/removal of the molten material and debris from the IEG. As contrary to the conventional grinding, there is no direct physical contact (except during short circuit, if so) between the tool and the workpiece, hence, fragile and thin sectioned specimens can be easily machined. During this process, the material removal takes place due to melting and/or vaporization but not by mechanical action like shearing by abrasive particle/cutting tool.

Because of a small IEG maintained between the tool and workpiece in EDM with a rotary disk electrode, the compensation for the overcut on the workpiece should be taken care of while designing the wheel. This m/c may have current in the range of 0.5 A–200 A, voltage in the range of 40–80 V, and pulse frequency in the range of 50–250 kHz. To achieve the desired MRR and surface integrity, optimum machining parameters should be chosen. The upper limit for wheel surface speed is about 180 m/min. *Peak* (1982) has suggested an empirical relationship to find the process cycle time.

This process is *capable to machine* extremely hard materials, say, carbides at 2 to 3 times faster than the conventional diamond grinding. Good surface finish (0.2 to 0.3 μm) and high accuracy (± 2.5 μm) are achieved by this process.

ELECTRIC DISCHARGE GRINDING (EDG)

EDM gives much lower material removal rate, lower machining efficiency, resolidified white layer or recast layer infested with microcracks, residual stresses etc. Further, while grinding hard materials by conventional grinding process, normal force is high, abrasive wear is fast (or, time between two dressings becomes lower) and grinding wheel clogging enhances. In order to overcome some of the problems of both, grinding and EDM, and to take the advantage of the characteristics of both these processes, a hybrid process has been developed by combining both of these in a single process that is electric discharge diamond (or abrasive) grinding abbreviated as EDDG (or, EDAG), or simply Electric Discharge Grinding (EDG). It was anticipated that the combination of EDM + Grinding would yield reduction in forces, better accuracy and surface integrity, and also improved dressing quality by automatic in-process dressing. Figure 7.38 shows a schematic diagram of electric discharge diamond grinding setup.

Electric Discharge Diamond Grinding is a hybrid machining process in which a metal bonded diamond wheel removes material by mechanical erosion, and it also acts as an electrode for electric discharge (spark) action, which results in thermal erosion. Adjusting wheel speed and electric voltage pulse parameters, one can control the percentage of mechanical grinding and thermal (electric discharge) erosion. Diamond grinding has many constraints such as high cost of installation, high operation cost, high specific material removal energy, high workpiece strength degradation, high wheel wear and its frequent dressing. As discussed earlier, Electric Discharge Abrasive Grinding process uses abrasive particles embedded in the grinding wheel for removing a part of material by grinding operation and remaining by EDM action simultaneously.

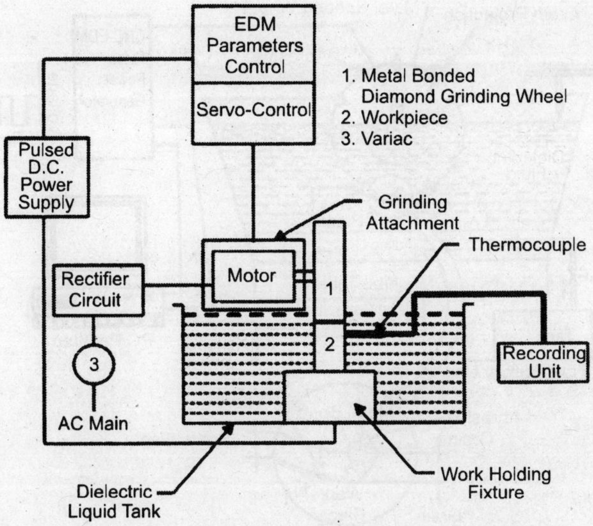

Fig. 7.38: General schematic diagram of electro discharge diamond grinding.

Working Principle

Electric Discharge Diamond Grinding is a hybrid process which involves machining of hard materials making use of positive features of both EDM as well as conventional diamond grinding. EDDG process permits the use of water or water based cutting fluid as dielectric. Sparking takes place between metallic bonding material of the grinding wheel and workpiece. Heat generated during sparking softens the work material ahead of an abrasive particle, and hence machining by abrasive particles becomes easier, Figure 7.39(a). As can be seen in schematic diagram of EDAG in Figure 7.39(a), a spark takes place between the wheel and the workpiece. By melting and/or vaporization, it removes material from the wheel as well as workpiece. Apart from the removal of the material by such thermal erosion, it also softens the workpiece material and tool material both by the heat conduction from the locations of sparks. The softened molten material is easily scooped out by the chains of the abrasive particles coming behind, due to the rotation of the grinding wheel. Thus, the mechanism of material removal in EDDG process involves material removal by thermal erosion as well as mechanical abrasion. Further, the softened workpiece material helps in decreasing the grinding forces, and increasing MRR. On the other hand, softened wheel bonding material facilitates the detachment of the abrasive particles from the grinding wheel. It is more true specially when the enough wear of the abrasive particle has taken place or depth of cut is very high resulting in increased required force to remove material from the workpiece. This detachment of the abrasive particles from the grinding wheel leads to its *auto dressing or self dressing of the grinding wheel.* Figure 7.39(b) shows the schematic illustration of on-machine Electro-Discharge (ED) trueing/dressing method. Precise trueing, dressing and profile generation on grinding

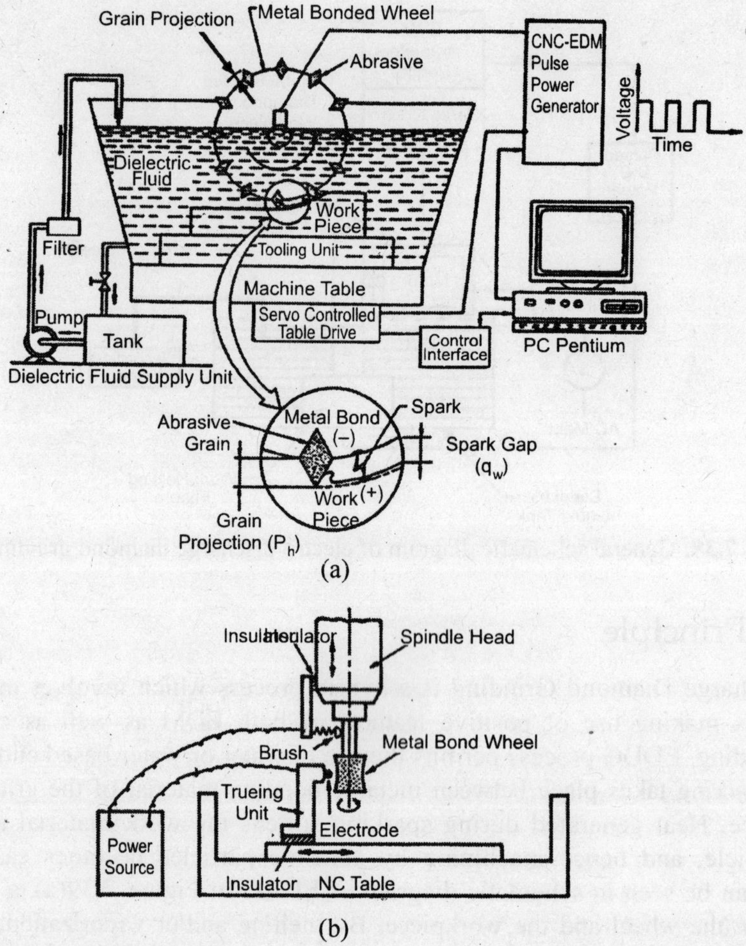

Fig. 7.39: (a) Numerically controlled electric discharge diamond grinding setup, (b) Schematic diagram of on-machine electric discharge trueing/dressing operation [*Suzuki et al.*, 1986].

wheels of various grain sizes is possible just by changing electro-discharge conditions. The dressing of the diamond wheels using tubular and solid electrodes has also been done. Rotary speed of the diamond wheel and amount of the swarf (very fine chips) generated during EDDG also influence the process performance [*Koshy et al.*, 1993].

The EDDG has also been named as electro-contact abrasion finishing, and has been employed for cutting blanks of various materials, profiles and dimensions. Figures 7.40(a) and 7.40(b) show wheel clogging/loading and spherical shaped deposited debris between the grinding wheel abrasive particles. Presence of spherical debris clearly indicates that the mechanism of material removal is melting and vaporization due to EDM effect. Material removed by the mechanical grinding process is in the form of micro-chips as in diamond grinding, Figures 7.40(a) and 7.40(b).

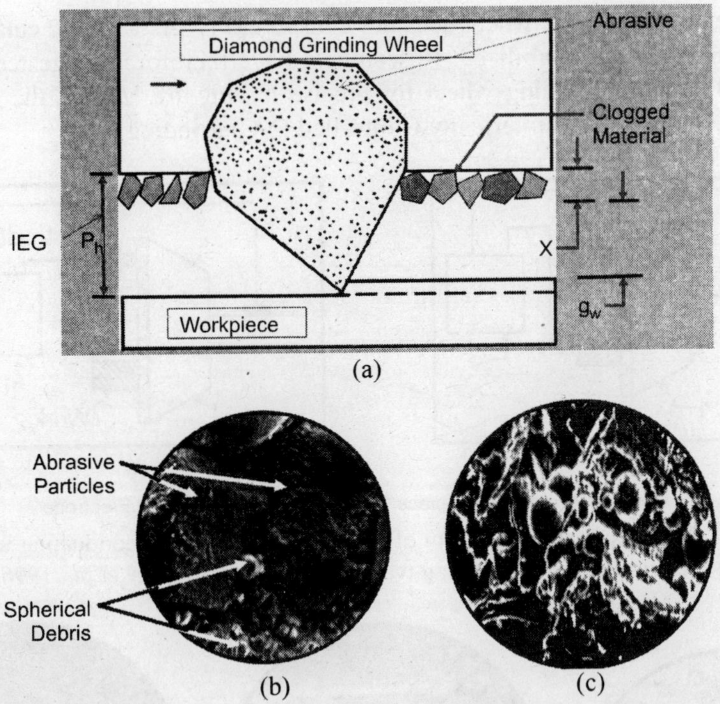

Fig. 7.40: (a) Schematic diagram of the loaded grinding wheel, (b) Abrasive particles of the grinding wheel showing grinding wheel loading with debris, (c) Micrograph of debris collected at 5 A (Speed 4.5 m/s) [*Koshy et al.,* 1993].

Figure 7.41(a) shows the application of EDG of an electrically conducting workpiece by making it as an electrode. Figure 7.41(b) shows that it can be used for grinding electrically non-conducting workpiece (2) also. Here, a dummy electrode (4) is used to complete the electrical circuit. Here, the use of a dummy electrode helps in auto dressing of the EDG wheel during the grinding process. In this case, there is no need to stop the grinding process for dressing of the grinding wheel. EDG, can also be used to grind electrically non-conducting workpiece still making use of auto dressing of the grinding wheel thus saving machine down time for dressing and maintaining comparatively higher MRR. It also reduces wheel clogging as in case of diamond grinding, Figure 7.40(b). Declogging takes place by EDM effect between the dummy workpiece and the grinding wheel.

The mechanism of material removal at different current conditions in EDG can be understood from Figure 7.42. Figure 7.42(a) shows pure grinding (current = 0 A) which is obvious from the grinding marks on the machined workpiece. Figure 7.42(b) shows mainly grinding and partly EDM (craters in the figure are evidence of EDM, but scratches indicate that the grinding is dominating). Figure 7.42(c) shows material removal mainly by EDM and partly by grinding at high current (EDM dominated as evidenced by craters) [*Koshy et al.,* 1993]. Sparking in the IEG during EDG, also results in continuous dressing of the grinding wheel and hence the wheel does not clog so easily. As a result, cutting properties of the grinding wheel are stabilized. Researchers have therefore, suggested to use EDG with diamond abrasive grinding wheel for precise *trueing/dressing* of the metal bonded grinding wheel employing Numerically Controlled (NC) grinding m/c.

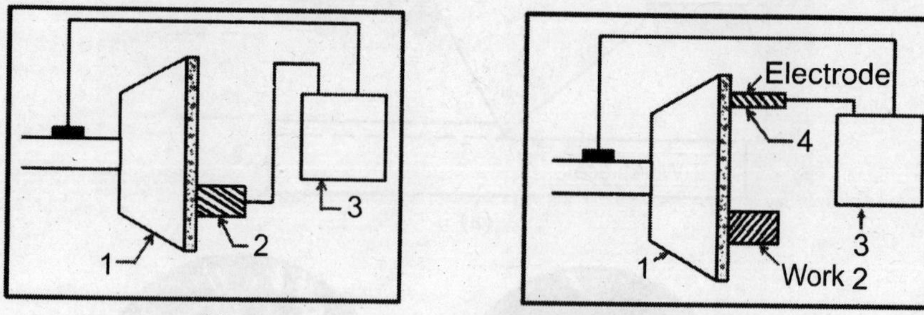

Fig. 7.41: A Simple schematic diagram of EDG of (a) Electrically conducting workpiece, (b) Electrically non-conducting workpiece material [*Koshy et al.,* 1996].

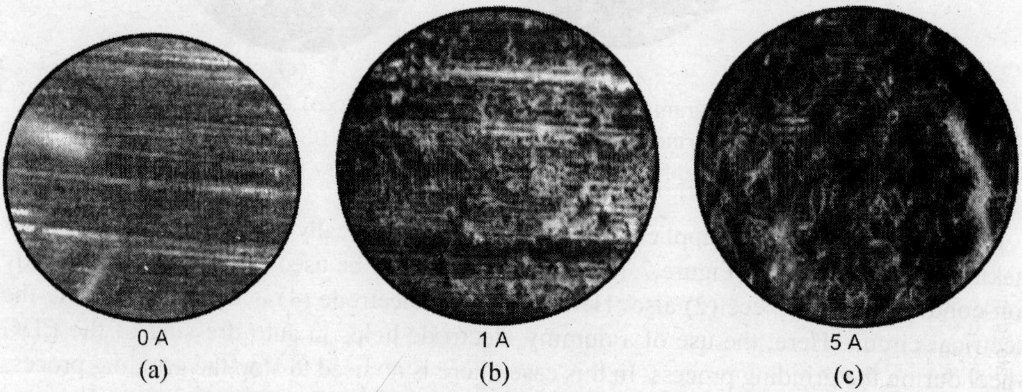

Fig. 7.42: Micrographs of surface ground at a current of (a) 0 A, (b) 1 A, (c) 5 A, at Wheel speed of 4.5 m/s.

PARAMETRIC ANALYSIS

In this section, the effects of some important process variables on the material removal rate have been discussed, in very brief.

Effect of Pulse Current

In EDG, as the current increases, MRR_g increases (Figure 7.43a). The reason is that, as the current increases, EDM action dominates over grinding. EDM is assisted by grinding in removing molten material/debris from the craters as per the model presented in Figure 7.43(b). It also decreases the chances of resolidification of molten material (also called as recast layer or white layer).

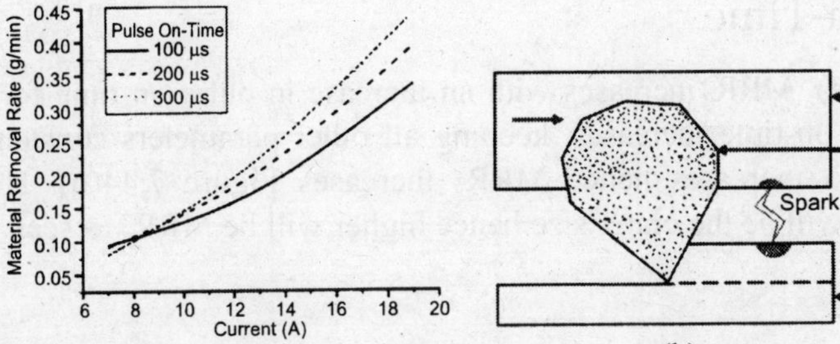

(a) (b)

Fig. 7.43: (a) Effect of pulse current on material removal rate for different pulse on-times, (b) Schematic representation of the wheel-workpiece interface in EDG [*Koshy et al.*, 1993]

Effect of Wheel Speed

MRR_g slowly increases as wheel speed increases (Figure 7.44(a)), for different current values. Improvement in MRR_g can be attributed to the increased abrasive action due to the increased wheel speed. Also, for a given current and pulse-on time (*i.e.*, for given discharge energy), molten material is being rapidly thrown away by the abrasive particles from the working gap (more correctly from the craters) as the wheel speed increases. Hence, MRR increases as the wheel speed increases.

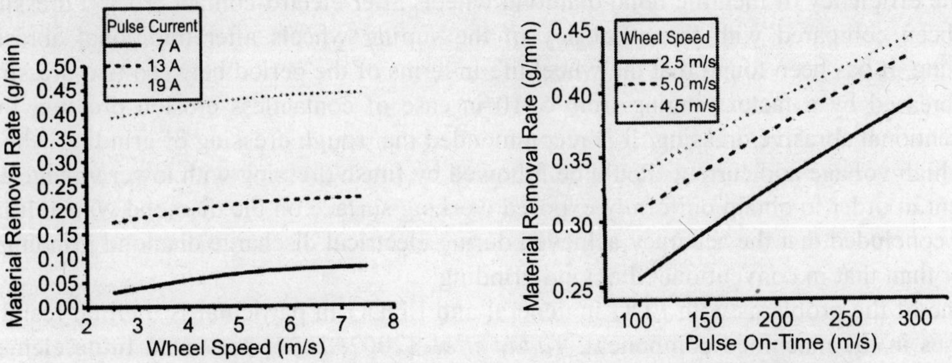

(a) (b)

Fig. 7.44: (a) Effect of wheel speed on material removal rate, (b) Effect of pulse on-time on material removal rate for different wheel speeds in EDDG [Koshy *et al.*, 1993].

MRR_g depends on the degree of interaction between the wheel and the workpiece. Hence, dynamics of the gap width (or interelectrode gap) is important from MRR point of view.

With increasing wheel speed, the reduced contamination of the dielectric in the gap results in decreasing gap width, which would increase the underformed chip thickness. Thus, an increase in wheel speed would increase MRR_g.

Effect of Pulse On-Time

As seen in Figure 7.44(b), MRR_g increases with an increase in pulse on-time for different wheel speeds. As pulse on-time increases, keeping all other parameters constant, energy input per unit time also increases hence MRR_g increases Figure 7.44(b). Higher the energy per spark, larger will be the crater size hence higher will be MRR_g as seen in Figure 7.44(b).

Capabilities and Applications

This process is also being used for machining of cremates, super alloys, and other metal matrix composites. EDG has been employed to machine tungsten free carbides with reverse polarity using fixed plunge feed rate. It has been reported that EDG gave MRR_v as 260 mm^3/min, *wheel life* as more than 120 min, and good surface finish (Ra value as good as 0.32 to 0.15 μm) without cracking. EDG is recommended over Electrochemical Discharge Grinding (ECDG) and Electrochemical Grinding (ECG) because of its much less corrosive effects.

It has been concluded that during electro-discharge diamond grinding of cemented carbide drill bits, the diamond wheel has about 25% higher wheel life as compared to conventional diamond grinding. It also gives higher MRR_v and lower operating costs.

The efficiency of metallic bond diamond wheels after electro-contact erosion dressing, has been compared with the efficiency of the similar wheels after traditional abrasive dressing. It has been found that the wheel life in terms of the period between two dressings is increased by a factor ranging from 5–10 in case of contactless erosion dressing than conventional abrasive dressing. It is recommended that rough dressing of grinding wheels with high voltage and current should be followed by finish dressing with lower voltage and current in order to obtain uniformly exposed working surface on the diamond wheel. It has been concluded that the accuracy achieved during electrical discharge diamond grinding is better than that in conventional diamond grinding.

One of the problems with EDG in general and EDDG in particular is thermal residual stresses in the machined component. *Yadav et al.* [2002, 2004] presented finite element analysis of EDDG and analyzed thermal (residual) stresses in the machined components. Effects of process parameters (current, pulse-on time, and duty cycle) on temperature distribution and thermal stresses in the workpiece are studied. The temperature gradient along the depth is found to be steeper than the temperature gradient along its radial distance from the center of a spark. It is also noted that the temperature gradient from the top surface is very high up to 0.63 mm depth, for all currents.

From the above discussion of the capabilities of EDDG/EDAG, it can be concluded that it is an attractive process for grinding electrically conductive materials and to some extent electrically partially conductive and non-conducting materials also. Optimization of input process parameters would enhance its economic viability. It is also feasible to grind very hard but electrically conducting materials with comparatively softer abrasive particles like Al_2O_3.

BIBLIOGRAPHY

[1] Polvani, R.S. and Evans, C.J. (1994). "Electrically Assisted Grinding of Advanced Ceramic Materials", (Personal correspondence).

[2] Koshy, P., Jain, V.K. and Lal, G.K. (1993). Experimental Investigations into EDM with rotating disc electrode, *Precision Engineering*, Vol. 15, No. 1, pp. 6–15.

[3] Koshy, Philip (1996). "Electrical Discharge Diamond Grinding: Mechanism of Material Removal and Modeling", *Ph.D. Thesis,* I.I.T. Kanpur (India).

[4] Suzuki, K., Uematsu, T. and Nakagawa, T. (1986). "On Machine Trueing Dressing of Metal Bond Grinding Wheels by Electro Discharge Machining", *Annals of the CIRP,* Vol. 36, No. 1, pp. 115–118.

[5] Yadav, Vinod, Jain, V.K. and Dixit, P.M. (2004). "Theoretical Analysis of Thermal Stress in Electro Discharge Diamond Grinding", *Mach. Sci. Tech.*, Vol. 8, No. 1, pp. 119–140.

[6] Yadav, Vinod, Jain, V.K. and Dixit, P.M. (2002). "Temperature Distribution in Electro Discharge Abrasive Grinding (EDAG)", *Mach. Sci. Tech.*, Vol. 6, No. 1, pp. 97–127.

SELF TEST QUESTIONS

Q1. Write True (T) or False (F).

1. EDM with rotary disc electrode wheel is a hybrid process.
2. EDM with rotary disc electrode wheel does not have the abrasive particles.
3. EDG without abrasive particles in the grinding wheel is more close to EDM than EDG.
4. Abrasive particles used in EDAG are electrically conducting.
5. Mechanism of material removal in EDG is exactly the same as that in EDM.
6. EDDG process is *capable to machine* extremely hard materials, say, carbides at 2–3 times faster than the conventional diamond grinding.
7. By changing the IEG, you can partly control the contribution in % by mechanical erosion and thermal erosion.
8. The magnitude of forces acting during the EDDG is higher than traditional diamond grinding.
9. Material removed by mechanical erosion during EDDG process is in the form of globules dispersed in the inter electrode gap.

10. With an increase in current during EDAG, MRR_v increases but surface finish deteriorates.
11. During electro-discharge diamond grinding of cemented carbide drill bits, the diamond wheel has about 25% lower wheel life as compared to conventional diamond grinding.
12. It can be concluded that EDDG/EDAG is an attractive process for grinding electrically non-conductive material.

Q2. Make the most appropriate choice(s).

1. Which of the following are (is) not hybrid processes:
   (a) Wire-EDM (b) Electric discharge abrasive grinding
   (c) Laser assisted EDM (d) Shaped tube electro-machining
   (e) Traditional grinding
2. Voltage used during EDG is
   (a) 0 to 30 V (b) 40 to 80 V (c) 100 to 220 V
3. During EDDG, sparking takes place between
   (a) Abrasive particle and workpiece
   (b) Abrasive particle and bonding material
   (c) Workpiece and bonding material (d) None of these
4. With a decrease in pulse-off time, surface finish value, R_a
   (a) Increases (b) Decreases (c) No effect on surface finish

Q3. Review Questions.

1. Why hybrid machining processes are needed? Explain with the help of an example.
2. What are the three different types of electric discharge grinding processes? Explain their mechanisms of material removal.
3. With the help of a neat sketch, explain the working of EDAG process, their merits and demerits, if any.
4. What is the difference between "truing" and "dressing" of a grinding wheel? Explain with the help of a figure.
5. What do you understand by wheel clogging and de-clogging during EDAG process?
6. With an increase in current, why surface finish deteriorates? Explain with the help of a figure.

ACRONYMS

ED	Electric-Discharge
EDAG	Electric Discharge Abrasive Grinding
ECG	Electrochemical Grinding
EDG	Electric Discharge Grinding
EDDG	Electric Discharge Diamond Grinding
EDM	Electric Discharge Machining
IEG	Inter Electrode Gap
kHz	Kilo Hertz
MRR	Material Removal Rate
MRR_g	Malarial Removal Rate in g/s
MRR_v	Volumetric Material Removal Rate in mm^3/s
m/c	Machine
NC	Numerical Control
USG	Ultrasonic Grinding
w/p	Work Piece

CHAPTER 7C
AT-A-GLANCE
ELECTRIC DISCHARGE GRINDING (EDG)

Hybrid Machining Process

- Combination of Two or More Machining Processes
- Exploits Potential Advantages and Restricts Disadvantages of the Constituent Processes
- Grinding: Conventional and Hybrid (EDG, EDAG, ECG, etc.)
- EDG–EDAG and EDDG (Both Come under EDG)
- Same Principle as that of EDM
- Electrically Conductive Grinding Wheel (No Abrasive Particles)
- Rotating Wheel → Proper Circulation of Dielectric in IEG
- Material Removal → Melting and Vaporisation Only

Working Principle

- Hybrid Procss → EDM + Grinding
- Water/Water Based Cutting Fluid → Dielectric
- Metal Bonded Diamond Grit Wheel
- Sparking → Between Bonding Material and W/P
- Material Removal by Both → EDM + Grinding ← Easier
- Self Dressing of Grinding Wheel
- Possible to Grind Both Electrically Conducting and Non-Conducting Materials
- Minimum Wheel Clogging/Loading Chances
- At High Current, EDM Dominates
- EDM Softens WP Material. Possible to Machine Very Hard Material by Comparatively Softer Abrasive.
- With Increase in Pulse Current, Pulse On-Time and Wheel Speed, MRR_g Increases

Capabilities and Applications

- Machining of Cermates, Super Alloys, Metal Matrix Composites
- High MRR_V, High Wheel Life, Good SF, Better Surface Integrity

Chapter Eight

LASER BEAM MACHINING (LBM)

PRODUCTION OF LASERS

Most of us are familiar with a common experiment performed for fun during the school days, in which sun rays are focused by a lens to burn a piece of paper. The energy density achieved in this experiment is about 1 W/mm^2. In the same way, let a laser beam be focused at the diamond surface. The energy density at the diamond surface may be achieved so high (about 1000 W/mm^2) that it can (melt and) even vaporize the diamond.

Such a tremendous amount of energy release is achieved due to collision of oscillating high energy-level atoms with electromagnetic waves having resonant frequency. These waves absorb energy from the atoms and become highly powerful, and they are called **MASER** (**M**icrowave **A**mplification by **S**timulated **E**mission of **R**adiation). Later, **LASER** (**L**ight **A**mplification by **S**timulated **E**mission of **R**adiation) was invented by amplifying ordinary light waves based on similar principle (*i.e.*, to transmit light waves with constant frequency and wavelengths without interference).

Einstein hypothesized that under appropriate conditions, light energy of a particular frequency can be used to stimulate the electrons in an atom to emit additional light with exactly the same characteristics as the original stimulating light source. An atom, initially in any of the excited state does not remain forever in that state (or energy level). Einstein proposed that when an atom at '*q*' energy level has light of right frequency acting on it, then this *atom absorbs* photons of that light and the transition takes place from lower energy level '*q*' to higher energy level '*p*'. This phenomenon of the movement of an atom in the higher energy level is called **absorption** (Figure 8.1). On the other hand, transition of an atom from the higher energy level '*p*' back to the lower energy level '*q*' is known as **emission**. The emission could be one of the two kinds, *viz.*, spontaneous emission (independent of light intensity) and stimulated emission (influenced by the intensity of light).

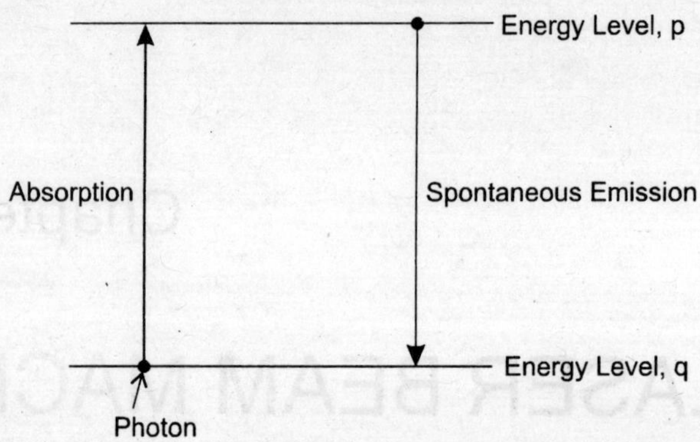

Fig. 8.1: Transition by absorption and spontaneous emission between two given energy levels of an atom [*Bhattacharyya*, 1973]

Suppose each horizontal line in Figure 8.2 indicates the allowed value of energy of an atom at that energy level. Let an atom (or molecule) be brought to higher energy level (say, E_3 in Figure 8.2) by an outside energy source (say, heat, light, or chemical.). Now, if it is allowed to decay back to its ground state energy level (E_o), a photon (unit of light) is released (Figure 8.2). If this *photon* comes in contact with another molecule or atom at high energy level (E_3), then this atom will also decay back to ground state releasing another

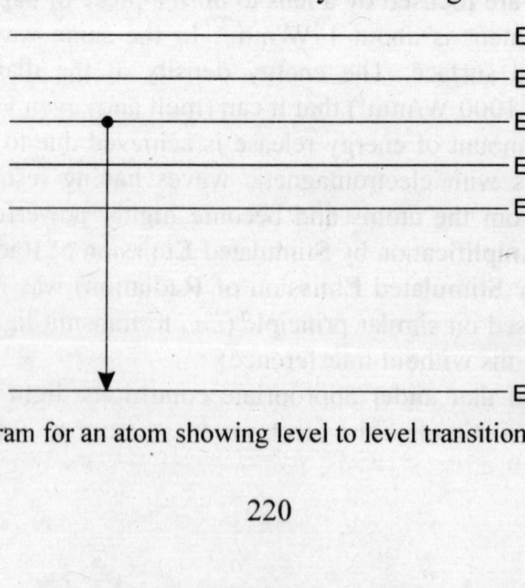

Fig. 8.2: Energy diagram for an atom showing level to level transition [*Bhattacharyya*, 1973]

'photon'. This chain of events would produce photons having same characteristics (*viz.*, wavelength, phase, direction and energy). This sequence of triggering clone photons from stimulated atoms (or molecules) is known as ***stimulated emission***. Stimulated emission forms the basis of laser operation. This process is reverse of the one in which photons (or electromagnetic waves) are absorbed by atomic system. Further, to produce a working laser, the energy source should be so powerful that most of the atoms (or molecules) of the lasing material are at their higher energy level. It is known as ***population inversion*** and refers to the population of atoms/molecules in the lasing material.

Feedback mechanism is an essential element of the laser production system. It captures and redirects a part of the coherent photons back into the active medium. These photons further stimulate the emission of some more photons of the same frequency and phase. This mechanism also permits a small percentage of coherent photons to exit the system in the form of laser light. This laser light is utilized for various useful purposes as discussed in the following sections. Rest of the photons remain in the system, and they are responsible to maintain the amplification process through stimulated emission.

WORKING PRINCIPLE OF LASER BEAM MACHINING

The objective of any machining process is to remove material from the workpiece and to get its desired shape and size within the prescribed tolerances. Photons of the laser beam interact with the electrons of the substrate and transfer momentum. As a result, the amplitude of vibration of the lattice points of the substrate increases within a period of 0.1 millisecond (ms). This results in almost instantaneous heating of the exposed area. This heat diffuses by conduction in all directions, and temperature of the exposed area increases. In Laser Beam Machining (LBM), a laser beam heats the substrate and material is removed by either or combination of melting, vaporization and ablation [*Bharti and Sivakumar*, 1996]. This molten and vaporized material from the machining zone can be easily removed by flow of high pressure assist gas jet. The major **LBM configurations** are: drilling (1-D), cutting (2-D) and grooving, turning and milling (3-D), and micromachining of different workpiece materials [*Dubey and Yadav*, 2008]. Laser beam turning and milling are 3-D operations that require two simultaneous laser beams to get the desired profile created in the workpiece. **Laser light** is *monochromatic*, (*i.e.*, its wavelength occupies a very narrow part of the spectrum). Hence, a simple lens is able to focus and concentrate laser light to a spot of much smaller diameter and much higher intensity than that obtained by other types of light. Laser light is *coherent* in nature (it travels in phase). Hence, it gives higher focused intensity than the normal light which is incoherent in nature. The low divergence rate of lasers is also responsible for high intensity of light. The advantage of using laser as a thermal energy source for material processing is due to its efficient control of depth and high magnitude of energy.

Thus, laser beam is a light source having unique **properties** like high monochromaticity, high degree of coherence, high brightness, high peak power, high energy per pulse, and very small size of the focused spot. Wave length of commonly used lasers lies between

0.21 μm–11 μm (Ruby = 0.7 μm, Nd:YAG ≈ 1.0 μm, CO ≈ 2.7 μm and CO_2 ≈ 10.6 μm). Figure 8.3 shows three important elements of any laser device, *viz.*; a laser medium (a collection of atoms, molecules, or ions), a pumping energy source required to excite these atoms to higher energy level, and optical feedback system. Consider a gas laser consisting of a thin tube filled with gas at low pressure. There are electrodes placed at both ends of the tube. Electric current when passed through provides sufficient energy to stimulate atoms/ molecules of the gas in the tube. As shown in Figure 8.3, the feedback mechanism for laser resonator consists of parallel mirrors kept at the ends of the tube. One of these mirrors is fully reflective (HR mirror in Figure 8.4) while the other one is partially transparent (output mirror) to provide the laser output. It allows a beam of radiation to either pass through, or bounce back and forth repeatedly through the laser medium.

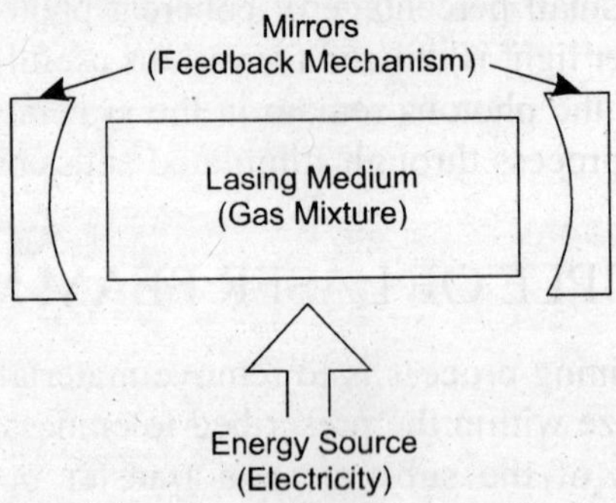

Fig. 8.3: Components of a gas laser [*Benedict*, 1987]

To make a laser beam useful for **processing of materials**, its power density should be high. The power density of laser beam and its interaction with the workpiece material determine whether the beam is able to perform the function of welding, cutting, heat treatment, or marking. To perform a machining operation, laser beam power density should lie between 1.5×10^6 to 1.5×10^8 W/cm^2, and the workpiece should be kept very close to the prime focus. However, for welding, lower power densities of the order of 1.5×10^4 to 1.5×10^5 W/cm^2 are adequate.

As the laser beam falls on the workpiece surface, reflection and transmission of electromagnetic waves at the interface of air-workpiece material take place. Reflection and transmission of electromagnetic waves of the given wavelength depend on its reflectivity and absorption coefficient. Lasers of different wave lengths, power and beam sizes have been developed for different kinds of applications, such as, excimer lasers, semiconductor lasers, dye lasers, and free electron lasers. Based on frequency, there are nanosecond lasers, femto second lasers, and picoseconds lasers.

Depending upon the intensity of the laser beam, one of the following events takes place:

1. In case of low intensity beam, there may be no phase change of the irradiated work material (or, no material removal would take place).

2. In case of high intensity beam, the work surface temperature would rise up to or above its boiling point and vaporization would take place (that is, it can be tuned for removal of material, or machining).

TYPE OF LASERS

Basically lasers are of two types, *viz.*, solid state laser and gas laser.

Solid State Lasers

Because of poor **thermal properties of solid state lasers** (*viz.*, ruby and Nd:glass), they can't be used for heavy duty work. Such lasers do not operate faster than 1 or 2 Hz. They are used only for low energy applications like spot welding, drilling, etc. However, Nd:YAG laser, most powerful in solid state lasers, is also used for manufacturing operations like cutting. Its average power (< 1000 W) is much lower than CO_2 laser. Hence, it is usually employed for light works.

Many materials having lasing property have been developed, *viz.* calcium fluoride crystals doped with neodymium ($Ca + F_2 \rightarrow Nd$). The round crystal rods with reflective ends are used. Crystalline ruby is another material used for laser action. It is made of aluminum oxide with chromium ion impurities distributed through the aluminium lattice sites ($Al_2O_3 + Cr_2\,(0.05\%)$). Flash lamp surrounding a ruby rod produces light. Flash lamp and ruby rod are enclosed in a cylinder. This cylinder has highly reflective internal surfaces. These surfaces direct light from the flash lamp into the rod. This light excites the chromium ion of ruby crystal to high energy levels. While on return journey to the normal state, these excited ions at high energy levels release the photons (or energy). Thus, the desired energy is obtained in the form of short duration pulses.

Gas Lasers

In a gas laser, CO_2, He, or N_2 acts as a lasing medium. These gases are recirculated and replenished to reduce the operating cost. Direct electrical energy is used to provide energy for stimulating lasing medium. Axial flow CO_2 laser is shown in Figure 8.4, and its output power capacity is usually 100 W for each meter length of the tube. For higher powers up to 1500 W and reduced floor space, folded resonator axial flow CO_2 lasers are used. For very high power (several thousand Watts) and still very compact CO_2 laser is known as transverse flow, or gas transport laser. Some of the details of the gas laser set-up are as follows.

Large amount of gas volume is used. The resonator mirrors are positioned to reflect the beam several times before it escapes through the output mirror. Most of the lasers are computer controlled to take advantage of their high speed processing. During processing of materials, motion can be given either to the workpiece or beam, or both depending upon the set-up design.

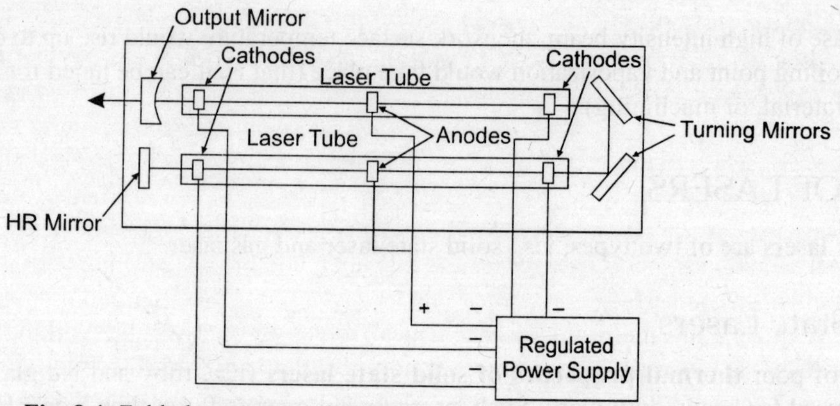

Fig. 8.4: Folded resonator axial flow carbon dioxide laser [*Benedict*, 1987]

PROCESS CHARACTERISTICS

Figure 8.5 shows various processes in which the laser power is utilized during LBM [*Yeo et al.*, 1994]. The relative magnitudes of **heat consumption** as losses and absorption by workpiece material depend upon thermal and optical properties of the work material, and intensity and pulse duration of the laser beam. It is also to be noted that a part of the material being expelled from the work surface stays in the path of the laser beam in the form of small droplets and continues to absorb energy of the laser beam.

High capital and operating costs and low machining efficiency (around 1% or so) prevent LBM from being competitive with conventional machining techniques. **Industrial lasers** operate either in Continuous Wave (CW) mode or impulse mode. CW lasers are used for processes like welding, Laser Chemical Vapor Deposition (LCVD), surface hardening, etc. which require uninterrupted supply of energy for melting and phase transformation. Further, LBM results in a **Heat Affected Zone (HAZ)**. Controlled pulse

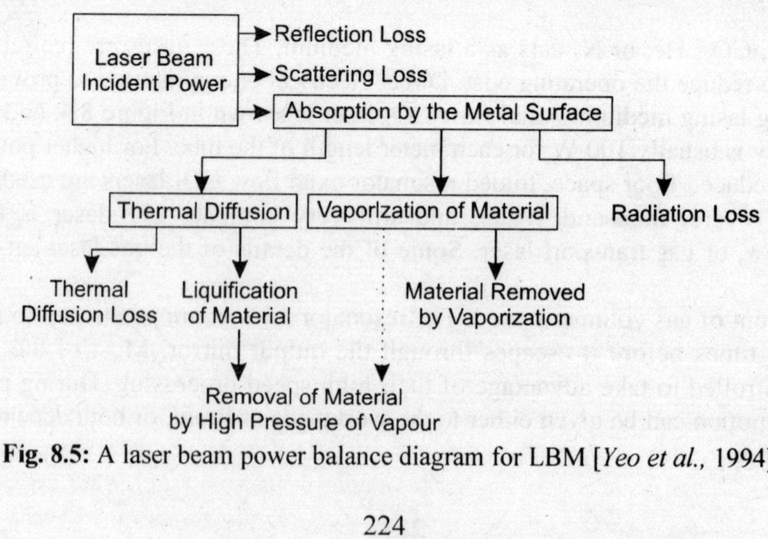

Fig. 8.5: A laser beam power balance diagram for LBM [*Yeo et al.*, 1994]

energy is desirable for the processes like cutting, drilling, marking, etc., so that the HAZ is minimum possible. It has also been found [*Huber and Marx*, 1979] that LBM technique reduces fatigue strength of the machined component as compared to the fatigue strength of the component when machined by the conventional machining process. It has also been found that as the feed rate in LBM increases, the thickness of the HAZ goes down. The thickness of the HAZ is also governed by the type of assisting gas and its pressure (in case of gas assisted laser cutting), gas nozzle diameter, and the distance between the nozzle tip and the workpiece surface.

In LBM, practically there are no mechanical forces exerted on the workpiece. LBM process is capable of easily machining refractory, brittle, hard, metallic, and non-metallic materials, *viz.*, cast-alloy, tungsten, titanium, alumina, and diamond. It can machine through any optically transparent material (say, glass). As long as the beam path is not obstructed, it can be used to machine in otherwise **inaccessible areas**. The laser beam can operate through transparent environment like air, gas, vacuum, and in some cases even liquids. However, LBM cannot be applied to the materials which have high thermal conductivity and/or high optical reflectivity (*viz.*, aluminum, copper, and their alloys). Because of this property, table made of aluminum is used to hold the workpiece during LBM process. It should also be noted that LBM systems are quite inefficient.

The least diameter to which a laser beam can be focused depends upon the laser beam divergence, which is a function of the quality of the lasing material and depth at which machining is being done. Using LBM, holes of high aspect ratio (= hole depth/diameter of hole) and a very small diameter can be made. Figure 8.6 shows a schematic diagram of cross section of a **drilled hole** using LBM process. The taper angle of a drilled hole reduces with an increase in the depth of the hole [*Scott*, 1976]. Geometry of Figure 8.6 can be used to evaluate taper angle, α (= (e-c)/d). **Recast layer** (*i.e.* any molten or vaporized material

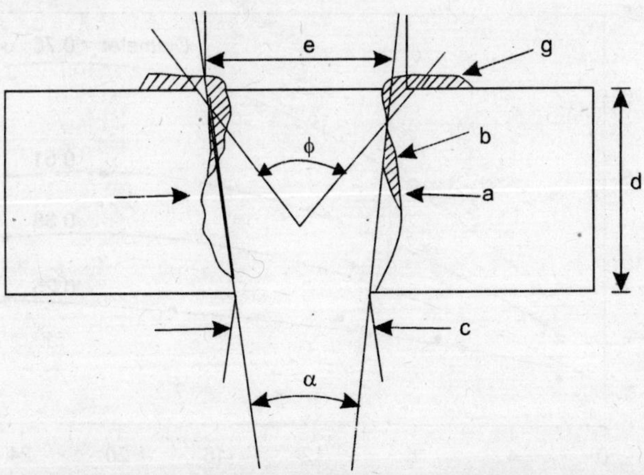

Fig. 8.6: Geometry of a drilled hole using LBM process: **a** - diameter of mid span, **b** - thickness of the recast layer, **c** - exit diameter; **d** - hole depth, **e** - inlet diameter; **g** - thickness of the surface debris, ϕ - inlet cone angle, α - taper angle [*Garcia et al.,* 1989].

that resolidifies and deposits on the machined surface) has micro/nano cracks and it is loose enough to be scraped off. It is also easy to drill holes at an angle other than 90° to the surface (not less than approximately 10°).

For a good quality drilled hole, high peak power (or ratio of high pulse energy and short pulse duration) and high power density are recommended. The recommended range of pulse duration suitable for deep hole drilling is 0.1 to 2.5 ms. It is reported [*James and Mike*, 1989] that the time required to drill a given size of hole is proportional to the amount of energy delivered per pulse (Figure 8.7). Figure 8.8 shows a relationship between thickness of the workpiece and time taken to drill the holes of different diameters.

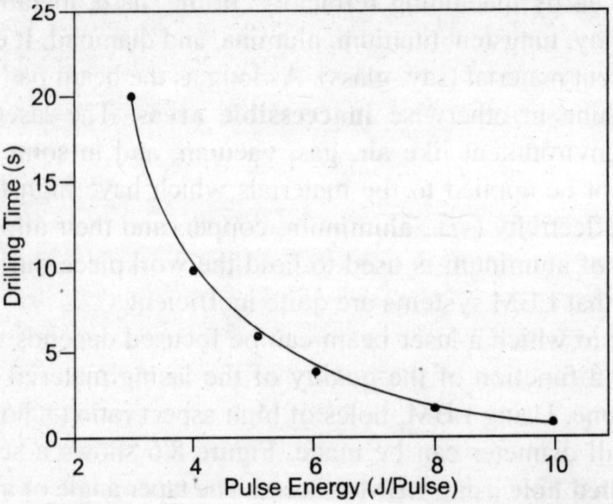

Fig. 8.7: Effect of pulse energy on the time required for drilling a hole of constant depth (6.355 mm) in steel alloy [*James and Mile*, 1989]

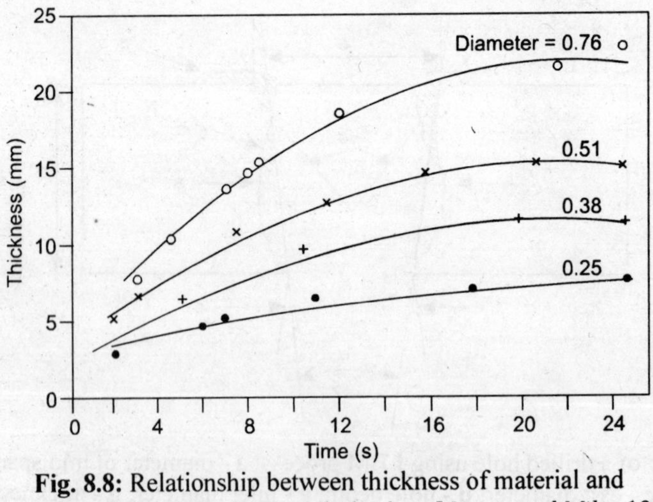

Fig. 8.8: Relationship between thickness of material and drilling time for various diameter hole sizes [*James and Mike*, 1989]

LASER ASSISTED HYBRID MACHINING PROCESSES

Hybrid machining processes are those processes, which are combinations of two or more than two machining processes. These constituent processes can be traditional or advanced machining processes in any combination. These hybrid processes are developed to exploit the potential advantages and to restrict the disadvantages associated with an individual constituent process. Usually, the performance of a hybrid machining process is better than the sum of the performance of their constituent processes with the same parameters settings. In some of these processes, besides the performance from individual component process, an additional contribution may also come from the interaction of the component processes [*Yadav et al.*, 2002]. Many of the hybrid machining processes have been developed by combining conventional or advanced machining processes with LBM or USM. Laser and non-laser hybrid machining processes are becoming more popular in industries in recent time as compared to the past.

APPLICATIONS

Laser beam energy has been favorably employed for machining difficult-to-machine materials such as hardened steels, composites, ceramics, etc. irrespective of whether they are electrically conductive or not. However, the process is employed to those materials which have favorable thermal and optical properties. Laser beam energy has been utilized for operations like drilling, cutting, micromachining, trepanning, trimming, marking, welding, soldering, brazing, etc. [*Kobayashi*, 1984].

Drilling

LBM process is extensively used for making small holes (microhole dia. < 1 mm; small hole dia. 1.0 mm–3.2 mm [*Yeo et al.*, 1994]), and also known as ***laser percussion hole drilling***. The workpiece is placed at or near the focal point of the laser beam. The localized high intensity heat results in melting of a part of the work material and a small part may vaporize also depending upon the temperature in that area. Escaping of vaporized material out of the hole results in the removal of a part of the molten material also as a spray of the droplets.

Super alloys due to their properties like toughness, creep strength, and hot corrosion resistance at high temperature, are commonly used materials for a turbine components such as blades, guide vanes, after burners, casings etc. where temperature as high as 2000°C can reach. A large number of **cooling holes** are required in some of these components. Laser beam drilling is also used [*Yeo et al.*, 1994] for cooling hole drilling apart from other processes, say, electrochemical drilling.

LBM process has been used to drill small diameter holes (diameter, 0.125 to 1.25 mm) with length to diameter ratio as high as 100:1 only under very special conditions. It has been used for drilling miniature holes in diamond dies for wire drawing, in sapphire and ruby bearings for watches, in turbine blades for cooling, etc. The drilled holes are found to have a taper, rough shape with low degree of roundness, recast layer and heat affected zone (≈ 0.0025 to 0.01 mm). The actual values of these characteristics (taper, roundness, etc.) depend upon the workpiece material properties, thickness of the workpiece and machining input parameters. Diametral repeatability of this process is about ± 0.025 mm or ± 10% of the diameter (whichever is greater). It is also reported [*Sona*, 1987] that a better quality hole and improved performance of the process can be achieved if low energy high frequency pulses are used in place of high energy single pulse.

The beam divergence (*i.e.,* beam spread, or an increase in beam diameter for unit distance of beam travel) depends on the average output power of the laser [*Bolin*, 1982]. Laser beam with low divergence should be recommended for deep hole drilling. Apart from the low divergence, laser beam should also have long focal length. To have straight sided hole, focal point of the laser beam should be located just below the surface of the material being machined. The beam is always focused to a "waist" rather than infinitely small spot due to the divergence and diffraction. The depth of focus (Figure 8.9) is the length of the beam over which the waist diameter does not change appreciably. A lens of larger focal length gives a larger depth of focus. The depth of focus (Δf) can be calculated using Eq. (8.1).

$$\Delta f = (2Wf)/(Di^*) \qquad \ldots(8.1)$$

where, f is the focal length of lens (mm), W (in mm) is the focused beam diameter $(= f.\,\theta)$, θ is the total beam divergence angle (radians), and Di^* is the beam diameter (mm).

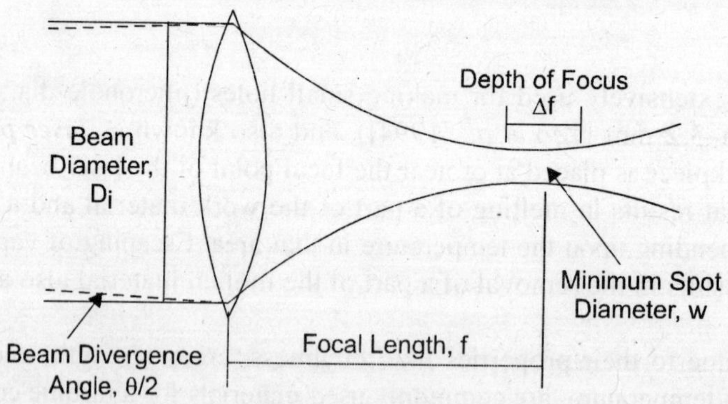

Fig. 8.9: Focal pattern of a converging lens [*James and Mike*, 1989]

Cutting

Larger sized holes (>1.2 mm diameter) can't be drilled by this process because of low power density of the focused beam. Hence, larger sized through holes are trepanned (or cut) rather than drilled. Laser cutting does not involve any mechanical type of forces. Cutting is done omni-directionally at high speed and it is capable to pierce the workpiece at any location. Cut shapes can be produced either by ***trepanning or CNC cutting.*** Trepanning is achieved by opto-mechanically sweeping a laser beam in the desired path on the workpiece. Large sized through holes using laser beam are made by trepanning operation in which either laser beam or workpiece is moved according to the profile of the hole or cavity to be produced. The size of the trepanned hole does not depend on the size of the laser beam spot. This technique has been employed to drill holes in ceramics also. Cycle time ranges from a fraction of a second to few seconds depending upon thickness of the workpiece material. Noncircular holes can also be easily made by CNC contour cutting machine using laser beam. Any NC part programmable shape can be machined by this technique.

In a ***gas assisted laser beam machining system***, the laser beam travels towards the optical assembly of the system having a focusing lens and a coaxial gas jet system. The laser beam and the gas jet are directed towards the workpiece. The gas jet assists in removing molten material from the cut, and also to keep debris away from contaminating the focusing lens [*Mishra and Yadav*, 2015]. Use of Oxygen gas as a jet is often recommended for oxidizable materials (say, carbon steels). Oxygen gives higher cutting speed compared to other gases namely, air and nitrogen but oxidized edges of the machined components result in larger HAZ. Argon gas gives very good cut edges, hence it is used for cutting workpieces which are to be welded or brazed at later stage.

Nd:YAG laser gives low mean power beam but its intensity is quite high ($>10^8$ W/cm^2) mainly due to better focussing behavior (beam divergence as 0.001 radian and focussed beam diameter as 0.08 mm). Due to low thermal load, Nd:YAG laser has been employed for cutting of brittle materials like SiC ceramics without crack damage [*Tonshoff et al.*, 1989] with smaller kerf width and HAZ, as compared to CO_2 laser. Commercially available laser units of about 800 W capacity can cut most of the metal plates up to 3 mm thickness or so, with cutting accuracy of about 0.8 mm.

Marking

The marking system is used to engrave letters, numerals and symbols on metal and nonmetal workpieces. The system is made up of *pulsating laser* with a computer controlled beam scanning system. As the laser beam scans the workpiece, the localized area in the form of overlapping blind holes is vaporized to produce blind grooves of maximum width and maximum depth as 0.25 mm. The depth can be as low as 5 μm (0.005 mm) which depends on the selected machining parameters. To obtain high quality marking (blind features) with minimum surface damage, a high power density pulsating laser beam lasting only for nanoseconds is used. To control the position of laser beam and timings of the laser pulses, a microcomputer is used. The necessary information to generate different characters is stored in the computer. As the system is put on, the beam scans the workpiece in the form of the desired patterns at the rate as high as 30 characters per second. The quality of the characters so obtained is quite high.

Miscellaneous Applications

LBM is being employed for both micromachining as well as macromachining. **A 3-D laser beam machining** process, utilizing two laser beams, for operations like threading, turning, grooving (Figure 8.10a), milling, etc. has been employed [*Chryssolouris et al.*, 1989]. During 3-D cutting, two independent laser beams are simultaneously used to cut two grooves which are moving closer to each other. These two grooves converge, and a volume is cut off without being melted/vaporized (Figures 8.10a and 8.10b).

Laser energy has also been employed to fracture in a controlled fashion, delicate items. Absorption of laser energy by a workpiece results in high thermal gradient leading to the mechanical stresses. Finally, it results in a controlled fracture. Further, laser beam thermal energy is also used to assist traditional machining processes, for example, laser beam assisted turning, laser beam assisted milling, etc.

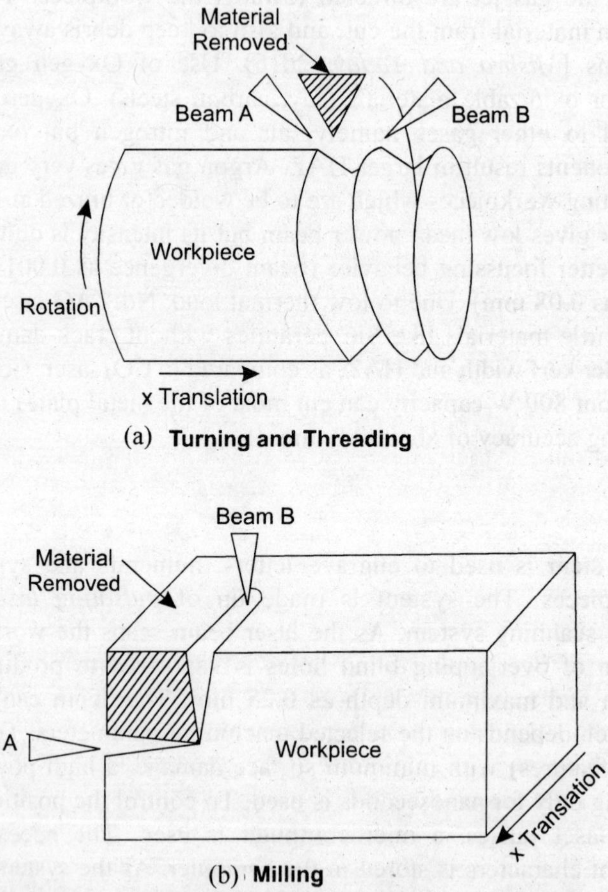

Fig. 8.10: Schematic diagram of (a) Turning/threading, and (b) Milling using two laser beams [*Chryssolouris et al.*, 1989]

LASER BEAM MICROMACHINING (LBMM)

LBM is one of the successful processes used for micromachining metallic as well as non-metallic workpieces. Basic principle behind the laser beam microcutting, microdrilling, microstructuring etc. is briefly discussed here. In Laser Beam Micromachining (LBMM), both short and ultrashort laser pulses are used for selective machining. The absorption coefficient (α) of the workpiece material depends on the temperature of the workpiece and wave length of the laser beam [*Mishra and Yadav*, 2015], and it controls the decay of laser intensity with the depth of machining inside the material. The laser beam is inertia less high energy density point heat source (*assumed for simplification*) which is useful for material processing; laser beam micromachining is one of them.

High-intensity short and ultra short laser pulses are powerful thermal energy sources for creating micro-features in wide range of materials with little or no thermal damage. Now-a-days, manufacturing industries are using LBMM in various fields like micro-optics, micro-electronics, micro-biology and micro-chemistry. It can also be used to fabricate 3-D submicron size structures, miniature photonic components, read only memory chips and hollow channel wave guides used in optical communication network, optical data memory and biological optical chips.

Working Principle of Laser Beam Micromachining

Laser Beam Micromachining (LBMM) can be performed by (i) direct writing, (ii) mask projection patterning and (iii) patterning by interference of laser. In direct writing, usually laser beam irradiates the substrate and the desired pattern is obtained by translating the laser beam on the substrate. The mask projection technique uses excimer laser to illuminate a given feature on the mask. Resolution of the features created by this technique depends on the mask material and the projection system. In interference technique, the primary laser beam is splitted into several single beams which are then superimposed to create interference pattern. The interference pattern is projected on the substrate and the micromachined pattern corresponding to the intensity profile of the interference pattern is obtained with structure size of 50–300 nm [*Gower*, 2000].

Depending on the characteristics of a particular material and laser processing parameters such as wavelength, fluence* and pulse length; a number of mechanisms are responsible for laser ablation. In LBMM, the input parameters affect the characteristics (size, shape, accuracy, heat affected zone, etc.) of the features created on the workpiece. Wavelength, pulse width, beam energy, focal length of the lens and beam shape (Gaussian, square, point or some other shape) are some input parameters and beam characteristics which affect the ablation rate, uniformity of micromachining, size of recast layer, etc.

---

* In general physics, the fluence is defined as the time-integrated flux of some radiation or particle stream. Specifically in optics, the fluence F *e.g.* of a laser pulse, is the optical energy delivered per unit area. Its most common units are J/cm^2 (joules per square centimeter).

Quick rise in temperature kinetically suppresses the solid to liquid transformation. Boiling point of the substrate is reached before appreciable amount of liquid is formed. Substrate that would normally transform to liquid before transforming to vapour starts sublimating under very high power density laser beam. If material removal is restricted to vaporization mode then the shape of the interface between vapour and condensed substrate determines the overall precision/tolerances. It is because of this reason that short high energy pulses (nanosecond, femtosecond and picoseconds) are preferred for micro-machining. The focal spot characteristics (spot size and energy distribution within the spot, say, Gaussian, uniform, point, etc.) will be governing factor for the degree of precision.

In a broad sense, **ablation** is removal of material from metals or glasses/crystals by vaporization. In polymers, the removal can be by photochemical changes. Ablation depth is determined by the heat absorption depth of the material and the heat of vaporization of the work material. The depth is also a function of beam energy density (Laser energy per unit area on the work material is also known as laser fluence), the laser pulse duration, and the laser wavelength. The laser-material interaction regime is significantly influenced by laser intensity and interaction time.

Figure 8.11 shows a schematic of laser-workpiece material interaction [Vasa, 2014]. The absorbed light energy in the material is converted into heat and its conduction produces a temperature distribution in the material (Figure 8.11(a)). Depending on the magnitude of the temperature, the material experiences heating, melting (Figure 8.11(b)) and vaporization (Figure 8.11(c)). The laser beam removes material through melting and vaporization, and the process is known as ablation. The ablation of metals is caused by the absorption of laser energy and it involves absorption of photons by electrons (time taken as approximately 10^{-15} s), and the energy is transferred to the lattice (time taken as approximately 10^{-12} s). The power density influences the vaporization rate, and high vaporization rate results in a shockwave that can reach speed of a few km/s. High pressure is created during the melting and superheating of liquid. As a result, the laser pulse ejects the material at a high speed. Sometimes, if the material is not completely vaporized, particularly if it is expelled from a depth, then it cools quickly and forms a recast layer as shown in Figure 8.11(e).

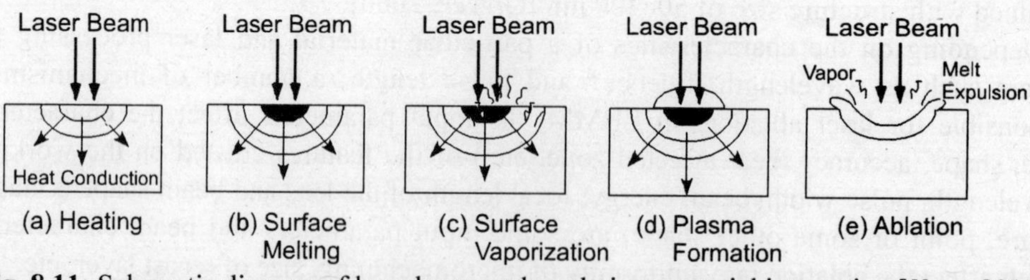

Fig. 8.11: Schematic diagram of laser beam and workpiece material interaction [*Vasa*, 2014]

Polymers are transformed in a different way than metals. If there is sufficient energy in ultraviolet light, then some bonds break, however some energy is used to create molecular vibration and heat. Since this is a lower percentage than in metals, it is said that ultraviolet light produces a cooler excitation in polymers. Infrared, on the other hand, causes mostly molecular vibration and material changes by thermal processes. Coupling efficiency in metals is quantified by lack of reflectivity, in polymers it is quantified by lack of transmission (absorption).

Applications of LBMM

Excimer laser has short wavelength (193 nm, 248 nm and 308 nm) in ultraviolet region. The spot diameter is small hence gives good accuracy. In many solids (particularly organic solids), the interaction is termed as photo ablation or simply ablation, where ultraviolet photons directly break the molecular bonds holding the material intact. As a result, the surrounding material remains unaffected and the process remains non-thermal. Hence, it is also called as ***cold machining***. The excimer lasers are commonly used in masking mode machining. Using this property, Figure 8.12 shows a human hair on which a word EXCIMER LASER has been written. Figure 8.13 shows many parts within a 2 mm × 2 mm frame which can be made using LBM.

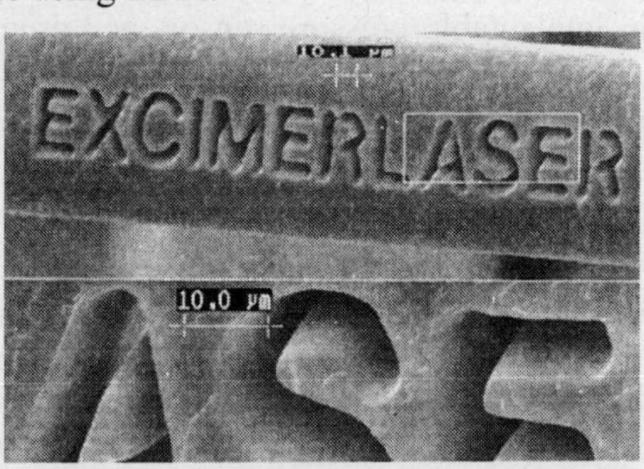

Fig. 8.12: Excimer laser written on human hair using excimer laser
[Ref: https://www.photonics.com/Articles/Excimer_Lasers_Photonic_Stamps_with_Micron/a25164]

Fig. 8.13: 2 mm × 2 mm size frame containing many parts.
It gives an idea of the kind of miniaturization that can be done using laser beam micromachining.

Laser beam ***microdrilling*** is usually performed by either single pulse drilling or precussion drilling. In the earlier case, a single laser pulse penetrates the material and creates a hole having its diameter larger or smaller than the laser beam diameter depending on the spatial intensity distribution of the laser pulse. Workpiece material thickness, thermal diffusivity and its boiling temperature also affect the hole diameter and its precision. The drilled hole taper is also affected by pulse energy and pulse duration both. Hole diameter to depth ratio can vary up to 1:25.

LBMM is used for drilling inter-connected holes for electronic packaging, drilling precise nozzle arrays for inkjet printer heads and many other similar applications. Multi-beam approach divides a beam into many individual beams (say, up to 1000 small beams) each one of which drills individual holes. It requires a well designed beam delivery system. LBMM can also machine sub-micrometer holes. Further, higher precision machining is possible by using ultra short pulse (femto second and pico second) laser beam system.

LBMM can be used for the precision machining of fluoropolymers which are used for the development of miniature lab-on-chip technologies. LBMM is used for fabrication of stents which are basically hollow tubes with mesh like walls, and micromachining of holes in catheters and patient monitoring probes. Micro-Electro-Mechanical-Systems (MEMS) which are widely used in implantable pressure sensors and drug delivery system can be fabricated by LBMM (*Gower*, 2000; *Rizvi*, 2003). Precision cutting of very thin and delicate metal foils is gaining popularity due to its wide applications in micro- and nano-technology applications.

Embedded 3-D micro fluidic channels can be fabricated by exposure of photo- sensitive glass to femto second laser. LBMM can perform patterning of high resolution electronic circuits in metal layers on variety of substrates. High speed drilling of meshes in metal foils with holes size as small as 2 μm, in situ trimming of machined and pre-assembled devices without damage to the adjacent parts can be performed with ultra-fast laser used in LBMM. *Hu et al.* [2010] reported micro-hole drilling and micro-structure machining in metals, alloys and ceramics using pico second laser. Microtexturing is one of the applications of LBMM.

Laser Beam Microtexturing

In surface microtexturing, a particular type of roughness pattern (micro or nano in size) is created on the surface in order to change its functional properties. Such textured surfaces are also referred as functional surfaces which have many new applications *e.g.*, wetting, tribological characteristics, optical properties and so on (*Patel et al.,* 2016). Some natural surfaces have inherent vibrant multifunctional properties, for example, a lotus leaf has water repellent property. Figure 8.14a shows a lotus leaf which has hydrophobic surface and it contains two levels of surface roughness that is micro protrusions and nano-hairs over microprotrusion (Figure 8.14b). Such surface containing two levels of roughness results in hydrophobicity and makes the surface to have self-cleaning property.

Arrangement of micro protrusions (pillars) and nano-hairs over the micro pillars provides very low surface-energy which is dominated by the surface tension of the water droplets, and the droplets try to maintain their spherical shape. Micro-protrusions (height ranges from 5–10 μm) present random gaps, support the water droplets and don't allow the droplets to get punctured. These protrusions provide large contact angle.

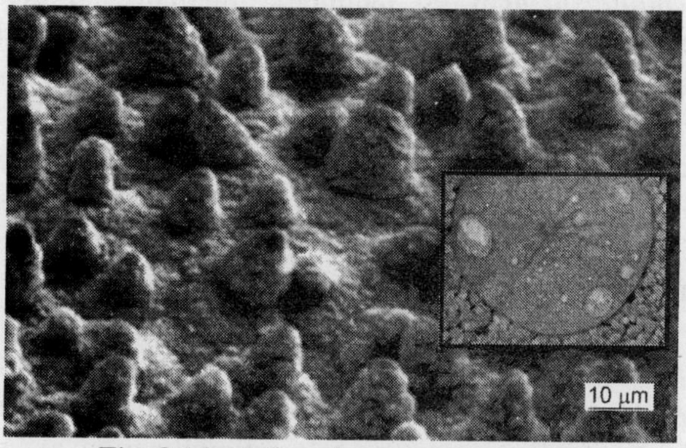

Fig. 8.14a: Micro features of lotus leaf

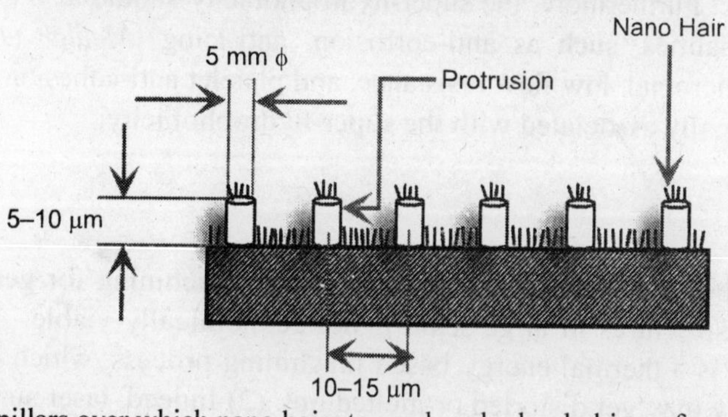

Fig. 8.14b: Micro pillars over which nano hairs to produce hydrophobic surface [Patel *et al.*, 2016]

Production of Hydrophobic Surfaces and their Potential Applications

There are two ways of generating hydrophobicity on metallic surfaces (i) Hydrophobic nano coating, and (ii) Surface texturing (*Patel et al.,* 2016). A large number of researchers are currently working in the challenging area of **mimicking natural surfaces** like lotus leaf. Hydrophobic surfaces have property to repel the water droplets and during rolling of these droplets, dirt and foreign particles adhered to the leaf are flushed out. Two levels of hierarchical structure (roughness) provide contact angle of >110° (with water) which results in *super hydrophobicity.* Surfaces having low contact angle (<65°) and high surface energy lead to hydrophillicity

Tang et al. [2011] produced super-hydrophobic surface on brass substrate using pulsed UV laser ablation. Surface roughness (Ra) increases from ~400 nm to ~5 µm. Meanwhile, plenty of nano-structures (200~600 nm) are also machined and scattered on the surface of the micro-structures. Such type of metal surfaces result in water contact angle lying between 110° to 161°. *Vorobyev et al.* [2015] experimentally investigated the creation of ***multifunctional metallic surfaces*** on platinum, titanium and brass by generating a hierarchical nano/micro-structure with the help of femto second laser pulses. Micro projection dimensions are: height ≈ 20 µm, width ≈ 30 µm and distance between two projections ≈ 30 µm.

Nowadays, hydrophobicity is highly demanded on the surfaces of ***critical components***, say, an impeller casing and blades of a centrifugal pump that can result in longer life of the impeller and reduced fouling by repelling the fluid being pumped by the impeller. Textured super-hydrophobic metal surfaces can be used to protect aerofoils and aircraft engines against the accumulation of ice during high altitude fly, as well as to make more weather proof infrastructures [*Alhart*], [*Sanchez*, 2015]. The areas of applications of hydrophobicity are quite wide, including power transmission lines, pipes of air conditioners and refrigerators drag reduction, radar or telecommunication antennas, weather proof infrastructures, etc. Furthermore, the super-hydrophobicity should also enable other highly desirable functionalities, such as anti-corrosion, anti-icing [*Meuler et al.,* 2010], anti-biofouling, anti-microbial, low flow resistance, and platelet anti-adhesion [*Sun, et al.,* 2005] which are intrinsically associated with the super-hydrophobicity.

REMARKS

However, it can be commented that: (1) Using laser machining for generation of micro/nano features or structures in large scale is not economically viable. Also, laser surface texturing process is a thermal energy based machining process, which can generate nano structures but they may get distorted or melted out. (2) Indeed, laser surface texturing is a good option for polymer but it is difficult to do so on metallic surfaces especially curved surfaces.

BIBLIOGRAPHY

[1] Singh, Abhilasha; Patel, D.; Ramkumar, J. and Balani, K. (2016). Micro-channel fabrication through pulsed Nd:YAG laser on Ti6Al4V. *Proc. Int. Conf. on Better Living* 3:120–120 doi: 10.3850/978-981-09-7519-7nbl16-rps-120.

[2] Singh, Abhilasha; Patel, D.S.; Ramkumar, J. and Balani, K. (2018). Single step laser surface texturing for enhancing contact angle and tribological properties. *Int. J. Adv. Manuf. Technol*. doi: 10.1007/s00170-018-1579-8.

[3] Alhart, T., "Super-hydrophobic Metals for Aircraft Applications", GE Global Research, http://www.growthconsulting.frost.com.

[4] Benedict, G.F. (1987). "Non-traditional Manufacturing Processes", *Marcel Decker Inc., New York.*

[5] Bharti, A. and Sivakumar, R. (1996). "The Mechanism of Material Removal in Laser Fusion Cutting", *Lasers in Engineering*, Vol. 5(1), pp. 87–105.

[6] Bhattacharyya, A. (1973). "New Technology", The Institution of Engineers (I), Calcutta, (India).

[7] Bhattacharyya, A. and Brain, K. (1989). "Drilling: Working through the Benefits", *Machining Prord. Eng.*, 147(3765), pp. 73–83.

[8] Bolin, S.R. (1982). "The Effect of Focus and Beam Divergence on Quality of Holes Drilled with Pulsed Nd:YAG Laser", *Proc. Mat. Process. Sympos., Laser Inst. of America,* Vol. 31, pp. 135–140.

[9] Bolin, S.R. (1983). "Nd:YAG Laser Application Survey", Laser Material Processing (Editor: M. Boss), North Holland Amsterdam, pp. 407–438.

[10] Chryssolouris, G. (2013). Laser Machining: Theory and Practice. Published by Springer Verlag, Berlin.

[11] Chryssolouris, G., Bredt, J., Kordas, S. and Wilson, E. (1988). "Theoretical Aspects of a Laser Machine Tool", *Trans. ASME, J. Engineering for Industry,* Vol. 110, pp. 65–70.

[12] Chryssolouris, G., Choi, W.C., Kyi, S.B. and Sheng, P. (1988). "Investigation of the Effect of a Gas Jet in Laser Grooving", *Proceedings 16th NAMRC*, pp. 217–223.

[13] Chryssolouris, G., Schonbeck, J., Choi, W. and Sheng, P. (1989). "Advances in three Dimensional Machining", *Presented in WAM of ASME (PED-VOL), held at San Francisco,* p. 17.

[14] Chryssolouris, G. and Yablon, A. (1993). "Depth Prediction in Laser Machining with the Aid of Surface Temperature Measurement", *Annals of the CIRP,* Vol. 42(1), pp. 250–257.

[15] Dubey, Avanish Kumar and Yadava, Vinod (2008). "Laser Beam Machining—A Review", *International Journal of Machine Tools & Manufacture*, Vol. 48, pp. 609–628.

[16] Gracia de Vicuna, G.E., Beitialarrangoitia, J.C. and Ghosh, S.K. (1989). "Defects Arising from Laser Machining of Materials", in: A. Niku-Lari and B.L. Mordike (Eds.) High Power Lasers, pp. 227–249.

[17] Gower, M.C. (2000). Industrial Application of Laser Micromachining. *Opt. Express*; Vol. 7(2), pp. 56–67.

[18] Hu, W., Shin, Y.C. and King, G.B. (2010). Micromachining of Metal Alloys and Ceramics by Pico-second Laser Ablation. *J. Manuf. Sci. Eng.,* 2010; 132:0110091–7.

[19] Huber, J. and Marx, W. (1979). "Production Laser Cutting", *ASM Conf. on Applications of Lasers in Materials Processing,* Washington DC, pp. 273–290.

[20] James, J. and Mike, P. (1989). "Deep Hole Drilling with Lasers", *Modern Machine Shop,* Vol. (62)1, pp. 53–63.

[21] Jiang, C.Y., Lau, W.S., Yue, J.M. and Chiang, L. (1993). "On the Maximum Depth and Profile of Cut in Pulsed Nd:YAG Laser Machining", *Annals of the CIRP*, Vol. 42(1), pp. 223–226.

[22] Kobayashi, A. (1984). "Recent Progress in Laser Processing in Japan", *Proceedings 5th Int. Conf. Prod. Engg., Held at Tokyo*, pp. 459–465.

[23] Lei, S., Shin, Y.C. and Incropera, F.P. (2001). Experimental Investigations of Thermo-mechanical Characteristics in Laser-Assisted Machining of Silicon Nitride Ceramics, *ASME Journal of Manufacturing Science and Engineering,* Vol. 123, pp. 639–646.

[24] Matsuzawa, T. (2002). "State of the Art of Micromachining", *Annals of the CIRP*, Vol. 49, pp. 473–488.

[25] McGeough, J.A. (1988). "Advanced Methods of Machining", *Chapman and Hall,* London.

[26] Mishra Sanjay, Yadava Vinod (2015), "Laser Beam Micro Machining (LBMM)—A Review", *Optics and Lasers in Engineering,* Vol. 73, pp. 89–122.

[27] Meuler, A.J., McKinley, G.H. and Cohen, R.E. (2010). "Exploiting Topographical Textures to Impart Ice Phobicity", *ACS NANO*, Vol. 4, pp. 7048–7052.

[28] Patel, Divyansh; Jain, V.K. and Ramkumar, J. (2016). Micro Texturing on Metallic Surfaces: State of the Art, *Proc. IMechE Part B: J. Engineering Manufacture,* 1–24; DOI: 10.1177/0954405416661583, IMechE 2016.

[29] Reddy, J.F. and Farion, D.F. (2001). LIA Handbook of LMP", *LIA MPI*, pp. 481–482.

[30] Rizvi, N.H. (2003). Femto Second Laser Micromachining: Current Status and Applications. *RIKEN Rev*, Vol. 50, pp. 77–82.

[31] Sanchez, N. (2015). "Laser Water-Repellent Metal Much More Slippery than Teflon", http://www.newsmax.com/TheWire/laser-water-repellent-metal

[32] Scott, B.F. (1976). Laser Machining and Fabrication—A Review, *Proc. 17th Int. Mach. Tool Des. Res. Conf., Birmingham,* Macmillan Press, pp. 335–339.

[33] Sona, S. (1987). Metallic Material Processing: Cutting and Drilling, in ODD Soares and M. Perez Amor (Eds) Applied Laser Tooling, Nijhoff Boston, pp. 105–113.

[34] Sun, T., Tan, H., Han, D., Fu, Q. and Jiang, L. (2005). "No Platelet can Adhere-Largely Improved Blood Compatibility on Nanostructured Super-Hydrophobic Surfaces", *Small,* Vol. 1 (10), pp. 959–963.

[35] Tang, M., Shim, V., Pan, Z.Y., Choo, Y.S. and Hong, M.H. (2011). "Laser Ablation of Metal Substrates for Super-Hydrophobic Effect", *Journal of Laser Micro/Nano engineering*, Vol. 6(1), pp. 6–9.

[36] Tonshoff, H.K. and Emmelmann, C. (1989). "Laser Cutting of Advanced Ceramics", *Annals of the CIRP*, Vol. 38, pp. 219–222.

[37] Vasa, N.J. (2014). Laser Micromachining Techniques and their Applications, Introduction to Micromachining (Ed. Dr. V.K. Jain), Narosa Publishers, New Delhi (India).

[38] Vorobyev, A.Y. and Guo, C. (2015). "Multifunctional Surfaces Produced by Femtosecond Laser Pulses", *Journal of Applied Physics*, Vol. 117, pp. 033103.

[39] Yadava, V., Jain, V.K. and Dixit, P.M. (2002). "Temperature Distribution during Electro-Discharge Abrasive Grinding", *Machining Science and Technology*, Vol. 6(1), pp. 97–127.

[40] Ye, O.C.Y., Tan, S.C., Jan, S. and Michael, W.S. Lau (1994). A Technical Review of the Laser Drilling of Aeroplane Materials, *J. Materials Process Tech.*, Vol. 42, pp. 15–49.

SELF TEST QUESTIONS

Q1. Make the correct choice(s).

1. LBM system cannot efficiently machine:
   (a) Refractory materials (b) tungsten carbide
   (c) copper (d) M.S.
2. The efficiency of LBM system is about:
   (a) 3 to 5% (b) 13 to 15% (c) 0.3 to 1.0% (d) 30 to 50%
3. While drilling deep hole by LBM process, it is (with reference to top surface):
   (a) Straight sided (b) Converging (c) Diverging (d) None of these
4. You can use LBM for machining soda lime glass:
   (a) Yes (b) no (c) insufficient information
5. The unsuitable process(es) for making 5 mm deep blind complex contoured shape in M.S. sheet of 20 mm thickness is (are):
   (a) LBM (b) EDM (c) ECM (d) AFM
6. Commonly used process for making straight sided cooling holes in a turbine blade is:
   (a) USM (b) AWJM (c) LBM (d) EBM (e) EDM
7. During LBM, one does not get
   (a) recast layer (b) thermal residual stresses
   (c) straight sided hole (d) none of these
8. Laser beam trepanning can be easily used for making a blind hole in a thick sheet of MS.
   (a) yes (b) no (c) insufficient information
9. Oxygen gas assisted laser beam machining helps in
   (a) removal of molten material (b) flushing out of debris
   (c) getting larger HAZ (d) all of these
10. Laser beam micromachining uses
    (a) CW laser beam (b) pulsed laser beam (c) any one of these two
11. During laser beam drilling, the drilled hole diameter compared to laser beam diameter can be
    (a) smaller (b) bigger (c) any of these two.
12. Super hydrophobic surface has contact angle with water as
    (a) $>65^\circ$ and $<100^\circ$ (b) $<65^\circ$ (c) $>110^\circ$ (d) any of these

Q2. Answer in brief.

(a) Write the full form of following acronyms: LASER, MASER and HAZ.
(b) Explain the effect of 'focusing' on the performance of LBM.
(c) LBM and EDM both are thermal processes. Which one results in more thermal damage to the machined component? Justify your answer.
(d) What do you understand by Laser assisted hybrid machining process?
(e) Explain, how can you grind a material harder than the abrasive particles used in a grinding wheel, using laser assisted grinding process?
(f) Will you get recast layer in case of laser assisted grinding process? Justify your answer.
(g) Is there any difference between trimming and trepanning operations, or they are the same? Explain your answer with the help of an example.

Q3. Write short notes on the following.

(a) Explain the production of laser beam and working principle of LBM.
(b) Laser beam diameter is 120 μm and the hole to be made has the diameter of 10 mm. Can you use this laser beam for making this hole? Justify your answer using figures etc., if required.
(c) What do you understand by "depth of focus"? Explain with the help of a figure.
(d) Explain with the help of a figure, how can you achieve 100 holes concurrently using one single laser beam?
(e) What are the two unique characteristics of laser light?
(f) What do you understand by mask projection micro patterning?
(g) Explain the term "Ablation" of metals and organic materials.
(h) What are the factors which govern the degree of precision in micromachining?
(i) What are the differences in normal LBM and *cold machining* using laser beam? Explain clearly.
(j) Sketch a stent and suggest the methodology you will use to make a stent from a solid wire. Make the necessary sketches.
(k) What do you understand by hydrophobic and hydrophilic surfaces? Can you create hydrophobic surface having contact angle around 150 degree, by using LBMM process? Explain in detail.

NOMENCLATURE

Di*	Beam Diameter
E, p, q	Energy levels
f	Focal length
Δf	Depth of focus
W	Minimum spot diameter
θ	Beam divergence angle

ACRONYMS

CNC	Computer Numerical Control
CW	Continuous Waves
HAZ	Heat Affected Zone
HR	Highly Reflecting
LASER	Light Amplification by Stimulated Emission of Radiation
LBM	Laser Beam Machining
LBMM	Laser Beam Micromachining
MASER	Microwave Amplification by Stimulated Emission of Radiation
MEMS	Micro Electro Mechanical System

CHAPTER 8

AT-A-GLANCE

LASER BEAM MACHINING (LBM)

Production of Laser

- Hypothesis of Einstein → possible to use light energy to stimulate electrons → emit light of same characteristics.
- An atom brought to high energy level '$E2$' by outside energy source (light, heat, etc.).
- If this atom is allowed to decay back to ground state energy level ($E1$) → a photon is released.
- This photon contacts other atom ← this will also decay back to ground state energy level → chain reaction → called as stimulated emission.
- External energy source → so powerful → most of the atoms of lasing material are at higher energy level.

Feedback Mechanism

- Feedback mechanism captures and redirects a few coherent photons back into the active medium.
- Redirected photons stimulate the emission of some more photons.
- Feedback mechanism also permits a small percentage of coherent photons to exit as laser light from partially transparent end.
- Thus it is **l**ight **a**mplification by **s**timulated **e**mission of **r**adiation.
- Electrodes at both ends of the tube.
- Electric current stimulates gas atoms/molecules in the tube.
- Feed back mechanism → parallel mirrors at the ends of the tube.
  - One mirror → fully reflective.
  - Another one → partially transparent → laser output.
- Laser light → monochromatic.
- Wavelength → very narrow portion of the spectrum.
  - A simple lens focuses & concentrates → very small diameter & high intensity.
  - Laser light coherent (travels in phase) in nature & very low divergence rate.
- Use of laser in materials processing → high power density.
- Machining operation → $1.5 \times 10^6 - 1.5 \times 10^8$ W/cm^2.
- W/P kept very near to focus.

Working Principle

- Photons and electrons of substrate interact and transfer momentum.
- Instantaneous heating leads to melting and/or vaporization resulting in material removal by removal of molten material, or vaporization/ablation.
- 3-D operations require two laser beams cutting simultaneously.
- Can be used for both macromachining as well as micromachining.
- Laser light is monochromatic.
- Laser beam can be focused in a small diameter and high heat intensity beam.
- Laser light source → high degree of coherence, high brightness, high peak power, high energy per pulse, and very small size of the focused spot.
- Laser beam power density decides the type of operation it can perform.

Types of Laser

- **Solid state lasers** → Ruby, Nd:YAG.
  - Can't be used for heavy duty work.
  - Used for spot welding, drilling, etc.
- **Gas lasers** → CO_2, H_e, N_2, etc.
- Gases recirculated & replenished.
  - Folded resonator axial flow CO_2 laser → 1500 W or more.
  - For very high power → transverse flow/gas transport laser:
    - Large amount of gas volume.
    - Beam reflected several times before it escapes.
    - Computer control to take advantage of high speed processing.
    - Motion → W/P or beam or both.

Process Characteristics

- Heat consumption as losses and absorption = f (WP material, laser intensity & pulse duration) + droplets in the path of laser beam.
- High capital and operating costs, and low machining efficiency: not competitive with traditional machining methods.
- Continuous mode/pulsed mode. Low fatigue strength. Minimizes HAZ.
- Thickness of HAZ → feed rate, gas pressure, and gas nozzle diameter.
- No mechanical forces.
- It can machine optically transparent material & in inaccessible areas.
- Can not be applied to materials with high heat conductivity & high reflectivity → Al, Cu. Hence machining table → aluminium.

- Quite inefficient system (<5%).
- The least diameter achievable: quality of the lasing material and depth at which machining is being done. Also can make tapered hole.

Laser Assisted Hybrid Machining Processes (HMPs)

- Laser beam + one or more traditional or/and advanced machinig process(es).
- Exploits potential merits and restricts demerits of the individual/constitutent processes.
- HMPs performance is better than the constituent processes.

Applications

- Employed for machining hardened steel, composites, ceramics, etc.
- Workpiece material should have favourable thermal and optical properties.
- Operations like drilling, cutting, milling, trpanning, trimming, micromachining, marking, etc.

Drilling Holes

- W/P placed at/near focal point.
- Mechanism of material removal: melting and/or vaporization.
- Hole diam. → 0.125–1.25 mm.
- Aspect ratio → under very special conditions 100:1.
- Usually < 20. Tapered, low degree of roundness, recast layered walls.
- HAZ (2.5×10^{-3}–10×10^{-3} mm).
- Diameter repeatability ± 0.025 mm, or ± 10% of the diameter (whichever is greater).
- Deep hole drilling → low divergence, large focal length.
- For sraight sided hole: focal point just below the surface being machined.
- Depth of focus: within 'waist'.
- Large sized holes → cut (trepanned) rather than drilled.

Cutting

- No mechanical forces → high speed cutting.
- Gas jet assists → clearing the cut material.
  - Oxygen for oxidizable materials → higher cutting speed.
  - But oxidized edges → larger HAZ.
- Argon → good cut edges.
- Macrocutting, microcutting and 3-D cutting.

- CNC contour cutting any programmable shape → trepaning optomechanically sweeping beam → desired path.

Marking

- Cycle time → fraction to few seconds.
- To make letters, numerals, and symbols → metals & non-metals both.
- Pulsating laser system and computer controlled beam scanning system.
- Blind grooves of very small depth (0.005–0.25 mm). Selected machining parameters.
- For high quality marking → laser beam lasts for neno-seconds.
- Microcomputer to control → position and timings.
- 30 characters/second.

Laser Beam Micromachining and Working Principle

- Short and ultrashort pulses create microfeatures → wide range of materials.
- Applications: optics, electronics, communications, chips, etc.
- LBMM 3 ways: direct writing, mask projection patterning, interference laser patterning.
- Input parameters (wavelength, fluence, pulse length, etc.) → affect feature properties, recast layer size, etc.
- High power density laser → quick rise in temperature suppresses solid to liquid transformation → sublimation.
- For micromachining: short high energy pulses. (nano-, femto-, pico-second pulses).
- Ablation: material removal by vaporization or breaking molecular bonds.

Chapter Nine

PLASMA ARC MACHINING (PAM)

WORKING PRINCIPLE

A gas molecule at room temperature consists of two or more atoms. When such a gas is heated to a high temperature of the order of 2000°C or so, the *molecules separate out as atoms*. If the temperature is raised to 3000°C, the electrons from some of the atoms dissociate and the gas becomes ionized consisting of ions and electrons. This state of gas is known as plasma.

Thus, **plasma** is the glowing, ionized gas that results from heating of a material to extremely high temperature. It is composed of free electrons dissociated from the main gas atoms. A gas in plasma state becomes electrically conductive as well as responsive to magnetism. Because of such behaviour, plasma is also known as a *fourth state of matter*. The plasma is encountered in electrical discharges. The source of heat generation in plasma is the recombination of electrons and ions into atoms, or recombination of atoms into molecules. This liberated bonding energy is responsible for increased kinetic energy of the atoms or molecules formed by recombination.

The temperature of plasma can be of the order of 33,000°C. When such a high temperature source reacts with work material, the work material melts out and may even vaporize, and finally is cut into pieces. Many materials (say, aluminium, stainless steel, etc) have high thermal conductivity, large heat capacity, and/or good oxidation resistance. As a result, such materials cannot be cut by conventional techniques like oxyfuel cutting. But these materials can be easily cut by **plasma arc cutting (PAC)**.

PLASMA ARC CUTTING SYSTEM

PAC system uses DC power source. PAC systems operate either on non-transferred arc mode or transferred arc mode (Fig. 9.1). In the earlier case, the thermal efficiency is low (65–75%) and power is transferred between the electrode and the nozzle. This **non-transferred arc** ionizes a high velocity gas that is streaming towards the workpiece. The workpiece may be electrically conductive or non-conductive.

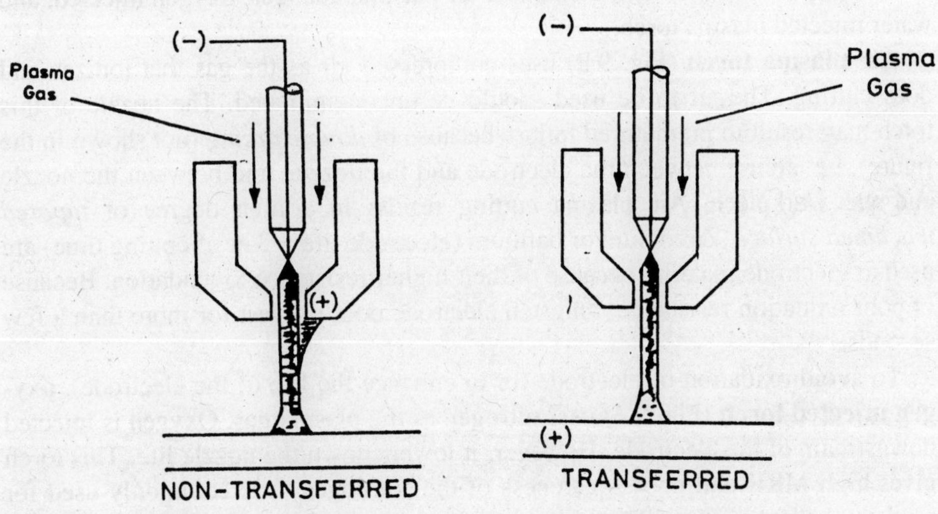

Fig. 9.1 Schematic diagram for non-transferred and transferred arcs.

In case of a **transferred arc mode**, the arc is maintained between the electrode (negative polarity) and the electrically conductive workpiece (positive polarity). Note that only *electrically conductive workpiece* can be machined or cut by transferred arc system. The arc heats a coaxial-flowing gas and maintains it in a plasma state. The electrothermal efficiency is up to 85–90%. PAC system can deliver up to 1000 A at about 200 V (DC). The flowing gas pressure may be up to 1.4 MPa resulting in a plasma velocity of several hundred metres/second. Higher the gas flow rate, more will be momentum of the plasma jet. It will ease out removal of the molten material from the machining zone. The plasma jet is constricted by the flowing gas which acts as a cooling agent sandwiched between the nozzle wall and the plasma jet.

In case of PAC, the material may be removed either by melting, or by melting and vaporization both. In either case, the material (in molten state or vaporized state) is blown off from the machining zone by high velocity plasma jet.

ELEMENTS OF PLASMA ARC CUTTING SYSTEM

The important **elements** of a PAC system are *power supply, gas supply, cooling water system, control console and plasma torch.* There are many torch designs which are practically used, for example air plasma, dual gas, oxygen injected, and water injected plasma torch.

Air plasma torch (Fig. 9.2) uses compressed air as the gas that ionizes and does cutting. The air to be used should be uncontaminated. The nozzle of this torch may result in prematured failure because of *double arcing* (not shown in the figure), i e arcing between the electrode and the nozzle, and between the nozzle and the workpiece. Air plasma cutting results in a high degree of *tapered machined surface*. Zirconium or hafnium (electrode life ≈ 2 hr of cutting time) are used as electrode material because of their higher resistance to oxidation. Because of poor oxidation resistance, tungsten electrode does not last for more than a few seconds.

To avoid oxidation of electrode (or to enhance the life of the electrode), **oxygen injected torch** (Fig. 9.3) uses nitrogen as the plasma gas. Oxygen is injected downstream of the electrode. However, it lowers down the nozzle life. This torch gives high MRR and poor squareness of the cut edges. It is commonly used for mild steel plate cutting. The presence of oxygen in the air helps in increasing MRR in case of oxidizable materials like steel. In case of certain ferrous metals, cutting speed is increased by about 25%. It is possible because oxygen backs up

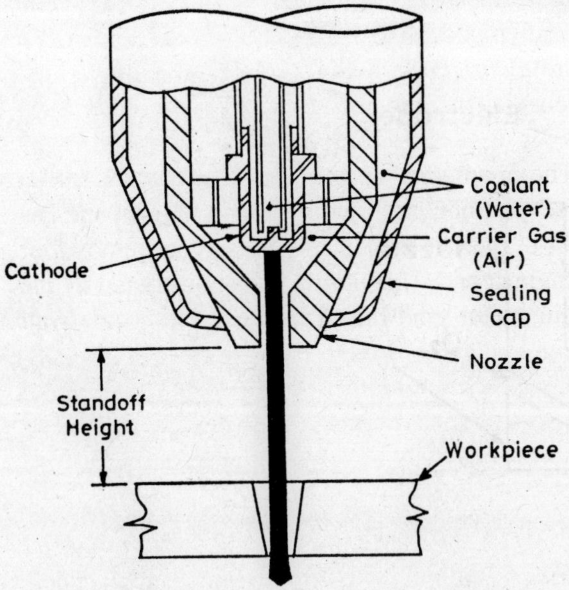

Fig. 9.2 Details of air plasma torch construction [*Benedict*, 1987; Courtesy: W.A. Whitney Corp. Rockford, III.]

the exothermic burning of oxidizable metals.

Dual gas system (Fig. 9.4) uses one gas (nitrogen) as the plasma gas while another gas as the shielding gas (O_2, CO_2, argon-hydrogen, etc). Secondary or shielding gas is chosen according to the material to be cut. Secondary gas system helps in maintaining sharp corners on the top side of the cut edges.

In **water injected torch** (Fig. 9.5), water (pressure ≈ 1.2 MPa) is injected (radially or swirling vertically) to constrict the plasma. A small quantity (about 10%) of water vaporizes. This thin layer of steam constricts the plasma [*McGeough*, 1988] and also insulates the nozzle. Nitrogen at about 1 MPa is used as the plasma gas. To avoid double arcing, the lower part of the nozzle is made of ceramic. Water constriction helps in reducing smoke, enhancing nozzle life, reducing HAZ, and limiting formation of oxides on the cut edges of the workpiece. In some systems, shielding gas or water is given swirling motion. As a consequence of this, plasma is also forced to swirl. This results in one edge of the cut being almost perfectly straight.

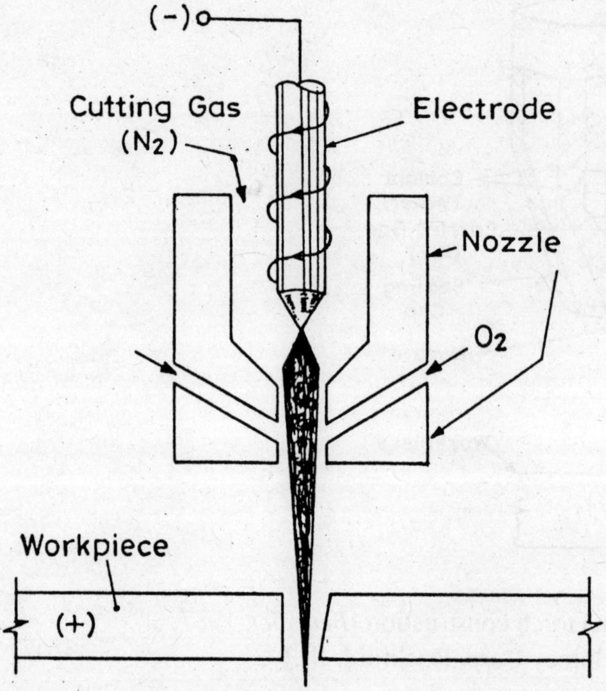

Fig. 9.3 Schematic diagram of oxygen-injected torch construction [*Benedict*, 1978; Hypertherm Inc., Hanover, N.H.]

Water muffler (a device that produces a covering of water around the plasma torch and extends down to the work surface) helps in reducing smoke and noise. Water mixed with a dye also absorbs part of the ultraviolet rays produced in PAC. In some cases, a water table is also used to reduce the level of noise and extent of sparks. Water below the workpiece quenches sparks and damps sound level. **Underwater PAC systems** are also available which effectively reduce the noise and smoke levels.

PROCESS PERFORMANCE

As the surface speed during PAC is increased, volumetric material removal rate is found to attain a maximum value and then starts decreasing. PAC has been employed to cut materials as thick as 150 mm. Cutting speed is affected by

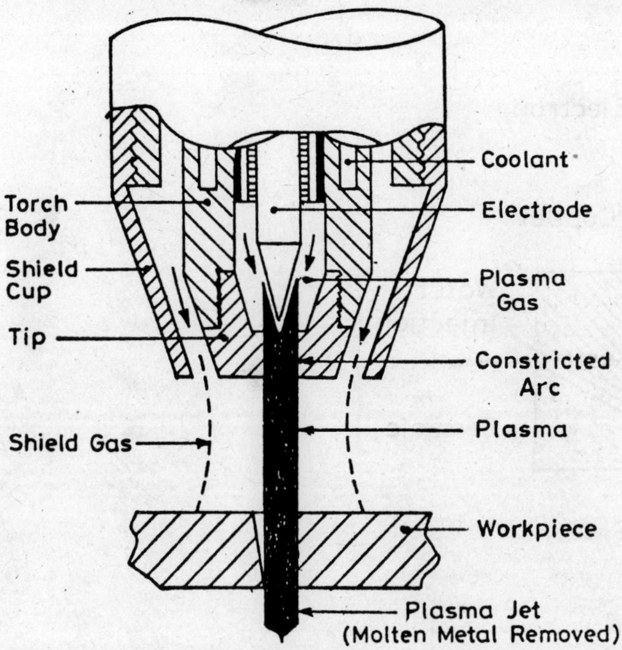

Fig. 9.4 Constructional details of dual gas plasma torch [*Benedict*, 1978: Courtesy: W.A. Whitney Corp., Rockford 111.]

the thickness (t) of the material being cut, type of the material being cut, and the current (I) being used during cutting. Cutting speed ('S' in m/min) is found to decrease with an increase in thickness ('t' in mm) of the material being cut.

$$S = 25.4/t \quad (\text{for } I = 500 \text{ A}). \qquad \ldots(9.1)$$

The tolerances obtained are poor and depend on the thickness of the material being cut. For example, the tolerance value of ± 0.8 mm may be achieved if $t <$ 25.0 mm, and ± 3.0 mm if $t >$ 25.00 mm. Width of cut is also quite high (2.5–9.5 mm) and taper depends on whether swirling is used or not (without swirling 5–7°). Surface finish of the cut edges ranges from 5–75 µm. Thickness of HAZ is governed by the material thickness, PAC system, type of material being cut and cutting conditions. It ranges from 0.75–5.0 mm.

It should be noted that for an appropriate performance of PAC system, an increase in power requires a corresponding rise in the gas flow rate.

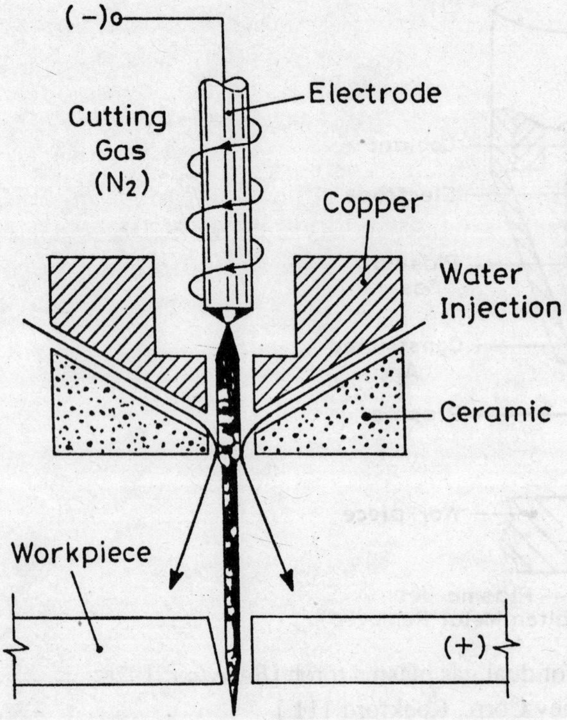

Fig. 9.5 Constructional details of water-injected plasma torch [*Benedict*, 1978, Courtesy: Hypertherm Inc., Hanover, N.H.]

APPLICATIONS

Multiple torch system is used for simultaneously cutting varieties of shapes from one plate. It is commonly used for preparing ends of a pipe section before welding. To achieve a bevel cut on the end of a pipe, mount the torch at a fixed angle and rotate the pipe underneath the torch.

Computer numerical control (**CNC**) PAC systems are also available in the market. They are useful for punching type of operations as well as for shape cutting on light duty plates made of stainless steel, aluminium and copper which are difficult to machine by oxy-fuel system.

Oxy-fuel system can cut the plates up to a definite thickness beyond which only PAC system can do the job. In case of thick plates, PAC is much more economical than oxy-fuel system. In case of cutting of a groove, its shape and size are governed by arc power, traverse speed, angle and height of the plasma torch. In case of electrically non-conducting materials (using non-transferred arc), MRR is low.

PAM has also been employed for turning of difficult-to-machine materials. Plasma torch is mounted tangentially (at about 30°) to the workpiece. Attempts are being made to develop successful **underwater plasma machining** systems although associated with lower speeds.

BIBLIOGRAPHY

1. Benedict G.F. (1987), *Non-traditional Manufacturing Processes*, Marcel Dekker Inc., New York.
2. Lucey J.A. and Wylie F.S. (Oct. 1967), Plasma Machining, *Proc. Conf. Machinability*, The Iron and Steel Institute, London, pp. 235-238.
3. McGeough J.A. (1988), *Advanced Methods of Machining*, Chapman and Hall, London.
4. Pandey P.C. and Shan H.S. (1980), *Modern Machining Processes*, Tata-McGraw Hill, New Delhi.
5. Troian J. (April 1982), How to Apply Water Injected Plasma Cutting to a CNC Punching and Nibbling Machine, *Fabricator*, Apr., pp. 22-24.

REVIEW QUESTIONS

1. What do you understand by 'fourth state of matter'? Explain in brief.
2. Can you machine electrically non-conductive materials by PAM?
3. How the molten material ejects out of the machining zone in PAC?
4. Why tungsten is not used as electrode material?
5. Write the specific applications of PAC.

CHAPTER 9
AT-A-GLANCE
PLASMA ARC MACHINING
(PAM)

PLASMA

- GAS HEATED TO 2000°C → MOLECULES SEPARATE OUT AS ATOMS (IF HAVING 2 OR MORE)
- AT 3000°C OR MORE, ELECTRONS DISSOCIATE FROM ATOMS AND GAS IONIZES
- THE GAS CONSISTING IONS & ELECTRONS RECOMBINES → AS ATOMS AND / OR ATOMS INTO MOLECULES
  + CHARACTERISTICS OF PLASMA → GLOWING GAS, ELECTRICALLY CONDUCTIVE & RESPONSIVE TO MAGNETISM (FOURTH STATE OF MATTER)
  + HEAT GENERATION → RECOMBINATION OF IONS & ELECTRONS AS ATOMS AND / OR RECOMBINATION OF ATOMS INTO MOLECULES

  ↓

  RELEASE OF BONDING ENERGY

  ↓

  TEMPERATURE AS HIGH AS 33000° C

  ↓

  CAN CUT ANY METERIAL INCLUDING THOSE HAVING HIGH THERMAL CONDCTIVITY, HIGH HEAT
  CAPACITY, AND/OR HIGH OXIDATION RESISTANCE

  ↓

  MATERIALS THAT CAN'T BE CUT BY CONVENTIONAL METHODS
  → PLASMA ARC CUTTING (PAC) CAN CUT THEM

- PAC USES DC POWER
- MATERIAL REMOVAL MECHANISM → MELTING AND / OR VAPORIZATION
  → MOLTEN MATERIAL IS BLOWN OFF BY HIGH VELOCITY GAS

(Contd...)

PAC SYSTEM

▽

ARC MODES

↓ NON-TRANSFERRED ARC

▽

- EFFICIENCY 65% - 75%
- ARC BETWEEN ELECTRODE & NOZZLE
- IONIZES HIGH VELOCITY GAS STREAMING TOWARDS W / P

↓ TRANSFERRED ARC

▽

- EFFICIENCY 85% – 90%
- ARC BETWEEN ELECTRODE & W/P
- W/P ELECTRICALLY CONDUCTIVE
- GAS PRESSURE (~1.4 MPa)
- VELOCITY SEVERAL HUNDRED m/s

▽

▽ REMOVES OUT MOLTEN MATERIAL FROM MACHINING ZONE.

▽ FLOWING GAS CONSTRICTS PLASMA JET & ACTS AS A COOLING AGENT

ELEMENTS OF PAC SYSTEM

1. POWER SUPPLY
2. GAS SUPPLY
3. COOLING WATER SYSTEM
4. CONTROL CONSOLE
5. PLASMA TORCH

∇ AIR AS PLASMA GAS → IONIZED & DOES CUTTING

∇ TORCH MAY FAIL PREMATURED → DOUBLE ARCING BETWEEN → ELECTRODE & NOZZLE, AND NOZZLE & W/P

∇ HIGH DEGREE OF TAPERED MACHINED SURFACE

∇ ELECTRODE MATERIALS
- ZIRCONIUM OR HAFNIUM (LIFE ~ 2 HR) → HIGH RESISTANCE TO OXIDATION
- TUNGSTEN ELECTRODE → LIFE ONLY A FEW SECONDS → POOR OXIDATION RESISTANCE

OXYGEN INJECTED TORCH

- HOW TO ENHANCE ELECTRODE LIFE?

  + TO AVOID OXIDATION OF ELECTRODE

    ↓

  + USE OXYGEN INJECTED TORCH

    ↓

  + USE NITROGEN AS PLASMA GAS
- THIS TORCH → HIGH MRR & POOR SQUARENESS

  ↓

  USED FOR M.S. (OXIDIZABLE) CUTTING
- PRESENCE OF OXYGEN IN AIR

  ↓

  + HIGHER MRR FOR OXIDIZABLE METAL (STEELS)
  + CUTTING SPEED MAY GO UP → 25%
  + OXYGEN BACKS UP EXOTHERMIC BURNING OF OXIDIZABLE METALS

DUAL GAS SYSTEM

PLASMA GAS (NITROGEN) & SHIELDING GAS (O_2, CO_2, ETC)

- ACCORDING TO THE MATERIAL TO BE CUT
- HELPS IN MAINTAINING SHARP CORNERS

WATER INJECTED TORCH

- WATER INJECTED (SWIRLING) → CONSTRICTS PLASMA
  10% VAPOURIZES
- VAPOUR LAYER CONSTRICTS PLASMA & INSULATES NOZZLE
- NITROGEN (1 MPa) → USED AS PLASMA GAS
- TO AVOID DOUBLE ARCING → LOWER PART OF NOZZLE → CERAMIC
- WATER CONSTRICTION → REDUCES SMOKE, HIGHER NOZZLE LIFE, SMALLER HAZ, LIMITED OXIDATION

- SWIRLING MOTION → SHIELDING GAS / WATER

  ↓

  SWIRLING OF PLASMA

  ↓

  ONE CUT EDGE → PERFECTLY STRAIGHT

WATER MUFFLER

- A COVERING OF WATER AROUND PLASMA TORCH & EXTENDS DOWN TO THE W/P SURFACE
- REDUCES SMOKE & NOISE
- DYE MIXED WITH WATER → ABSORBS ULTRAVIOLET RAYS FROM PLASMA ARC
- WATER TABLE → REDUCES LEVEL OF NOISE & DAMPS SOUND LEVEL
- UNDERWATER PAC SYSTEMS → REDUCE NOISE & SMOKE LEVELS

PERFORMANCE

- SURFACE SPEED & MRR → OPTIMUM
- MAX. CUT THICKNESS → 150 mm
- CUTTING SPEED = ϕ (MATL. THICKNESS (T) AND TYPE) → DECREASES WITH INCREASE IN "THICKNESS"
  S(m/min) = 25.4/T(mm) FOR I = 500 A
- POOR TOLERANCES → ± 0.8 mm (T < 25.0 mm); ± 3.0 mm (T > 25.0 mm)
- WIDTH OF CUT → 2.5 - 9.0 mm
- TAPER → 5 - 7°
- SURFACE FINISH → 5 - 75 μm
- HAZ (0.75 - 5.0 mm) = ϕ (T, PAC SYSTEM, CUTTING CONDITIONS, & MATERIAL TYPE)

APPLICATIONS

- MULTIPLE TORCH SYSTEM → VARITIES OF SHAPES ON A PLATE
- BEVEL CUT ON THE END OF A PIPE
- CNC PAC SYSTEM → PUNCHING TYPE OPERATION ON LIGHT DUTY PLATES → SS, Al, Cu, ETC ← CANNOT BE CUT BY OXY-FUEL SYSTEM
- OXY-FUEL CANNOT CUT BEYOND A THICKNESS → PAC DOES
- SHAPE & SIZE OF A GROOVE = ϕ (ARC POWER, TRAVERSE SPEED OF PLASMA, TORCH HEIGHT)
- NON-TRANSFERRED ARC → LOW MRR
- UNDERWATER PAC SYSTEM ← UNDER DEVELOPMENT

Chapter Ten

ELECTRON BEAM MACHINING (EBM)

WORKING PRINCIPLE

Electron beam machining (EBM) process is classified into two categories, viz '*Thermal type*' and '*Non-thermal type*'. In the thermal type EBM process, the surface of thermoelectronic cathode is heated to such a high temperature that the electrons acquire sufficient speed to escape out to the space around the cathode. The stream of these large number of electrons moves as a small diameter beam of electrons towards the anode. As a result, the workpiece is heated by the bombardment of these electrons in a localized area, to such a high temperature that it is melted and vaporized at the point of bombardment. In the second type (*non-thermal* EBM) process, the electron beam is used to cause a chemical reaction. Here, only the first type of EBM process (thermal type) is discussed.

The high velocity beam of electrons strikes the workpiece. **The kinetic energy of electrons** converts into heat which is responsible for melting and vaporization of workpiece material. This process can produce any shape of hole; however, round holes are usually drilled in metals, ceramics, plastics, etc. It can machine electrically conducting as well as non-conducting materials. Before machining starts, vacuum is created in the machining chamber.

The diameter of the electron beam focussed onto the work should be slightly smaller than the desired hole diameter. As the electron beam strikes the workpiece, the material gets heated, melted and partly vaporized. On the exit side of the hole, the *synthetic or organic backing material* is used. The electron beam after complete penetration into the workpiece, also partly penetrates in the auxiliary backing material. The backing material vaporizes and comes out of hole at a high pressure. The molten material is also expelled along with the vaporized backing material. In case of a *non-circular hole* to be produced, the electron beam is deflected with the help of the *computer control*, along the perimeter of the hole to be produced. As an alternative method, the beam can be kept stationary but the work-table can be moved in the desired path with the help of CNC.

ELECTRON BEAM MACHINING SYSTEM

There are three important **elements** of EBM system, viz vacuum system, electron beam gun and power supply.

(i) Electron Beam Gun

It is used to produce electron beam (Fig. 10.1) of the desired shape and to focus at the predetermined location. EBM gun is operated in the pulsed mode. A superheated cathode (tungsten filament type) generates the electrons cloud. Sometimes cathode may be used as a solid block indirectly heated by radiation emitted from a filament. Due to force of repulsion from the cathode, electrons move at a very high acceleration towards the anode which attracts them. The velocity with which electrons pass through the anode is approximately 66% that of light. On the path of electrons, there is a kind of **switch** (bias electrode) which generates the pulses.

A magnetic lens is used to shape the electron beam into a converging beam. This beam is passed through a variable aperture to reduce the diameter of the focussed beam by removing the stray electrons. Magnetic lenses are used to pin point the location of the beam, deflect it, and make it a round beam falling on the workpiece.

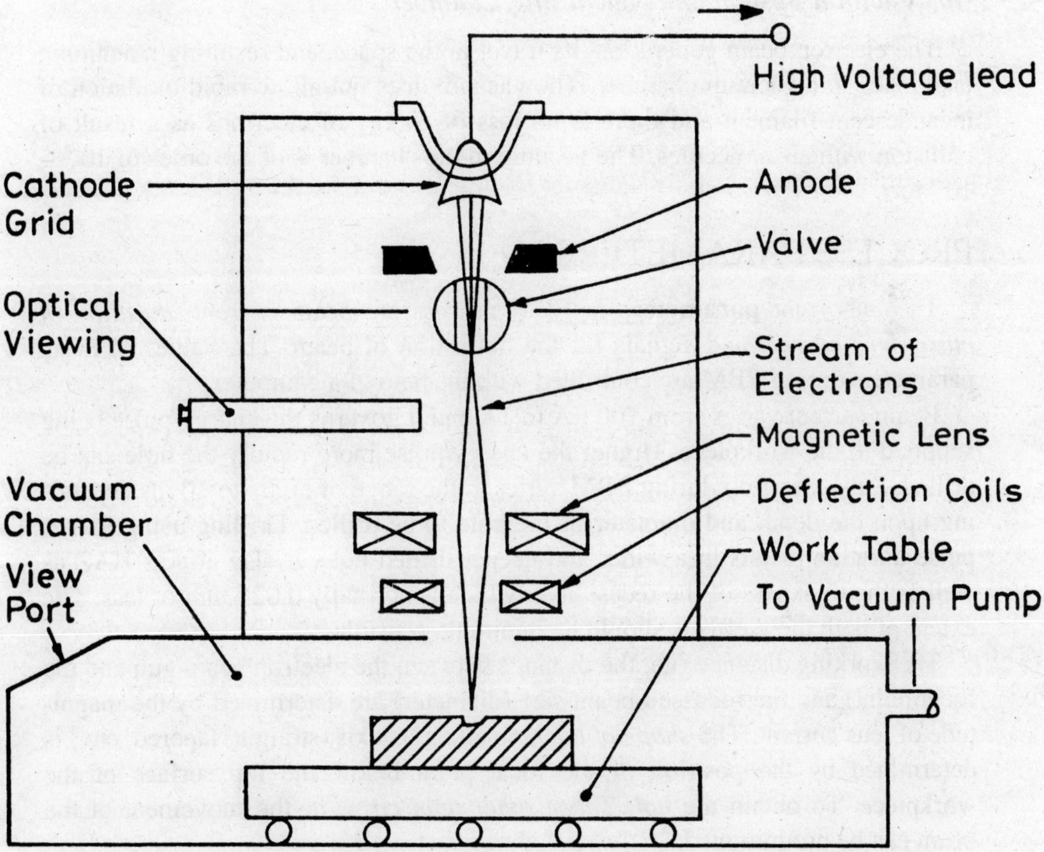

Fig. 10.1 Schematic diagram of electron beam machining system.

(ii) Power Supply

The power supply generates a voltage as high as 150 kV to accelerate electrons. The EBM gun of a powerful system is usually operated at about 12 kW and an individual pulse energy as 120 J/pulse. The *power density* at the work surface is too high that is why it is capable to melt and vaporize the workpiece material. Thus, material removal in EBM is basically due to vaporization.

(iii) Vacuum System and Machining Chamber

The electron beam generation, its travel in the space, and resulting machining take place in a vacuum chamber. The vacuum does not allow rapid oxidation of incandescent filament and there is no loss of energy of electrons as a result of collision with air molecules. The vacuum in the chamber is of the order of 10^{-4} – 10^{-5} torr.

PROCESS PARAMETERS

The *important* **parameters** in EBM process are *beam current, duration of pulse, lens current* and signals for the deflection of beam. The values of these parameters during EBM are controlled with the help of a computer.

Beam current varies from 100 μA to 1A and it governs the energy/pulse being supplied to the workpiece. Higher the energy/pulse more rapidly the hole can be drilled. Pulse duration during EBM varies in the range of 50 μs to 10 ms depending upon the depth and diameter of the hole to be drilled. Drilling using longer pulse duration results in a wider and deeper drilled hole. It also affects HAZ as well as the thickness of the recast layer which is normally 0.025 mm or less. The extent of both these effects should be minimum possible.

The working distance (i.e. the distance between the electron beam gun and the focal point) and the focussed beam size (diameter) are determined by the magnitude of lens current. The *shape of the hole* along its axis (straight, tapered, etc.) is determined by the position of the focal point below the top surface of the workpiece. To obtain the *hole shape other than circular*, the movement of the beam can be programmed.

The **material removal rate** (MRR) at which the workpiece material is vaporized can be calculated from Eq. (10.1).

$$MRR = \eta \frac{P}{W} \qquad \ldots(10.1)$$

where, η is the cutting efficiency, P is the power (J/s) and W is the specific energy (J/cm^3) (Eq. (10.2)) required to vaporize the work-material. Specific energy (W) can be calculated as follows:

$$W = C_{Ps}(T_m - T_i) + C_{Pl}(t_b - t_m) + H_f + H_v \qquad ...(10.2)$$

where, C_P is specific heat, T_m is melting temperature, T_i is initial temperature of workpiece, T_b is boiling temperature, H_f is latent heat of fusion, and H_v is the latent heat of vaporization. The cutting efficiency is usually below 20%. Here, C_P is assumed constant although it varies with temperature. Suffix s and *l* indicate solid and liquid states, respectively.

CHARACTERISTICS OF THE PROCESS

This process can be used to machine both electrically conductive as well as non-conductive materials, viz Ni, Cu, Al, ceramics, leather, plastics, etc. It has been observed that at the entry side of the electron beam, a small sized burr (a solidified layer) may be left out. In general, **performance** (viz MRR) of the EBM process is not significantly influenced by the properties (physical, mechanical and metallurgical) of workpiece material.

This process can machine small diameter holes (0.1 to 1.4 mm) to a large depth (say, 10 mm) or in other words, a hole with high *aspect ratio* (up to 15:1). The geometry of the hole and depth of the hole to be drilled, determine the average machining rate (or penetration rate). Further, no mechanical force is applied on the job; hence fragile (or brittle), thin, and/or low strength workpieces can be easily machined. *Off-the-axis holes* (or inclined holes) can also be machined by this process.

There is no distortion due to mechanical forces; however, very high temperature gradient would result in **residual thermal stresses**. Another limitation of the process is very high cost of the equipment. The operator also should be skilled one. The quality of the edges produced is determined by the thermal properties of the workpiece material and the pulse energy [Kaczmarek, 1976]. The heat affected zone (HAZ) depends upon pulse duration and the diameter of the hole being drilled.

APPLICATIONS

EBM is more popular in **industries** like *aerospace*, *insulation*, *food processing*, *chemical*, *clothing*, *etc*. It is very useful in those cases where number of holes (simple as well as complex shaped) required in a workpiece may range from *hundreds to thousands* (perforation of sheets, etc). This Process is also used for drilling thousands of holes (diameter < 1.00 mm) in very thin plates used for turbine engine combustor domes. Many thousand holes (diameter < 1.0 mm) in a cobalt alloy fibre spinning head of thickness around 5 mm are drilled by EBM. This drilling is claimed to be 100 times faster than EDM. Holes in the filters and screens used in food processing industries are also made by this process.

The *applications of EBM* also encompass the areas like making of fine gas orifices in space nuclear reactors, holes in wire drawing dies, cooling holes in turbine blades, metering holes in injector nozzles of diesel engines, etc. EBM is also being employed for *pattern generation* for integrated circuit fabrication.

In EBM, the beam power, focus, pulse duration and mechanical motion have been *controlled numerically*. This would permit more accurate and complex shaped machining using numerically controlled EBM system (NC-EBM).

BIBLIOGRAPHY

1. Bellows G. and Kohls J.B. (March 1982), Drilling without Drills, Special Report 743, *Am. Machinist*, pp. 173-188.
2. Bhattacharyya A. (1973), *New Technology*, The Institution of Engineers (I), Calcutta, India.
3. Kaczmarek J. (1976), *Principles of Machining by Cutting, Abrasion and Erosion*, Peter Pergrinus Ltd., Stevenge, pp. 514-528.
4. McGeough J.A. (1988), *Advanced Methods of Machining*, Chapman and Hall, London, pp. 11-33.
5. Pandey P.C. and Shan H.S. (1980), *Modern Machining Processes*, Tata-McGraw Hill, Publishing Co. Ltd., New Delhi.

PROBLEMS

(1) Explain the working principle of EBM process. Make the necessary sketch.

(2) Can you machine electrically non-conducting materials using EBM process?

(3) How the work-table is protected from getting damaged by the electron beam which has completely penetrated the workpiece?

(4) How a complex shape can be cut using EBM process?

(5) Write an equation to compute specific energy of vaporization.

(6) Write four specific applications where you feel that EBM should be the preferable choice.

NOMENCLATURE

C_{ps}	Specific heat of solid
C_{pl}	Specific heat of liquid
H_f	Latent heat of fusion
H_v	Latent heat of vaporization
P	Power
T_b	Boiling temperature
T_i	Initial temperature of the workpiece
T_m	Melting temperature
W	Specific energy
η	Cutting efficiency

CHAPTER 10

AT-A-GLANCE
ELECTRON BEAM MACHINING
(EBM)

ELECTRON BEAM MACHINING

↓

THERMAL TYPE

↓

CATHODE HEATED

↓

STREAM OF A LARGE NO. OF ELECTRONS
AS A SMALL DIAMETER BEAM

↓

MOVES TOWARDS WORKPIECE

↓

BOMBARDMENT OF ELECTRONS IN A LOCALIZED AREA

↓

VERY HIGH TEMPERATURE

↓

MELTING AND VAPORIZATION OF W/P MATL.
KINTTIC ENERGY OF BEAM CONVERTS INTO HEAT

↓

- ∇ CAN PRODUCE HOLES OF DIFFERENT SHAPES
- ∇ CONDUCTING AS WELL AS NON-CONDUCTING MATERIALS
- ∇ VACUUM REQUIRED IN MACHINING CHAMBER
- ∇ BEAM SIZE < DESIRED HOLE SIZE
- ∇ EJECTION OF MOLTEN WORKPIECE MATERIAL
  - VAPORIZED BACKING MATERIAL COMES OUT AT HIGH PRESSURE → EXPELS MOLTEN WORKPIECE MATERIAL
- ∇ NON-CIRCULAR HOLES → MOVE ELECTRON BEAM BY COMPUTER CONTROL ALONG HOLE PERIMETER

NON-THERMAL TYPE

↓

ELECTRON BEAM
CAUSES CHEMICAL
REACTION

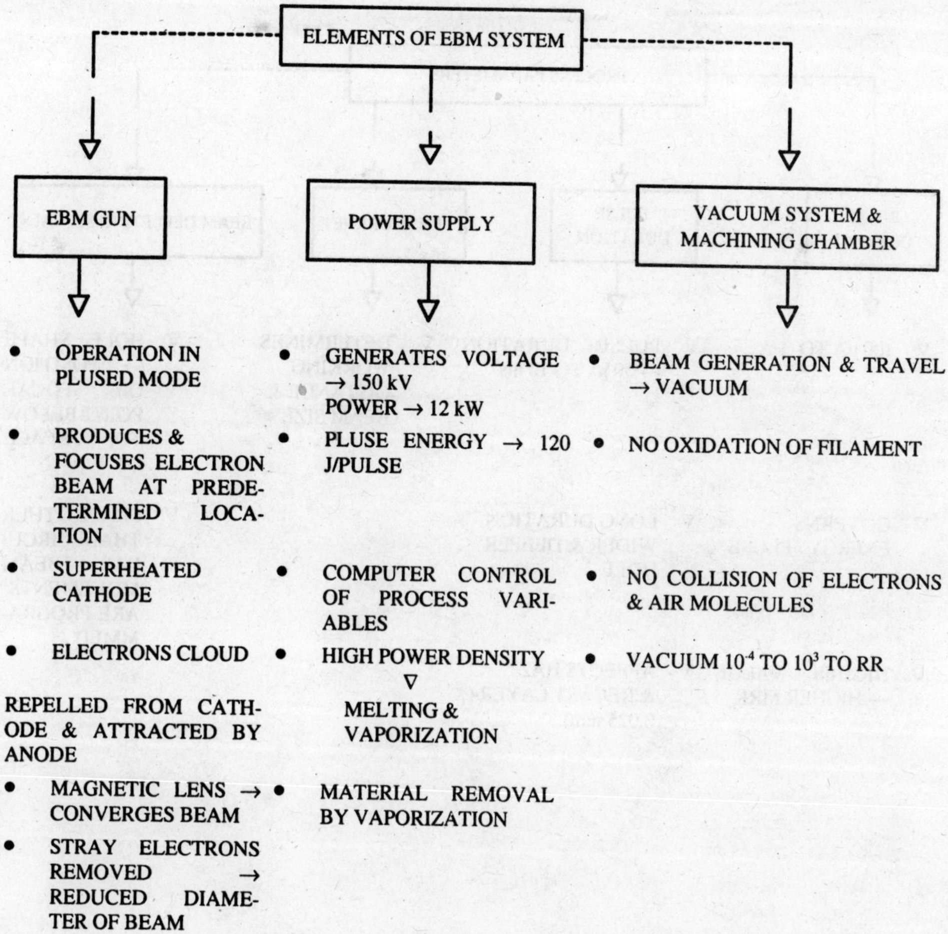
ELEMENTS OF EBM SYSTEM
EBM GUN
POWER SUPPLY
VACUUM SYSTEM & MACHINING CHAMBER
• OPERATION IN PLUSED MODE
• PRODUCES & FOCUSES ELECTRON BEAM AT PREDETERMINED LOCATION
• SUPERHEATED CATHODE
• ELECTRONS CLOUD
REPELLED FROM CATHODE & ATTRACTED BY ANODE
• MAGNETIC LENS → CONVERGES BEAM
• STRAY ELECTRONS REMOVED → REDUCED DIAMETER OF BEAM
• GENERATES VOLTAGE → 150 kV
POWER → 12 kW
• PLUSE ENERGY → 120 J/PULSE
• COMPUTER CONTROL OF PROCESS VARIABLES
• HIGH POWER DENSITY
MELTING & VAPORIZATION
• MATERIAL REMOVAL BY VAPORIZATION
• BEAM GENERATION & TRAVEL → VACUUM
• NO OXIDATION OF FILAMENT
• NO COLLISION OF ELECTRONS & AIR MOLECULES
• VACUUM 10^{-4} TO 10^{3} TO RR

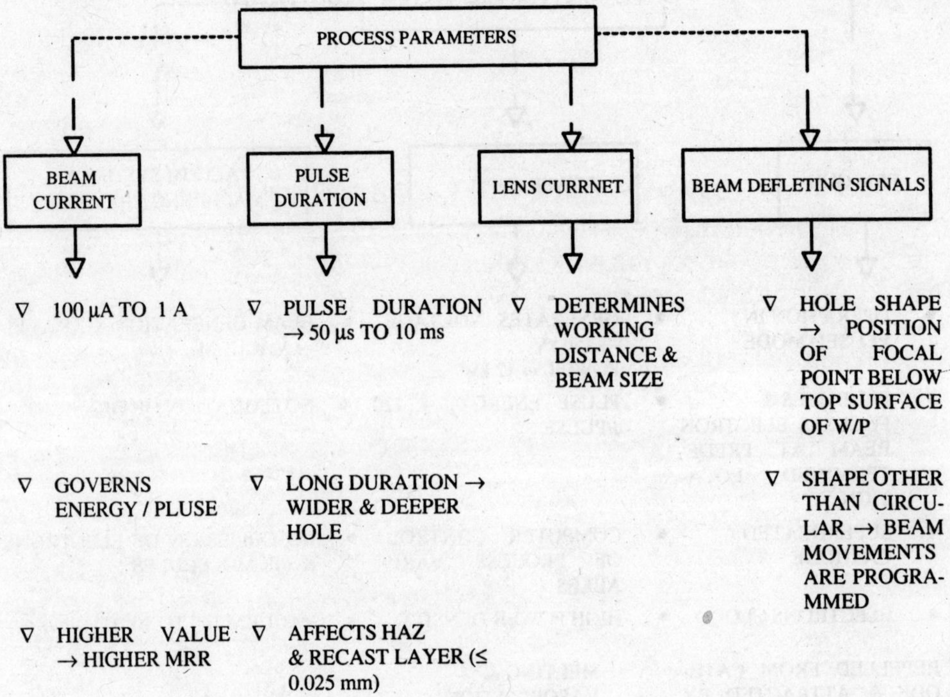
PROCESS PARAMETERS
BEAM CURRENT
PULSE DURATION
LENS CURRNET
BEAM DEFLETING SIGNALS
∇ 100 μA TO 1 A
∇ GOVERNS ENERGY / PLUSE
∇ HIGHER VALUE → HIGHER MRR
∇ PULSE DURATION → 50 μs TO 10 ms
∇ LONG DURATION → WIDER & DEEPER HOLE
∇ AFFECTS HAZ & RECAST LAYER (≤ 0.025 mm)
∇ DETERMINES WORKING DISTANCE & BEAM SIZE
∇ HOLE SHAPE → POSITION OF FOCAL POINT BELOW TOP SURFACE OF W/P
∇ SHAPE OTHER THAN CIRCULAR → BEAM MOVEMENTS ARE PROGRAMMED

- $MRR = \eta P/W$
- $W = C_{Ps} (T_M - T_i) + C_{Pl} (T_B - T_M) + H_F + H_V$

CHARACTERISTICS

- ∇ FOR CONDUCTIVE AS WELL AS NON-CONDUCTIVE MATERIALS
- ∇ AT ENTRY SIDE OF BEAM → A SMALL SIZED BURR
- ∇ W/P MATERIAL PROPERTIES DO NOT AFFECT PERFORMANCE
- ∇ SMALL DIAMETER HOLES (0.1 TO 1.4 mm)
- ∇ ASPECT RATIO = 15 : 1
- ∇ NO MECHANICAL FORCE → FRAGILE, THIN, LOW STRENGTH COMPONENTS → EASILY MACHINED
- ∇ OFF-AXIS HOLES EASILY MADE
- ∇ RESIDUAL THERMAL STRESSES ← HIGH TEMPERATURE GRADIENT
- ∇ VERY HIGH INVESTMENT COST
- ∇ SKILLED OPERATOR
- ∇ MACHINED EDGE QUALITY → THERMAL PROPERTIES OF W/P MATERIAL & PULSE ENERGY
- ∇ HAZ → ϕ (PULSE DURATION & HOLE DIAMETER)

APPLICATIONS

- ∇ MORE POPULAR → AEROSPACE, INSULATION, FOOD PROCESSING, CHEMICAL, CLOTHING, ETC
- ∇ HUNDREDS TO THOUSANDS OF HOLES IN A W/P ← COMPLEX-SHAPED, DIFFICULT-TO-MACHINE MATERIAL, PERFORATION, ETC
  - EXAMPLE : TURBINE ENGINE COMBUSTOR DOME
  - FIBER SPINNING HEAD (THICKNESS < 5 mm), MUCH FASTER THAN EDM
  - HOLES IN FILTERS & SCREENS ← FOOD INDUSTRY
  - COOLING HOLES IN TURBINE BLADES
  - METERING HOLES IN INJECTOR NOZZLES

PART–III

Electrochemical and Chemical Advanced Machining Processes

- Electrochemical Machining (ECM)
- Electrochemical Grinding (ECG)
- Electrostream Drilling (ESD)
- Electrochemical Deburring (ECDe)
- Shaped Tube Electrolytic Machining (STEM)
- Chemical Machining (ChM)
- Anode Shape Prediction and Tool Design for ECM Processes
- Electrochemical Spark Machining (ECSM)

Chapter Eleven

ELECTROCHEMICAL MACHINING (ECM)

INTRODUCTION

Electrolysis

Electrical energy is transported through metals by conduction of electric charges from one place to another. As opposed to the metallic conduction, where only electrons are the charge carriers, salt solutions conduct electrical energy by the migration of ions in the medium.

Fig. 11.1 shows an **electrolytic cell** in which DC battery sends electric current through the molten sodium chloride salt. Electrons from the battery enter the melt at the cathode, and when the circuit is complete, they leave the melt at the anode returning to the battery. Sodium ions (Na^+) from the medium combine with the electrons available at the cathode, and produce sodium metal which accumulates at the cathode region as

$$Na^+ + \bar{e} \rightarrow Na.$$

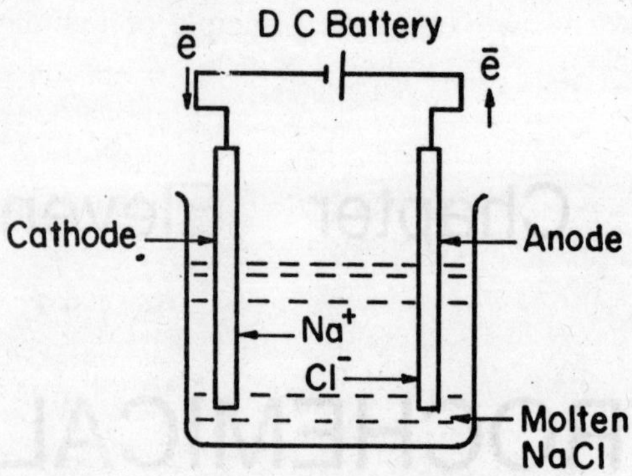

Fig. 11.1 The electrolysis of molten sodium chloride salt.

Thus, sodium ions are reduced (*reduction process* involves the addition of electrons) at the cathode. At the same time, chloride ions migrate towards the anode and are oxidized (in *oxidation process* electrons are released) to chlorine as:

$$Cl^- \rightarrow \frac{1}{2} Cl_2 + \bar{e}.$$

In order to get a sustained flow of current, and to avoid accumulation of ions at the electrode, reactions must keep occurring at the electrode, to maintain electrical neutrality. Thus, the external circuit feeds electrons into the medium at the cathode. These electrons are consumed at the cathode by the positive ions (or *cations* → since they migrate to the cathode), and are released by the negative ions (or *anions*) of the medium at the anode.

Electrochemical Machining (ECM)

Electrolysis has been successfully put to work in the areas like electroplating, electroforming and electropolishing. The process of metal removal by electrochemical dissolution was known as long back as 1780 AD but it is only over the last couple of decades that this method has been used to advantage. It is also known as *contactless electrochemical forming* process. The noteworthy feature of electrolysis is that the electrical energy is used to produce a chemical reaction, therefore, the machining process based on this principle is known as *electrochemical machining (ECM)*. This process works on the principle of Faraday's laws of electrolysis. In ECM, small electric DC* potential (5–25 V) is applied across the two electrodes, i e cathode and anode (anode is work and cathode is tool) immersed in electrolyte (Fig. 11.2). The transfer of electrons between the ions and the electrodes completes the electrical circuit. The metal is detached, atom by atom, from the anode surface and appears in the electrolyte as positive ions. In electrochemical machining, detached metal appears as precipitated solid of metal hydroxides. During the electrolysis of water, its molecules gain electrons from cathode so that they separate into free hydrogen gas and hydroxyl ions.

$$2H_2O + 2e^- \rightarrow H_2\uparrow + 2OH^-.$$

As the anode dissolves, negatively charged hydroxyl ions are electrically balanced by positively charged metal ions entering into the electrolyte.

$$Fe - 2e \rightarrow 2Fe^{2+}.$$

$$Fe^{2+} + 2(OH)^- \rightarrow Fe(OH)_2.$$

Metal ions do not remain as ions in the solution when neutral electrolytes are used, but combine with the hydroxyl ions to form **metal hydroxides**. These hydroxides are insoluble in water hence they appear as solid precipitates and no longer affect the electrochemical reaction.

---

*The application of direct current through a solution of electrolyte results in redox (reduction and oxidation) reactions, while the application of an alternating current (AC) leads to conduction only. The reason being that the electrodes change their polarities vary fast hence the electrode reaction occurring in the first half of the cycle is reversed in the later half of the cycle. Hence, only DC is used in this application.

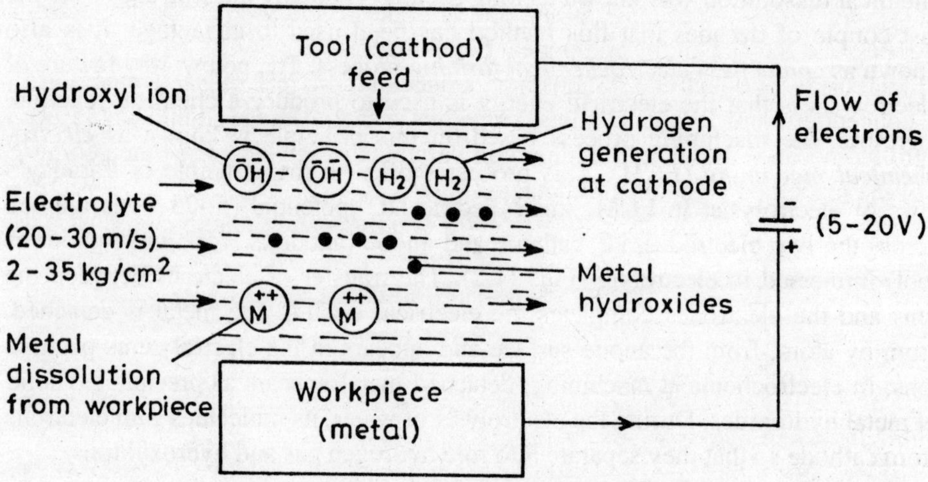

Fig. 11.2 The schematic diagram of electro-chemical machining.

$$Fe + 2H_2O \rightarrow Fe(OH)_2 + H_2 \uparrow .$$

Ferrous hydroxides may further react with water and oxygen to form ferric hydroxide:

$$4Fe(OH)_2 + 2H_2O + O_2 \rightarrow 4Fe(OH)_3.$$

Smaller the **interelectrode gap** (*IEG*) (the gap, y, between the two electrodes), greater will be the current flow because resistance (ρy/A) decreases and higher will be the rate of metal removal from the anode. High current density, in the small spacing (usually about 0.5 mm or less), promotes rapid generation of **reaction products**, viz hydroxide solids and gas bubbles. These reaction products act as a barrier to the flow of electrolyzing current and their effect is minimized by supplying the electrolyte at a pressure of 2 to 35 kg cm^{-2}, leading to the electrolyte flow velocity as high as 20–30 m/s.

Electrolytes used in ECM consist of either acids or, more generally, basic salts dissolved in water. Electrolyte flowing at high velocity in the IEG serves multifarious functions, viz dilutes the electrochemical reaction products and removes them out from the gap, dissipates heat at a faster rate, and limits the concentration of ions at the electrode surface to give higher machining rates. Electrolyte flow rate (volumetric) is determined based upon the desired flow velocity of electrolyte, *IEG*, and the size of the component being machined.

Electrolyte properties (composition, concentration, pH value, temperature and concentration of foreign elements) together with tool shape should be closely controlled because these are the important variables which determine the geometry of the machined component (anode profile).

Amount of hydroxides in the electrolyte is confined by continuous removal using large settling tanks, filters, and centrifuging pumps. Composition, concentration and pH value of electrolyte solution are controlled by adding water and salt solution. Their quantity to be added depends upon the periodic analysis of the check samples. Temperature is another important factor which governs the **electrical properties** of the electrolyte. It is controlled (within ± 1° C) by heating or cooling the electrolyte while in the tank.

The cathode is moved towards the anode at the same rate at which the work is being dissolved so that the gap (spacing) between the two electrodes remains constant. It really does not matter even if work is fed towards the tool. This will help in maintaining a constant **material removal rate** (MRR). Smaller gap at various points (Fig. 11.3a) between confronting surfaces of the electrodes-tool and work-will result in higher current density (J) and hence higher MRR. With the progressive movement of cathode towards the anode, ultimately the two surfaces will closely match (Fig. 11.3b).

Selection of electrolyte is quite important. However, inexpensive, easily available and commonly used electrolyte is sodium chloride (common salt). It is also necessary to pump the electrolyte at very high pressure through the IEG, so that the desired MRR can be maintained.

Thus, the anode profile to be produced depends upon so many factors but they can be narrowed down to only two, current density and cathode shape.

The principle of electrochemical machining (ECM) discussed above has been applied for performing a variety of **operations** as listed in Fig. 11.4. Some more operations can still be included in this list; however, the principle of working of all these operations remains the same, ie electrochemical dissolution of anode.

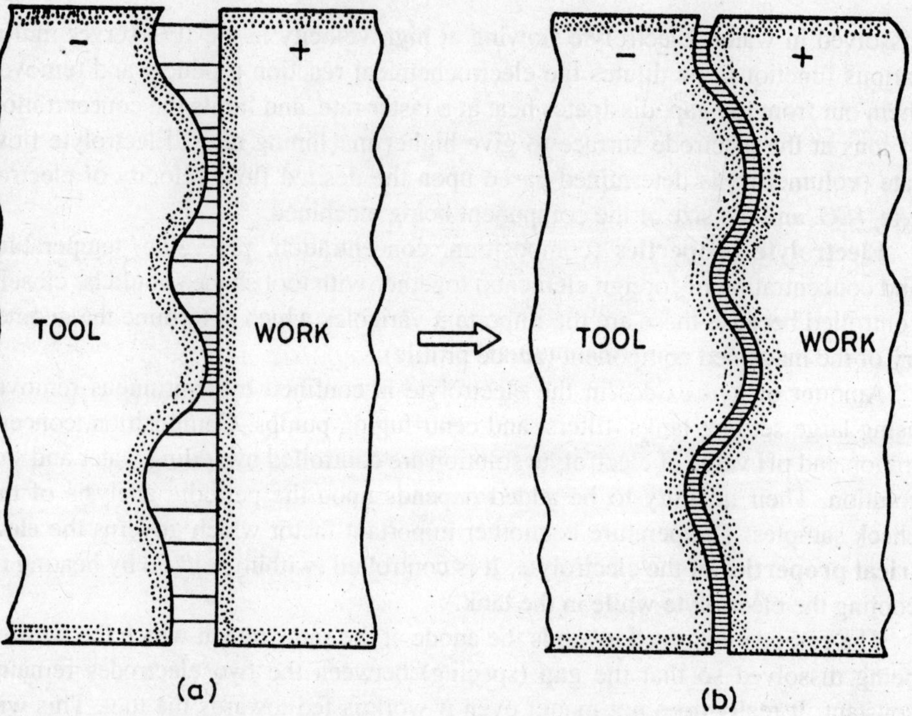

Fig. 11.3 Electrochemical machining: (a) Initial condition, (b) final condition.

ECM MACHINE TOOL

ECM machine tool (m/t) consists of mainly four sub-systems [*Jain, 1980*]:

(i) power source,
(ii) electrolyte supply and cleaning system,
(iii) tool and tool-feed system, and
(iv) workpiece and work-holding system.

A schematic diagram of the horizontal type ECM m/t is shown in Fig. 11.5.

(i) Power Source

During ECM process, a high value of direct current (may be as high as 40,000 A) and a low value of electric potential (in the range of 5–25 V) across the IEG is desirable. The highest current density achieved so far is around 20,000 A/cm^2.

- ECB (Electrochemical Boring) [15]
- ECB_r (Electrochemical Braoching) [18]
- ECB_z (Electrochemical Ballizing) [13]
- ECD (Electrochemical Drilling) [14]
- ECD_e (Electrochemical Deburring) [9]
- ECDS (Electrochemical Die Sinking) [7]
- ECG (Electrochemical Grinding) [6]
- ECH (Electrochemical Honing) [19]
- ECM (Electrochemical Machining) [16]
- ECM_l (Electrochemical Milling) [11]
- ECMM (Electrochemical Micromachining)
- ECS (Electrochemical Sawing)
- ECT (Electrochemical Turning)
- ECT_r (Electrochemical Trepanning)
- ECWC (Electrochemical Wire Cutting)
- ESD (Electrostream Drilling)
- STEM (Shaped Tube Electromachining)

Fig. 11.4 Various machining operations based on electrochemical dissolution of anode (*Jain and Pandey*, 1991).

Hence, with the help of a rectifier and a transformer, three phase AC is converted to low voltage, high current DC. Silicon controlled rectifiers (SCRs) are used both for rectification as well as for voltage regulation because of their rapid response to the changes in the process load and their compactness. Voltage regulation of ± 1% is adequate for most of the precision ECM works. However, lack of process control, equipment failure, operator's error, and similar other reasons may result in sparking between tool and work. The electrical circuitry detects these events and power is cut off (using devices like SCRs) within 10 μs to prevent severe damage to the work and tool. In case of precision works, even a small damage to an electrode is not acceptable. It may be minimized by using a bank of SCRs placed across the DC input to ECM m/c.

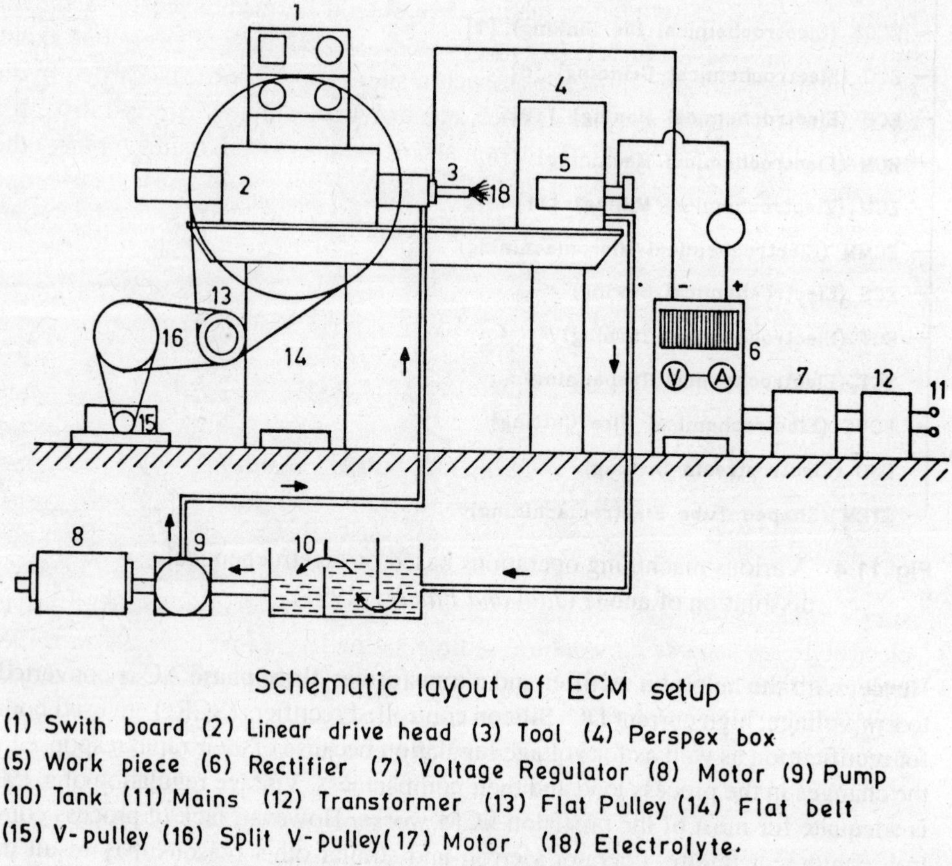

Fig. 11.5 A schematic diagram of horizontal type ECM m/t [*Jain*, 1980].

(ii) Electrolyte Supply and Cleaning System

The electrolyte supply and cleaning system consists of a pump, filters, pipings, control valves, heating or cooling coils, pressure gauges and a storage tank (or reservoir). Electrolyte supply ports may be made in the tool, work, or fixture, depending upon the requirements of the mode of electrolyte flow. Small inter-electrode gap, usually smaller than 1 mm, should be maintained for achieving high MRR and high accuracy. For this purpose, smooth flow of electrolyte should be maintained, and any blockade of such a small gap by particles carried by the electrolyte, should be avoided. Hence, electrolyte cleanliness is imperative. It is normally done with the help of filters made of stainless steel, monel, or any other anti-corrosive material. Filters should be periodically cleaned for their proper functioning. To achieve good results, filters can be placed in the supply pipe just prior to the work enclosure.

It should be ensured that the piping system does not introduce any foreign materials like corroded particles, scale, or pieces of broken seal material. Piping system is therefore made of stainless steel, glass fibre-reinforced plastics (GFRP), plastic lined mild steel, or similar other anti-corrosive materials. If the metallic piping is used, all the metallic parts should be earthed to prevent their anodic corrosion. Tables and fixtures should also be insulated from the anode. In case of heavy structures; if it is not possible to insulate from anode, they must carry a cathodic potential to prevent corrosion. The required minimum capacity of electrolyte tank is about 500 gallons for each 10,000 A of current. Fig. 11.5 shows a **single tank system** having three compartments. Reaction products from the electrolyte are separated by natural sedimentation in the first compartment. Electrolyte is then made to go to compartment 2 via a filter, and finally to the last compartment via a passage between the compartments 2 and 3. Electrolyte is further cleaned by a filter attached at the exit port of the tank.

ECM system is supposed to machine different metals and alloys at optimum machining conditions, and with varying requirements of accuracy, surface texture, etc. Under such situations, a single tank system is not recommended because of loss of time and wastage of electrolyte during draining, cleaning, mixing, or filling of new electrolyte in the tank. It results in higher cost and poor accuracy of electrochemically machined components and also poor control of operating conditions. *More than one tank* therefore, can be used and their number would depend upon the range of electrolytes needed to meet the work load. Use of more than one tank is justified when several ECM machines are in use. Further, concrete tank or swimming pool type of tank are pleaded because of their low cost, easy maintenance, and reliability.

(iii) Tool and Tool Feed System

Use of anti-corrosive **material for tools and fixtures** is important because they are required for a long period of time to operate in the corrosive environment of electrolyte. High thermal conductivity and high electrical conductivity of tool material are the main requirements. Easy machining of tool material is equally important because dimensional accuracy and surface finish of the tool directly affect the workpiece accuracy and surface finish. Aluminium, brass, bronze, copper, carbon, stainless steel and monel are a few of the materials used for this purpose. Further, those areas on the tool where ECM action is not required, should be insulated. For example, lack of insulation on the sides of a die sinking tool causes unwanted machining of work, and results in a loss of accuracy of the machined workpiece. However, under such situations use of **bit type of tools** can be recommended [*Jain and Pandey, 1982*]. Fig. 11.6 shows three categories of tools, viz bare tool, coated tool, and bit type of tool with the expected drilled hole profiles.

Use of non-corrosive and electrically non-conducting materials for making fixtures is recommended. Also, the fixtures and tools should be rigid enough to avoid vibration or deflection under the high hydraulic forces to which they are subjected.

(iv) Workpiece and Work Holding System

Only electrically conductive *work materials* can be machined by this process. The chemical properties of anode (work) material largely govern the material removal rate (MRR). *Work holding devices* are made of electrically non-conductive materials having good thermal stability, and low moisture absorption properties. For example, graphite fibres-reinforced plastics, plastics, perspex, etc are the materials used for fabricating the work holding devices.

ADVANTAGES AND LIMITATIONS

ECM offers impressive and long lasting **advantages**. It can machine highly complicated and curved shapes in a single pass. A single tool has been used to machine a large number of pieces without any loss in its shape and size. Theoretically, tool life in ECM is very high. The process offers several advantages, viz

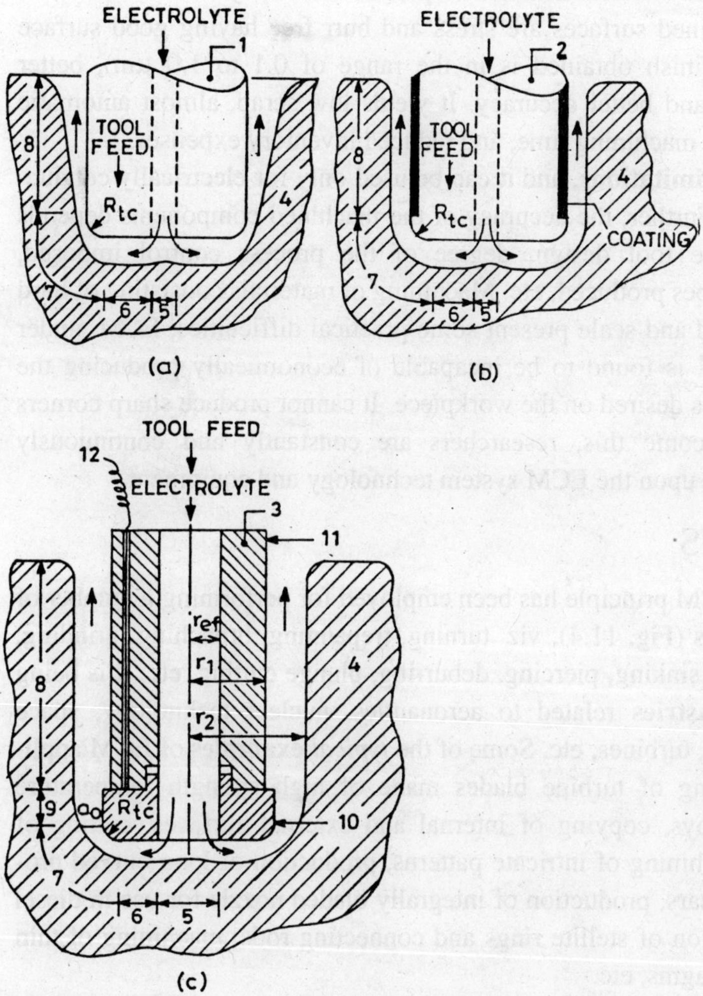

Fig. 11.6 Various types of tools: (a) bare tool, (b) coated tool, (c) bit type of tool[1] [*Jain and Pandey*, 1987].

[1] **1. bare tool, 2. coated tool, 3. bit type of tool, 4. workpiece, 5. Stagnation zone, 6. Front zone, 7. transition zone, 8. side zine, 9. tapered side, 10. tool bit, 11. perspex tool bit holder, 12. connecting wire**

the machinability of the work material is independent of its physical and mechanical properties. Machined surfaces are stress and burr free having good surface finish (best surface finish obtained is in the range of 0.1 to 1.0 μm), better corrosion resistance, and better accuracy. It yields low scrap, almost automatic operation, low overall machining time, and reduced inventory expenses.

ECM has its own **limitations**, and it can be used only for electrically conductive work-materials. Further, the accuracy of the machined components depends upon the factors like tool design, degree of the process control imposed, complexity in the shapes produced, etc. Machining of materials consisting of hard spots, inclusions, sand and scale present some practical difficulties. ECM, under certain circumstances, is found to be incapable of economically producing the dimensional tolerances desired on the workpiece. It cannot produce sharp corners and edges. To overcome this, researchers are constantly and continuously engaged in improving upon the ECM system technology and equipment.

APPLICATIONS

Over the years, ECM principle has been employed for performing a number of *machining* **operations** (Fig. 11.4), viz turning, trepanning, broaching, grinding, fine hole drilling, die sinking, piercing, deburring, plunge cutting, etc; it is being widely used in **industries** related to aeronautics, nuclear technology, space vehicles, automobiles, turbines, etc. Some of the *typical* **examples** of ECM applications are: machining of turbine blades made of high strength temperature resistant (HSTR) alloys, copying of internal and external surfaces, cutting of curvilinear slots, machining of intricate patterns, production of long curved profiles, machining of gears, production of integrally bladed nozzle for use in diesel locomotives, production of stellite rings and connecting rods, machining of thin large diameter diaphragms, etc.

This process has attracted the maximum attention of those involved in machining of hard and tough materials, specially with complex contours. However, the capabilities of this process have not been exploited to its fullest extent mainly because of the inherent problems associated with tool design [*Jain and Pandey, 1980*], and environment friendliness.

MECHANICAL PROPERTIES OF ECM'd PARTS

It is important to know the effects of ECM on the mechanical properties of electrochemically machined (ECM'd) components. This will greatly affect the acceptability of the process in different industries. Hardly any evidence is available about the hydrogen embrittlement of the ECM'd components. The basic reason being that hydrogen is evolved at the cathode while metal removal is taking place due to anodic dissolution at the anode. It has been reported that there is no effect of ECM on ductility, yield strength, ultimate strength, and microhardness of the machined component.

Surface layers damaged during conventional machining or by some other processes, may be removed by ECM and this may result in improvement in the properties of the work material. However, such removal of layers from the work surface reduces *fatigue strength* of a conventionally machined component. The conventionally machined surfaces have compressive residual stresses responsible for higher **fatigue strength**. This fact has been verified experimentally [*Evans et al*, 1971]. However, the required fatigue strength can be restored by further appropriate post mechanical finishing treatment. These subsequent mechanical treatments impart compressive stresses to the surface, so that the resulting workpiece can exhibit fatigue properties comparable to or better than those of mechanically finished parts.

The **surface finish** produced by ECM may also be the source of reduction in fatigue properties. The surfaces produced by ECM generally have better wear, friction, and corrosion resistance properties than those produced by mechanical means [*Gurklis*, 1965]. Table 11.1 gives and idea of the effect of type of finish on fatigue strength [*Evans et al*, 1971].

Table 11.1 Fatigue lives obtained [Evans et al, 1971] after ECM at a stress of ± 386 MN/m^2

Surface Finish	Fatigue Life, $10^5 \times$ cycles
Electrochemically polished	4.9
Etched	4.4
Intergranular attack	4.25
Hemispherical pits	3.3

Micro examination of the specimen has revealed that fatigue cracks usually originate from pits. Due to flowing electrolyte, these defects are more severe than those found in electropolishing. However, reduction in fatigue life is low and depth of intergranular attack measured is about hundredth of a millimeter or even less. Removal of compressive stress layer seems to be the main reason, and intergranular attack and other defects induced during ECM appear to be the secondary reasons for reduction in the fatigue strength. Improper combination of work and electrolyte, or inappropriate selection of operating conditions may result in non-uniform dissolution of metals and alloys. It would lead to selective etching, intergranular attack, or pitting. But, such defects can be minimized by employing proper heat treatment procedure, and also by developing an appropriate electrolyte and selecting proper operating conditions.

THEORY OF ECM

There are many *industrial processes* which work based on the principle of Faraday's laws of electrolysis. ECM is one of them. It is considered as the reverse of electroplating process. But, there are two major differences between ECM process and other electrolytic processes (viz, electropolishing, electrolytic pickling, etc). In ECM, it is not merely to remove metal from the workpiece but also change the shape and size of the workpiece in a controlled manner. Secondly, the magnitude of minimum current density employed in ECM is very high (8 A mm^{-2}) as compared to that used in electroplating or electrolytic pickling.

Faraday's Laws of Electrolysis

During ECM, metal from the anode (or workpiece) is removed atom by atom by removing negative electrical charges that bind the surface atoms to their neighbours. The ionized atoms are then positively charged and can be attracted away from the workpiece by an electric field. In an *electrolytic cell* (or ECM cell), material removal is governed by *Faraday's laws* of electrolysis given below.

(i) The amount of chemical change produced by an electric current (or the amount of substance deposited or dissolved) is proportional to the quantity of electricity passed.

(ii) The amounts of different substances deposited or dissolved by the same quantity of electricity are proportional to their chemical equivalent weights.

These laws can be expressed in mathematical form as follows:

$$m \propto I\,t\,E$$

or,
$$m = I\,t\,E\,/\,F \qquad \text{...(11.1)}$$

where, I is the current strength (amperes), t is the time (seconds), E is the gram equivalent weight of material (or A/Z, where A is atomic weight and Z is valency of dissolution), F is Faraday's constant (96500 As), and m is mass in grams. Eq. (11.1) is used to calculate MRR_g (= IE/F). Table 11.2 gives the values of 'A' and 'Z' for some metals, their density, and theoretical material removal rate.

This equation is based on the number of simplified assumptions and does not account for the effects of some of significant process variables, viz change in valency of electrochemical (EC) dissolution during the operation, gas evolution and bubble formation, electrolyte's electrical conductivity (k) and temperature variation along the electrolyte flow path, overpotential, presence of passivation film, etc. Passivity arises as a result of chemical and electrochemical behaviour of metals which results in the formation of protective film on their surfaces. Further, dissolution of iron in NaCl solution, depending upon the machining conditions, may be either in the form of ferrous hydroxide (Fe^{2+}) or ferric hydroxide (Fe^{3+}). Mode of dissolution during machining of alloys, is still more difficult to know. The preferential valency mode of dissolution has been found to depend upon the electrolyte flow rate, IEG and length of electrolyte flow path.

Electrochemical Equivalent of Alloys

Eq. (11.1) has been used for evaluating the amount of metal (m) removed from an anode made of single element. The evaluation of 'm' becomes difficult when the anode is made up of an alloy because the value of electrochemical equivalent of the alloy $(A/Z)_a$ is not known. However, two methods have been suggested to solve this problem, viz, '*percentage by weight method*' and '*superposition of charge*' method.

In the **first method**, the value of electrochemical equivalent of an alloy is calculated by multiplying the chemical equivalent of individual elements (A_i/Z_i) by their respective proportions by weight X_i, and then summing them up as given in Eq. (11.1a)

$$\left(\frac{A}{Z}\right)_a = \frac{1}{100}\sum_{i=1}^{n}\left(\frac{A_i}{Z_i}\right)X_i \qquad \text{...(11.1a)}$$

where, n is the number of the constituent elements.

Table 11.2 Theoretical material removal rates for a current of 1000 A [*McGeough*, 1974].

Metal	Atomic Weight (g)	Valency	Density (g/cm^3)	Removal rate (g/s)	(10^{-6} m^3/s)
Aluminium	26.97	3	2.67	0.093	0.035
Beryllium	9.0	2	1.85	0.047	0.025
Chromium	51.99	2	7.19	0.269	0.038
		3		0.180	0.025
		6		0.090	0.013
Cobalt	58.93	2	8.85	0.306	0.035
		3		0.204	0.023
Niobium	92.91	3	9.57	0.321	0.034
(Columbium)		4		0.241	0.025
		5		0.193	0.020
Copper	63.57	1	8.96	0.660	0.074
		2		0.329	0.037
Iron	55.85	2	7.86	0.289	0.037
		3		0.139	0.025
Magnesium	24.31	2	1.74	0.126	0.072
Manganese	54.94	2	7.43	0.285	0.038
		4		0.142	0.019
		6		0.095	0.013
		7		0.081	0.011
Molybdenum	95.94	3	10.22	0.331	0.032
		4		0.248	0.024
		6		0.166	0.016
Nickel	58.71	2	8.90	0.304	0.034
		3		0.203	0.023

(Contd.)

Table 11.2 (*Contd.*)

Silicon	28.09	4	2.33	0.073	0.031
Tin	118.69	2	7.30	0.615	0.084
		4		0.307	0.042
Titanium	47.9	3	4.51	0.165	0.037
		4		0.124	0.028
Tungsten	183.85	6	19.3	0.317	0.016
		8		0.238	0.012
Uranium	238.03	4	19.1	0.618	0.032
		6		0.412	0.022
Zinc	65.37	2	7.13	0.339	0.048

In the **second method** (superposition of charge method), the amount of electrical charge required to dissolve the mass contribution of individual constituent element of the alloy is calculated. The total electric charge required to dissolve 1g of the alloy is given by $(Z/A)_a$ F coulombs (from Eq. (11.1)).

Now, equate the electrical charge required by each element to dissolve their individual mass contribution to the total charge required for dissolving 1 gram of alloy,

$$\left(\frac{Z}{A}\right)_a F = F \sum_{i=1}^{n} \frac{X_i}{100}\left(\frac{Z_i}{A_i}\right),$$

or,

$$\left(\frac{A}{Z}\right)_a = \frac{100}{\sum_{i=1}^{n} X_i\left(\frac{Z_i}{A_i}\right)}. \quad \text{...(11.1b)}$$

The values of electrochemical equivalent of alloys calculated by these two methods are not exactly same. Hence, one should be careful while choosing a particular method for calculating $(A/Z)_a$.

All of the current (I) is not utilized only in dissolving metal from the workpiece. Hence, actual MRR depends on the **current efficiency**, η. It ranges from 75% to 100%. Theoretical MRR depends upon the current passing through the workpiece and its chemical composition. As soon as machining starts, the tool is fed towards the workpiece so that the gap can be maintained at a constant value. The feed rate is approximately equal to the linear material removal rate (MRR_l, i e penetration rate).

Example 11.1

An alloy consists (% by weight) nickel (72%), chromium (20%), iron (5%), titanium (0.5%), copper (0.5%), silicon and manganese (1.0% each). Determine chemical equivalent (using both methods) and density of the alloy. Use the lowest valency of dissolution (Table 11.2). Compare MRR_g in the two cases for a current value of 500 A at 100% current efficiency.

Solution

The density of an alloy can be determined as follows:
Consider one gram of alloy having density as ρ_a and volume as V_a. Then,

$$\rho_a V_a = 1$$

or,
$$\rho_a = \frac{1}{V_a} = \frac{1}{V_1 + V_2 + V_3 + .. + V_n}\ .$$

For constituent 1, $V_1\,\rho_1 = 0.72$ g

Therefore,
$$V_1 = \frac{0.72}{\rho_1} = \frac{72}{100\rho_1}$$

Similarly,
$$V_2 = \frac{20}{100\rho_2}$$

.

.

.

$$V_n = \frac{X_n}{100\rho_n}\ .$$

Therefore,
$$\rho_a = \frac{1}{\frac{X_1}{100\rho_1} + \frac{X_2}{100\rho_2} + \ldots + \frac{X_n}{100\rho_n}}$$

$$\rho_a = \frac{100}{\sum X_i/\rho_i}.$$

Substitute the values in the above equation.

$$\rho_a = \frac{100}{\frac{72}{8.90} + \frac{20}{7.19} + \frac{5}{7.86} + \frac{0.5}{4.51} + \frac{0.5}{8.69} + \frac{1}{2.33} + \frac{1}{7.43}}$$

$$= \frac{100}{8.09 + 2.78 + 0.64 + 0.11 + 0.058 + 0.43 + 0.13}$$

$$= 8.17 \text{g/cm}^3.$$

Now, chemical equivalent of the alloy using percentage by weight method is calculated as given below:

First Method (Eq. 11.1a)

$$\left(\frac{A}{Z}\right)_a = \frac{1}{100} \sum_{i=1}^{n} \left(\frac{A_i}{Z_i}\right) X_i$$

$$= 0.72 \times \frac{58.71}{2} + 0.20 \times \frac{51.99}{2} + 0.05 \times \frac{55.85}{2} + 0.005 \times \frac{47.9}{3}$$

$$+ 0.005 \times \frac{63.57}{1} + 0.01 \times \frac{28.09}{4} + 0.01 \times \frac{54.94}{2}$$

$$= 21.135 + 5.199 + 1.396 + 0.080 + 0.318 + 0.070 + 0.275$$

$$\approx 28.473 \text{ g}$$

Second Method

$$\left(\frac{A}{Z}\right)_a = \frac{100}{\sum_{i=1}^{n} \left(\frac{Z_i}{A_i}\right) X_i}$$

$$= \frac{100}{72 \times \frac{2}{58.71} + 20 \times \frac{2}{51.99} + 0.5 \times \frac{2}{55.85} + 0.5 \times \frac{3}{47.9} + 0.5 \times \frac{1}{63.57} + 1 \times \frac{4}{28.09} + 1 \times \frac{2}{54.94}}$$

$$\approx \frac{100}{2.453 + 0.769 + 0.018 + 0.031 + 0.008 + 0.142 + 0.036}$$

≈ 28.927 g.

Material removal rates (in grams – MRR_g) using $(A/Z)_a$ obtained from the two different methods are calculated as given below:

$$MMR_{g_1} = \frac{IE}{F} = 500 \times \frac{28.473}{96500}$$

$$\approx 0.1475 \text{ g/s.}$$

$$MRR_{g_2} = \frac{IE}{F} = \frac{500 \times 28.927}{96500}$$

$$\approx 0.1499 \text{ g/s.}$$

Material Removal Rate in ECM

It is always desirable to have minimum possible gap (usually ≤ 0.5 mm) between the two electrodes (tool and work) mainly to get accurate reproduction of tool shape on the workpiece. To simplify the analysis given in this section, the following assumptions are made:

(i) Electrical conductivity (k) of electrolyte in the IEG remains constant.

(ii) Electrical conductivities of tool and work materials are very large (1,00,000 Ω^{-1} cm^{-1}) as compared to that of electrolyte (less than 1.0 Ω^{-1} cm^{-1}). Hence, surfaces of the electrodes can be considered as equipotentials.

(iii) Effective voltage working across the electrodes is $(V - \Delta V)$, where ΔV is a small fraction of V and includes electrode voltage, overvoltage, etc. It is assumed to remain constant.

(iv) The anode dissolves at one fixed valency of dissolution.

For the simplicity of mathematical modelling, a case of plane parallel electrodes normal to feed direction [*Tipton, 1964*] is considered (Fig. 11.7). The equations derived in the following will be applicable mainly for such cases.

Electrolyte is assumed to flow in the direction of increasing X across the gap between two electrodes. It is also assumed that the properties of electrolyte remain unchanged in the Z direction, i e the problem is two-dimensional in nature.

The Eq. (11.1) can be used to calculate material removal rate (in grams MRR_g, or in terms of volume as MRR_v):

$$\dot{m}(MRR_g) = \eta I\, E / F, \qquad \text{...(11.1c)}$$

$$MRR_v = \eta I\, E / F\, \rho_a. \qquad \text{...(11.1d)}$$

where, η is current efficiency.

Leaving aside a few specific applications of ECM (for example, deburring), tool (cathode) is usually fed towards the workpiece (anode) at a constant rate. During equilibrium, feed rate (f) of the cathode is equal to the rate at which the thickness of the anode is being reduced, i.e. MRR_l. It is given by the following equation (divide both sides of the last equation by the common area 'A' through which current is flowing):

$$MRR_l = \frac{J\, E\, \eta}{\rho_a F}, \qquad \text{...(11.1f)}$$

where, $J = I/A$ (A/mm^2), and ρ_a is density (g/mm^3) of the anode material. In ECM, a properly shaped tool concentrates electric current on those areas of the work-piece from which preferential removal of the metal is desired. Metal removal rate (MRR) from an area on the workpiece is controlled by current density on that area. Current density during ECM is a function of shapes of the electrodes (work and tool, Fig. 11.3), their distance apart, voltage applied across them, and electrical conductivity of electrolyte flowing through the gap. Current density can also be written (Eq. 11.2a) as a multiplication of potential gradient ($(V-\Delta V)/y$) in the IEG, and electrolyte conductivity (k).

$$J = \frac{(V - \Delta V)k}{y}. \qquad \text{...(11.2a)}$$

Here, $(V-\Delta V)$ is the voltage available for driving current through the electrolyte. In Eq. (11.1f), MRR_l is nothing but penetration rate, or rate of change of IEG, i e dy/dt.

$$\frac{dy}{dt} = \frac{JE\eta}{\rho_a F}. \qquad \text{...(11.2b)}$$

During actual ECM, tool is usually moved towards the workpiece at a rate of 'f' units/s. Hence, effective rate of change of IEG is given by

$$\frac{dy}{dt} = \frac{JE\eta}{\rho_a F} - f. \qquad \text{...(11.2c)}$$

Under equilibrium conditions, $\frac{dy}{dt} = 0.$ ...(11.3a)

Therefore, under equilibrium condition, feed rate is given by

$$\frac{JE\eta}{\rho_a F} = f. \qquad \text{...(11.3b)}$$

Substitute the value of 'J' from Eq. (11.2a),

$$\frac{(V - \Delta V)k}{y_e} \frac{E\eta}{\rho_a F} = f \qquad \text{...(11.3c)}$$

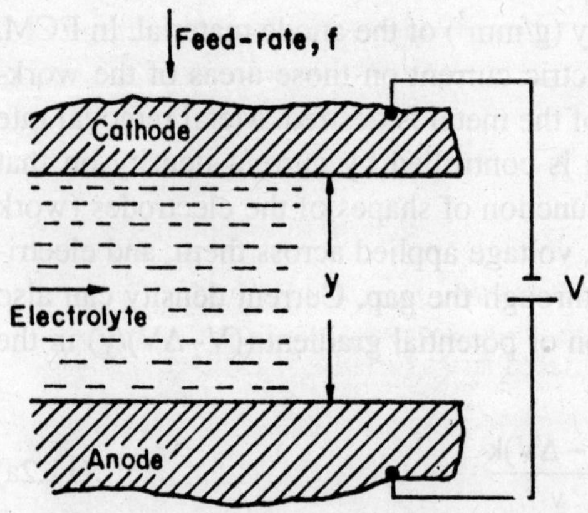

Fig. 11.7 Electrochemical machining with plane parallel electrodes at a constant voltage.

Inter-electrode Gap in ECM

Eq. (11.3c) can be used to evaluate equilibrium interelectrode gap (y_e)

$$y_e = \frac{(V - \Delta V)\eta k E}{\rho_a f F}. \qquad ...(11.4)$$

y_e can be achieved only if the feed rate (f) and applied voltage (V) are constant throughout machining. It is not important whether the work is stationary and the tool is fed towards it or vice versa. Since the tool surface during ECM remains unchanged, it is preferable to use it as a reference surface.

Example 11.2

During machining of iron (atomic weight = 55.85, valency = 2, density = 7.85 g/cm^3), the equilibrium gap is approximately 0.125 mm and the measured value of specific conductance of electrolyte = 0.2 Ω^{-1} cm^{-1}. Faraday's constant is 26.8 Ah, applied voltage is 10 V, and overvoltage is 1.5 V. Calculate the value of feed rate, f.

Solution

Substitute the values in Eq. (11.4)

$$f = \frac{(V - \Delta V)k E}{y_e \rho_a F} = \frac{(10 - 1.5)0.2(55.85/2)}{0.0125 \times 7.85 \times (26.8 \times 60)}$$

or $$f = 0.3 \text{ cm/min.}$$

$$\boxed{f = 0.05 \text{ mm/s.}} \text{ Ans.}$$

The rate of change of IEG (y) can be calculated from Eq. (11.5) [*Tipton*, 1964].

$$\frac{dy}{dt} = \frac{E(V - \Delta V)k}{F \quad \rho_a \quad y} - f. \qquad ...(11.5)$$

However, in case of equilibrium condition, dy/dt will attain the value as zero. In a more convenient form, above Eq. (11.5) can be written in terms of f, y and machining constant, C (mm^2 s^{-1}) ($C = E(V - \Delta V)k/F\rho_a$ is a constant for a particular work and electrolyte combination only if, (V–ΔV) and k remain constant during the process):

$$\frac{dy}{dt} = \frac{C}{y} - f. \qquad ...(11.6)$$

It is a basic differential equation of the system.

In actual practice, two cases arise for which Eq. (11.6) can be solved, zero feed rate ($f = 0$) and finite feed rate ($f \neq 0$).

Zero Feed Rate

For $f = 0$, Eq. (11.6) can be written as:

$$\frac{dy}{dt} = \frac{C}{y},$$

or,

$$y \; dy = C\, dt. \qquad ...(11.7)$$

Integrate both sides to get

$$y^2 = 2C\,t + K$$

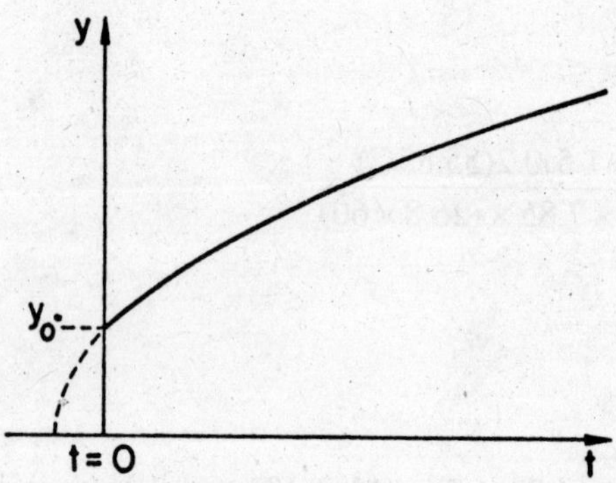

Fig. 11.8 The variation of the gap thickness (y) with time at zero feed-rate.

For the initial condition as $y = y_0$ at $t = 0$, above equation gives, $K = y_0^2$. Eq. (11.8) is obtained by substituting the value in the above equation.

$$y = (2Ct + y_o^2)^{1/2} \qquad ...(11.8)$$

It means that the gap increases in proportion to the square root of the time (Fig. 11.8).

Finite Feed Rate

For $f \neq 0$, the IEG during equilibrium condition will remain constant ($dy/dt = 0$) and from Eq. (11.6) we get

$$y = y_e = \frac{C}{f}, \qquad \ldots(11.9)$$

where, y_e is *equilibrium gap*. The IEG always tends to attain the equilibrium value during ECM. To make the analysis more generalized, some of the parameters (time and IEG) are made dimensionless, i.e.

$$\left.\begin{aligned} y' &= \frac{y}{y_e} = \frac{y\,f}{C}.\\ t' &= \frac{t}{(y_e/f)} = t\,\frac{f^2}{C}. \end{aligned}\right\} \qquad \ldots(11.10)$$

Here, y' indicates the ratio of the gap to the equilibrium gap, and t' indicates the number of times required to machine one 'y_e' distance.

Now, from Eq. (11.6) and Eq. (11.10), we get

$$\frac{dy'}{dt'} = \frac{1}{y'} - 1.$$

This equation can also be written as

$$\frac{dt'}{dy'} = \frac{y'}{1 - y'}. \qquad \ldots(11.11a)$$

or,

$$dt' = \left(\frac{y'}{1 - y'}\right) dy'$$

$$= \left(\frac{-1 + y' + 1}{1 - y'}\right) dy'$$

$$t' = -y' - \ln(y' - 1) + K \qquad \ldots(11.11b)$$

Solve the above Eq. (11.11a) to give Eq. (11.11b). Substitute the initial condition as $y' = y_0'$ at $t' = 0$ in Eq. (11.11b), to give the following:

$$0 = -y'_o - \ln(y_o' - 1) + K$$

$$\therefore \quad K = y_o' + \ln(y_o' - 1).$$

Substitute the value of K in Eq. (11.11b).

$$t' = -y' - \ln(y' - 1) + y_o' + \ln(y_o' - 1)$$

$$t' = y_o' - y' + \ln\frac{(y_o' - 1)}{(y' - 1)}. \qquad \text{...(11.12)}$$

In the above Eq. (11.12), only positive values of y' are possible, and $y' = 0$ implies a short circuit between the tool and workpiece.

For the case of $f \neq 0$, solution of Eq. (11.5) can also be obtained [*McGeough*, 1974] when parameters are not non-dimensionalised, as given below:

$$t = \frac{1}{f}\left[y_o - y_t + y_e \ln\left(\frac{y_o - y_e}{y_t - y_e}\right)\right] \qquad \text{...(11.13)}$$

where, y_t is IEG at any time 't'.

Self-Regulating Feature

To illustrate *self-regulating feature* of the ECM process, Eq. (11.12) has been plotted in Fig. 11.9 where y' is asymptotic to $y' = 1$ (ie MRR_l is equal to feed rate, or $y = y_e$).

Three relationships are possible between 'f' and 'MRR_l': $f = MRR_l$, $f > MRR_l$, and $f < MRR_l$.

(i) If $f = MRR_l$, stable machining will take place and IEG will be equal to the equilibrium gap ($y = y_e$).

(ii) If $f < MRR_l$, then initially the IEG will increase and it will attain a value greater than the equilibrium gap value ($y > y_e$). Because of this, current density will decrease as compared to the current density at the time when the gap is equal to the equilibrium gap. As a result, MRR_l will also decrease. Or the difference between f and MRR_l will decrease. Finally, the

IEG will start decreasing and it will attempt to attain the equilibrium gap ($MRR_l = f$).

(iii) If $f > MRR_l$, then initially the gap will be smaller than the equilibrium gap ($y < y_e$). As a result, current density will increase as compared to the situation when $y = y_e$, hence MRR_l will also increase. In other words, the difference between f and MRR_l will decrease. Finally, the gap will tend to attain the equilibrium gap (or $MRR_l = f$) value.

This characteristic has been utilized to illustrate the levelling up or shaping process.

Generalized Equation for Inter-electrode Gap

A quadratic Eq. (11.8) predicts an infinite value of the gap as 't' approaches infinity. However, in practice, as the gap increases, the current density decreases so that MRR gradually diminishes, and for a very large value of the gap the process would come to a *stand still*. This aspect is not well reflected in Eq. (11.8). Eqs. (11.12) and (11.13) are *implicit equations*, and can be solved *iteratively for* y_t (or y'). But, it consumes a lot of computer time. In order to evaluate the IEG at any instant of time t, solution of Eq. (11.13) is a bit tedious. Further, these equations are not convenient to handle specially in those cases where numerical methods like finite element method (FEM), finite difference method (FDM), or boundary element method (BEM) are applied to analyze the ECM process. Keeping this in view, a single equation has been derived [*Jain and Pandey*, 1981] which can be conveniently used in case of FEM, FDM, or BEM, and it is applicable simultaneously for both the cases, i e zero feed rate (f = 0) as well as finite feed rate ($f \neq 0$). Eq. (11.5) can be written as

$$\frac{dy}{dt} = \frac{EJ\eta}{F\rho_a} - f,$$

where, η is current efficiency.

$$dy = \left(\frac{EJ\eta}{F\rho_a} - f\right)dt. \qquad \text{...(11.14a)}$$

In the parentheses on the right hand side of the Eq. (11.14a), all the *factors are constant except J*. In FEM, BEM, and FDM, element size is quite small. Hence,

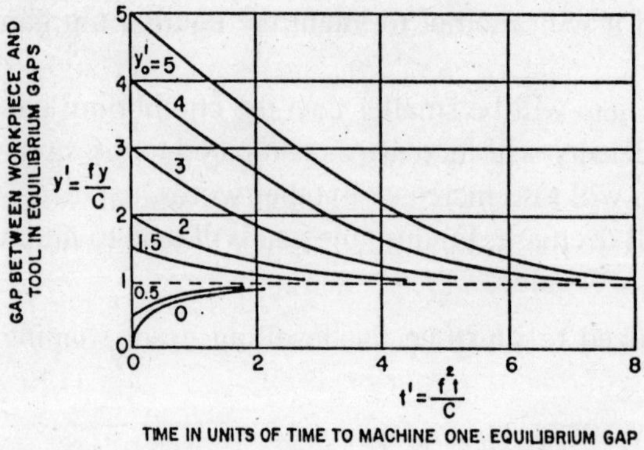

Fig. 11.9 y′ approaches to the equilibrium gap (y′ = 1) for a constant feed rate and for various initial gaps y'_o.

within an element, J can be assumed to remain constant over a small interval of time dt. Above equation therefore, can be written as

$$dy = (c - f)dt. \qquad ...(11.14b)$$

After integration, Eq. (11.14b) can be written as

$$y = (c - f)t + C'.$$

At time t = 0, $y = y_o$, then the above equation can be written as

$$C' = y_o.$$

Therefore,

$$y = y_o + (c - f)\ t. \qquad ...(11.15)$$

Eq. (11.15) is valid for a small interval of time Δt, hence, more conveniently, it can be written as

$$\boxed{y = y_o + (c - f)\ \Delta t.} \qquad ...(11.16)$$

Table 11.3 [*Jain and Pandey*, 1980] shows that the values of different parameters for three different problems obtained from the above discussed two different methods are very close to each other. It establishes the *validity* of a single and simple Eq. (11.16) applicable to both the cases of feed rate (f = 0 and f ≠ 0).

However, computer time required to solve ECM problems by employing the two Eqs. (11.13) and (11.16) and using numerical methods, would be very different.

One may come across a case when a plane parallel *gap is inclined* at an angle θ to the feed direction, Fig. 11.10. In such cases, feed rate 'f' should be replaced by its component ('f cos θ') in the above analysis. When θ = 90°, the feed rate normal to the work is zero and hence the condition f = 0 would apply. Note that here, angle θ is measured between a normal to the work surface and the feed direction.

Table 11.3 Comparison of Eqs. (11.13) and (11.16) with the data for anode shape (t = 300 s) after [*Jain and Pandey*, 1980].

x,	y (mm), 10^{-1}		Temperature rise (°C)		
mm	Eq. (11.13)	Eq. (11.16)	Eq. (11.13), $y_0 = 0.8$	Eq. (11.16), $y_0 = 0.8$	Experimental
7.0	7.8500	7.8477	5.2595	5.2623	5.0
17.0	9.0419	9.0452	13.7024	13.6954	11.625
22.0	9.7153	9.7192	18.4565	18.4422	15.0*
7.0	3.4808	3.3670	5.7189	5.3100	–
17.0	4.0584	4.0890	14.9660	15.0073	15.8687
22.0	4.3841	4.3980	20.1821	20.4389	22.45*
7.0	5.0018	5.0027	6.3056	6.3056	10.0875
17.0	5.9177	5.9176	16.6205	16.6170	–
22.0	6.4434	6.4426	22.5309	22.5336	23.6*

*x = 21.5 mm

MAXIMUM PERMISSIBLE FEED RATE IN ECM

Equation for volumetric material removal rate (MRR_v) can be written as

$$MRR_v = \frac{E\,I\,\eta}{96500\,\rho_a}. \qquad \text{...(11.17)}$$

The current efficiency (η) is close to 100% while using aqueous solution of NaCl as electrolyte. It is somewhat lower than 100% in case of nitrate and sulphate

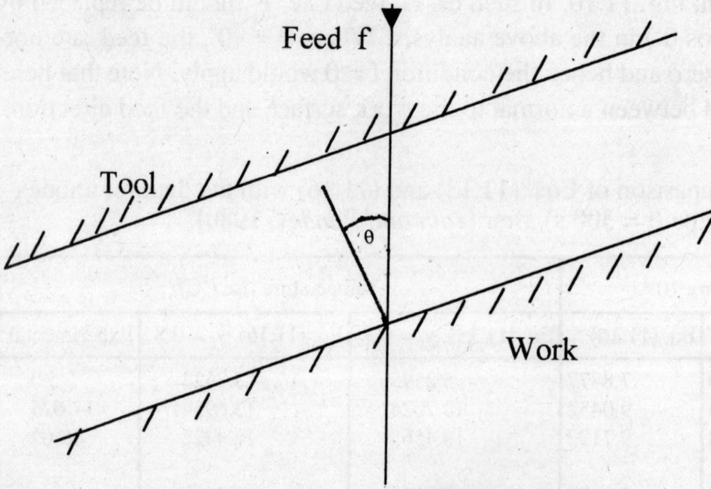

Fig. 11.10 Feed direction inclined at an angle θ measured between the normal to the work surface and feed direction.

electrolytes. In some cases, efficiency has been reported [*De Barr and Oliver, 1975*] to be more than 100% also specially when electrically non-conducting inclusions (Fig. 11.11) are there in the anode.

From Eq. (11.17), specific metal removal rate ($m^3 A^{-1} s^{-1}$) can be written as

$$MRR_s = \frac{\eta}{96500} \frac{E}{\rho_a} (m^3 A^{-1} s^{-1}). \qquad \text{...(11.18)}$$

Therefore, electrode feed rate (m/s) is given by

$$f = J.MMR_s$$

$$= \frac{(V - \Delta V)}{y} k \cdot \frac{E\eta}{96500\rho_a} \, ms^{-1} \text{ (y in metre)} \qquad \text{...(11.19)}$$

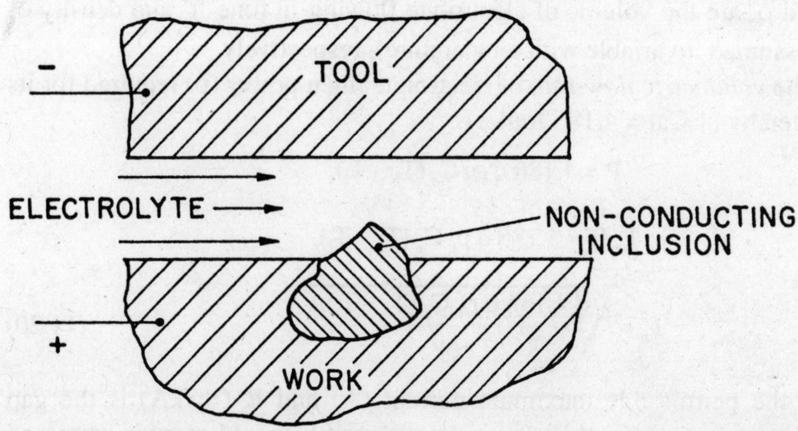

Fig. 11.11 Dissolution of anode with electrically non-conducting inclusion.

One of the important functions of the electrolyte is to carry away the heat out of the IEG. There is always an upper limit to the electrolyte flow rate that can be achieved from a given electrolyte flow system. This constraint in turn also poses a limit to the maximum feed rate that can be used during ECM without permitting the electrolyte to boil. In the worst case, it can be assumed that the electrolyte is *allowed to heat up to the boiling temperature*. Assuming that only Ohmic heating is significant, an approximate expression for maximum permissible feed rate can be derived [*Pandey and Shan*, 1980].

Using the law of **conservation of heat**, heat (H_o) required to raise the electrolyte temperature from T_i (temperature at inlet) to T_b (electrolyte boiling temperature) can be evaluated as:

$$H_o = m_e C_e (T_b - T_i),$$

where, m_e is the mass of electrolyte and C_e is the specific heat of electrolyte. It can also be written as

$$H_0 = \overline{V}_e \, \rho_e \, C_e \, (T_b - T_i),$$

or,
$$\frac{H_0}{t} = \frac{\overline{V}_e}{t} \rho_e \, C_e \, (T_b - T_i),$$

where, $\overline{V}_e$ and ρ_e are the volume of electrolyte flowing in time 't' and density of electrolyte (assumed invariable with temperature), respectively.

Let $\dot{q}$ be the *volumetric flow rate* of electrolyte; then power (p) required for its heating is given by (1 Cal = 4.186 Joules):

$$P = 4.186\, \dot{q}\, \rho_e\, C_e\, (T_b - T_i),$$

or,
$$I_m^2 R = 4.186\, \dot{q}\, \rho_e\, C_e\, (T_b - T_i),$$

or,
$$I_m = \sqrt{\frac{4.186\, \dot{q}\, \rho_e\, C_e\, (T_b - T_i) A}{y}}, \qquad \text{...(11.20)}$$

where, I_m is the permissible maximum current (A) and R (=y/kA) is the gap resistance. An increase in current (or current density) would require increased feed rate for machining under equilibrium conditions.

Using above equations (11.18-11.20), *maximum permissible feed rate* (f_m) can be calculated as:

$$f_m = J.MRR_s = \frac{I_m}{A} MRR_s$$

$$= \sqrt{\frac{4.186\, \dot{q}\, \rho_e\, C_e\, (T_b - T_i)\, A}{y}} \cdot \frac{1}{A^2} \cdot \frac{\eta\, E}{\rho_a\, 96500}$$

$$= \frac{\eta\, E}{\rho_a\, 96500} \sqrt{\frac{4.186\, \dot{q}\, \rho_e\, C_e\, (T_b - T_i)\, k}{yA}} \qquad \text{...(11.21)}$$

Eq. (11.21), after simplification (as given in the following), can also be used to calculate change in temperature ($\Delta T = T_b - T_i$) for the specified feed rate (f).

$$\Delta T = 2.23 \times 10^9 \left[\frac{yA}{k\, \dot{q}\, C_e\, \rho_e}\right] \left[\frac{f\, \rho_a}{E\, \eta}\right]^2$$

ELECTROLYTE CONDUCTIVITY (k)

In the above analysis, electrolyte's electrical conductivity has been assumed to remain constant throughout the gap. However, in actual practice, it is not so. Its conductivity is a function of local temperature, presence of bubbles, and contamination of the electrolyte by sludge. Size and distribution of hydrogen gas bubbles (ignoring oxygen, water vapour, and other gas bubbles) will affect the magnitude by which the electrolyte conductivity changes. Temperature increases the value of k while hydrogen gas bubbles decrease its value. As a result, the gap gets tapered, ie IEG varies along the electrolyte flow direction (Fig. 11.12). Fig. 11.13 shows the effect of variation in temperature and concentration on the conductivity of NaCl and HCl. The effects of temperature and gas bubbles on the electrolyte conductivity will be discussed now.

Effect of Temperature (T)

In the following analysis, it is assumed that heat generated due to viscous flow of electrolyte, heating due to overvoltage and chemical reactions, and heat transferred through the electrodes are all negligible.

Using the law of conservation of heat, temperature gradient along the path of electrolyte flow can be derived as follows:

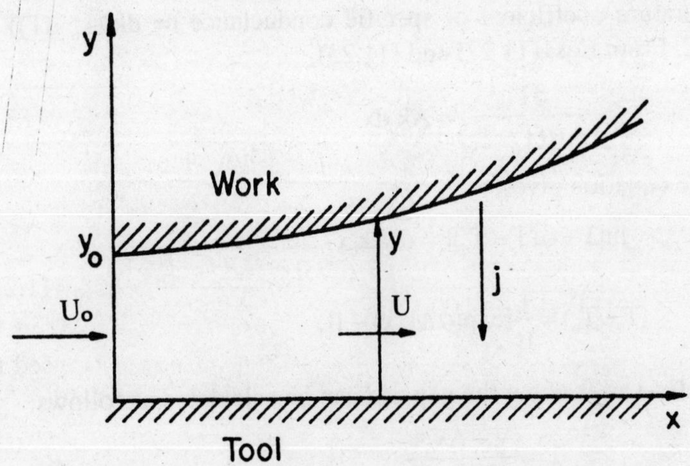

Fig. 11.12 The equilibrium gap with variable conductivity [*Tipton*, 1964].

$$I^2 R = U \rho_e C_e (\Delta T) y$$

(where, U is electrolyte flow velocity)

or, $$\frac{(V-\Delta V)^2}{y} k \,.\ dx = U \rho_e C_e (\Delta T)\, y$$

(Assuming unit width in z direction)

or, $$\frac{dT}{dx} = \frac{(V-\Delta V)^2}{U \rho_e C_e y^2} . k$$

where, ρ_e and C_e are assumed to remain constant in the IEG.

It is also safe to assume IEG and electrolyte flow velocity at inlet (y_o and U_o) to be constant. Then, initial temperature distribution in a plane parallel gap of thickness y_o is given by

$$\frac{dT}{dx} = \frac{(V-\Delta V)^2}{U_o \rho_e C_e y_o^2} = A.k. \qquad ...(11.22)$$

Conductivity of the electrolyte varies linearly with temperature

$$k = k_o(1 + \alpha(T - T_o)), \qquad ...(11.23)$$

where, α is the temperature coefficient of specific conductance (= $dk/(k_o\, dT)$) with the units as per °C. From Eqs. (11.22) and (11.23),

$$\frac{dT}{1 + \alpha(T - T_o)} = Ak_0 dx.$$

Integration of the above equation gives,

$$\ln[1 + \alpha(T - T_o)] = \alpha\, Ak_o x$$

or, $$(T - T_o) = \frac{1}{\alpha}[\exp(\alpha A k_o x) - 1].$$

The current density, J also varies along the gap and can be calculated as follows:

$$J = \frac{(V-\Delta V)k}{y}$$

$$= \frac{(V-\Delta V)}{y} k_o\, [1 + \alpha(T - T_o)]$$

$$= \frac{(V-\Delta V)}{y} k_o\, [\exp(\alpha A k_o x)]$$

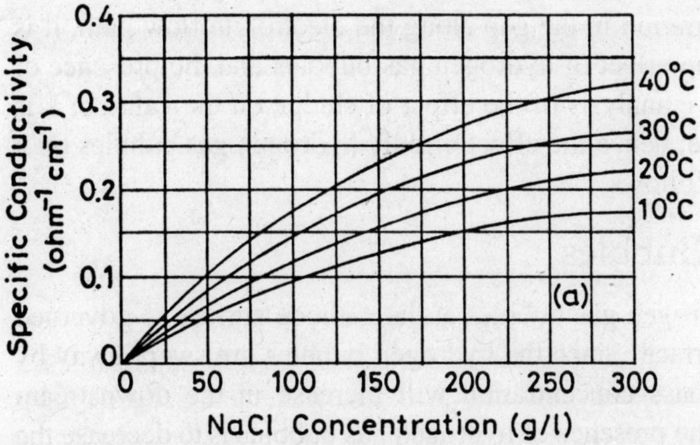

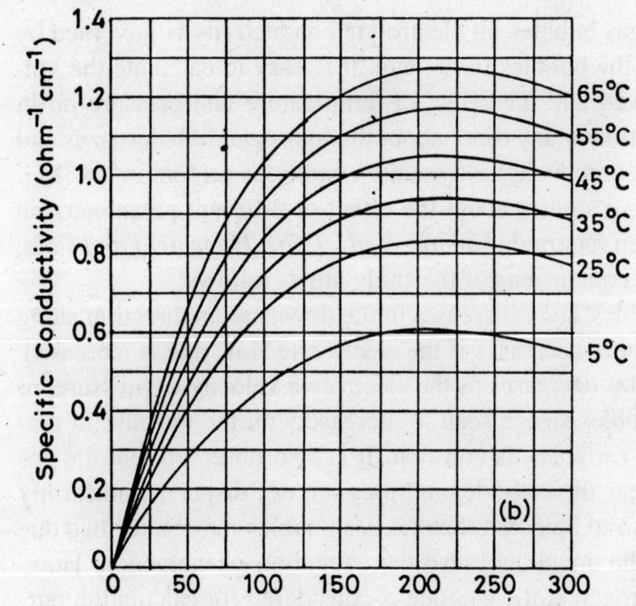

Fig. 11.13 Specific conductivity variation with electrolyte concentration of (a) NaCl, (b) HCl, at various temperatures.

To accurately predict variation in the gap along the electrolyte flow path, it is also desirable to consider the effect of hydrogen gas bubbles and the presence of sludge on k. The quantitative analysis of the effect of sludge on the value of k is not available in the literature; hence the effect of only hydrogen gas bubbles on k has been discussed in what follows.

Effect of Hydrogen Bubbles

Rate of evolution of hydrogen gas bubbles at the cathode surface is governed by current density on its surface. Since the hydrogen bubbles are swept away by flowing electrolyte, their mass concentration will increase in the downstream direction. The net effect of the presence of hydrogen gas bubbles is to decrease the electrolyte's electrical conductivity. Thus, it would decrease the local anodic dissolution rate.

The effect of hydrogen gas bubbles on electrolyte conductivity is governed by the size and distribution of the bubbles in the gap. It is easy to calculate the volume of hydrogen gas evolved and the effect of temperature and pressure on it. But, it is very difficult to estimate any thing about the hydrogen bubbles sizes and their distribution under the complex hydrodynamic conditions present in the IEG.

However, some qualitative studies about the effect of different parameters on the size of bubbles have been reported [*Landolt et al, 1970; Hopenfeld and Cole, 1966, 1969*]. The important conclusions of the study are as follows:

The thickness of the bubbles-layer increases in the downstream direction along the gap. However, its thickness decreases if the electrolyte flow rate is increased. The size of individual bubbles decreases as the electrolyte velocity or pressure, or both are increased. But, bubbles size is seen to increase with the increase in current density, keeping other variables as constant. It is also observed that the gas bubbles are concentrated near the cathode, and they are not dispersed uniformly throughout the IEG. It is also to be noted that the gas bubbles may be formed due to evolution of oxygen at the anode and also the cavitation phenomenon. However, only evolution of hydrogen at the cathode is considered for calculation purposes.

To incorporate the effect of temperature as well as hydrogen gas bubbles on electrolyte conductivity, the following equation has been suggested [*Hopenfeld and Cole,1969*; *Thorpe and Zerkle,1969*]:

$$k = k_o(1 + \alpha\Delta T)(1 - \alpha_v)^n \quad \ldots(11.24)$$

where, α_v is void fraction and n is an exponent. The value of exponent 'n' in Eq. (11.24) depends on the distribution of voids in the IEG. For uniform void distribution, *Hopenfeld and Cole* [1969] suggested a value of n = 1.5, and for non-uniform distribution when bubbles are concentrated near the cathode, *Thorpe and Zerkle* [1969] suggested a value of n = 2.0. But, during electrochemical drilling (ECD) and electrochemical bit drilling (ECBD), it is extremely difficult to know the distribution of voids in the IEG. Hence, an average value of n = 1.75 has also been used [*Jain et al*, 1987].

BIBLIOGRAPHY

1. Abdel Mahboud A.M. (1986), Improving accuracy and economics of ECM: Modelling, simulation and control,. *ASME Press*, Vol. 7(1), pp. 23-23.
2. Benedict G.. (1987, *Nontraditional Manufacturing Processes*, Marcel Dekker, New York.
3. Datta M., Romankiw L.T., Vigliotti D.R. and Von Gutfeld R.J. (1989), Jet and laser jet electrochemical micromachining of nickel and steel, *J. Electrochem. Soc.*, Vol. 136, No. 8, pp. 2251-2256.
4. DeBarr A.E. and Oliver D.A. (Ed) (1975), *Electrochemical Machining*, Macdonald & Co. Ltd., London.
5. Evans J.M., Boden P.J. and Baker A.A. (1971), Effects of surface finish produced during ECM upon the fatigue life of Nimonic- 80A, *Proc. 12th Int. Mach. Tool Des. & Res. Conf.*, p. 271.
6. Gurklis J.A. (1965), Metal removal by electrochemical methods and its effects on mechanical properties of metals, *Defense Metals Information Centre Report 213, Bettle Memorial Institute.*
7. Hewidy M.S., Rajurkar K.P., Royo G.F. and Fattouh M. (1989), Theoretical and experimental investigations of electrochemical ballizing, *Int. J. Prod. Res.*, Vol. 27, No. (8), pp. 1335-1347.
8. Hofstede A. and Berkel Van den (1970), Some remarks on electrochemical turning, *Annls CIRP*, Vol. 18, pp. 93-106.
9. Jain V.K. (1980), *An Analysis of ECM process for anode shape prediction*, Ph.D. Thesis, Roorkee University, India.
10. Jain V.K. and Pandey P.C. (1980), Tooling design for ECM, *Precision Engg.*, Vol. 2, pp. 195-206.

11. Jain V.K. and Pandey P.C. (1980), CAD of Tools in ECM, *Computer Aided Design*, Vol. 12, No. 6, pp. 309-315.
12. Jain V.K. and Pandey P.C. (1981), Tooling design for ECM: A finite element approach, *Trans. ASME, J. Engg. for Ind.*, Vol. 103, pp. 183-191.
13. Jain V.K. and Pandey P.C. (1982), Some investigations into the anode profile in the transition zone in an electrochemical hole sinking operation, *J. Appl. Electrochem.*, Vol. 12, pp. 497-500.
14. Jain V.K. and Pandey P.C. (1982), Investigations into the use of bits as a cathode in ECM, *Int. J. Mach. Tool Des. & Res*, Vol. 22, No. 4, pp. 341-352.
15. Jain V.K. Tandon S. and Kumar Prashant (1990), Experimental investigations into electrochemical spark machining of composites, *Trans ASME, J. Engg. for Ind.*, Vol. 112, No. 2, pp. 194-197.
16. Jain V.K., Jain Vinod K. and Pandey P.C. (1984), Corner reproduction accuracy in electrochemical drilling of blind holes, *Trans. ASME, J. Engg. for Ind.*, Vol. 106, pp. 55-61.
17. Jain V.K., Yogindra P.G. and Murugan S. (1987), Prediction of anode profile in ECBD and ECD operations, *Int. J. Mach. Tools Manuf.*, Vol. 27, No. 1, pp. 113-134.
18. Janseen J.W. (1989), New horizons for STEM drilling, *Proc. Nontraditional Machining Conf.*, Orlando, pp. MS89-81-8-1 to MS89-81-8-7.
19. Kawafune K., Mikoshiba T., Noto K. and Hirta K. (1967), Accuracy in cavity sinking by ECM, *Annls CIRP*, Vol. 15, pp. 443-455.
20. Krabacher E.J., W.A. Haggerty, Allison C.R. and Devis M.F. (1962), Electrical methods of machining, *Int. Conf. Production Research organised by ASME*, pp. 232-241.
21. Landolt D., Acosta R., Mullor R.H. and Tobias C.W. (1970), An optical study of cathodic hydogen evolution in high rate electrolysis, *J. Electrochem. Soc.*, Vol. 117, No. 6, pp. 839-845.
22. McGeough J.A. (1974), *Principles of ECM*, Chapman and Hall, London.
23. Merchant M.E. (1962), Newer methods for the precision working of metals: Research and present status, *Proc. Int. Conf. Prod. Res., ASME*, pp. 93-107.
24. Molley J. (1966), Electrochemical machining, *Machinery & Prod. Engg.*, pp. 13-17.
25. Pandey P.C. and Shan H.S. (1980), *Modern Machining Processes*, Tata-McGraw Hill, New Delhi, India.

26. Pourassamy A., Santosh Kumar K.J., Krishnan A. and Rajgopalan S.R. (1987), Investigations on the insulation of electrochemical trepanning tool, *Indian J. Technol.*, Vol. 25, pp. 617-620.
27. Ranganathan V. (1976), Electrochemical grinding of titanium, *Proc.* 7th *AIMTDR Conf.*, pp. 164-168.
28. Rolsten R.F. and Manning E.J. (1969), Electrochemical deburring, *Tool Manuf. Engr.*, pp. 4-16.
29. Thorpe J.F. and Zerkle R.D. (1969), Analytical determination of the equilibrium gap in electrochemical machining, *Int. J. Mach. Tool Des. Res.*, Vol. 9, pp. 131-144.
30. Tipton H. (1964), The dynamics of electrochemical machining, *In Advances in Machine Tool Design and Research*, Pergamon Press, Birmingham.
31. Wilson J.F. (1971), *Practice and Theory of Electrochemical Machining*, Wiley Interscience, New York.

SELF-TEST QUESTIONS

Q1. Write the correct answer(s). More than one may be correct.

(i) Which of the following workpiece properties affect MRR during ECM?
(a) Physical, (b) Mechanical, (c) Chemical, (d) None.

(ii) Desirable properties of electrolytes used in ECM are:
(a) high electrical conductivity, (b) high thermal conductivity,
(c) low specific heat, (d) high specific heat.

(iii) Theoretically the ratio of MRR to tool wear rate during electrochemical drilling is:
(a) Zero, (b) usually $\approx$ 1, (c) depends on the type of work material and electrolyte combination, (d) none of these.

(iv) Value of void fraction along the path of flow of electrolyte is increasing but temperature of the electrolyte remains approximately constant. Slope of the machined surface will be
(a) zero, (b) varying linearly, (c) varying nonlinearly, (d) unpredictable.

(v) Electrochemically machined surfaces have
(a) high residual stresses and improved fatigue strength,

(b) high residual stresses and reduced fatigue strength,
(c) insignificant residual stresses and reduced fatigue strength,
(d) insignificant residual stresses and improved fatigue strength.

Q2. Write whether the following statements are true or false:

(i) Electrolyte is not consumed in ECM. Type of electrolyte used during ECM, therefore, has no effect on MRR.
(ii) Electrochemically ground surface requires deburring.
(iii) Theoretically tool wear rate during ECM is zero.
(iv) IEG during ECM is usually in the range of 1-3 mm.
(v) Feed rate during ECD_e is usually low.

Q3. (i) Complete the following chemical reactions (ECM)

(a) $2Fe + 4\,H_2\,O + O_2 \rightarrow$
(For low valency of dissolution)

(b) $2Fe + 4\,H_2\,O + O_2 \rightarrow$
(For high valency of dissolution)

(ii) During ECM of iron using aqueous solution of NaCl as electrolyte, what are the possible reactions at anode and cathode?

(iii) Complete the following equations:

(a) $J = (V - \Delta V)$
(b) $\dot{m} = A$
(c) $f = J$

Q4. Write the correct answer from the given choices. Show all the calculations necessary to arrive at your answer.

(i) During ECM of an iron plate, ferrous hydroxide is formed and MRR_v obtained is 300 mm^3/s. What would be MRR_v (mm^3/s) if iron dissolves as ferric hydroxide in place of ferrous hydroxide.
(a) 200, (b) 450, (c) 300, (d) none of these.

(ii) During ECM, electrode feed rate is calculated as 2 mm/min. The electrolyte conductivity is 0.2 $\Omega^{-1}\,cm^{-1}$. It is now decided to change the feed rate to 3 mm/min without changing any other parameter except conductivity. What is the new value of conductivity ($/\Omega$–cm) for the same performance of the process?
(a) 0.4, (b) 0.3, (c) 0.133, (d) none.

(iii) In ECM, the maximum permissible feed rate is f_1 while using electrolyte 'A' having sp. heat as C_1 and electrolyte density as ρ_{e1}. In the second case, while using electrolyte 'B', the maximum permissible feed rate becomes f_2. In this case, specific heat of electrolyte 'B' is 1.125 C_1 and density as $2\rho_{e1}$. If the highest allowable temperature of the electrolyte and the values of other parameters remain the same, what is the ratio of f_1/f_2?
(a) 1.0, (b) 0.667, (c) 2.25, (d) 1.5.

PROBLEMS

Q1. (i) The composition (% by weight) of Nimonic 75 alloy is given in the accompanying table:

Element:	Ni	Cr	Fe	Ti	Si	Mn	Cu
% by weight:	72.0	20.00	5.00	0.50	1.00	1.00	0.50
Atomic weight:	58.71	51.99	55.85	47.90	28.09	54.94	63.50
Valency:	2, 3	2, 3, 4	2, 3	3, 4	4	2, 4, 6, 7	1, 2
Density:	8.90	7.19	7.89	4.51	2.33	7.43	8.96

(a) Calculate MRR (specific MRR and volumetric MRR) when a current of 2000 A is used. Use the lowest valency of dissolution in each case.

(b) Determine the equilibrium gap if applied voltage = 15 V, over potential = 1.5 V, k = 0.21/Ω-cm, and f = 1 mm/min.

(c) Calculate rise in temperature for the following machining conditions: feed rate = 0.05 mm/s, room temperature = 38°C, electrolyte density = 1.05 g/cm^3, η = 100%, electrolyte flow rate = 120 litres/min, flow velocity = 2000 cm/s and specific heat, C_e = 1.06 cal/g-°C.

(ii) Geometry of the workpiece surface with single curvature is given by the equation

$y = 10 + 0.20x - 0.05x^2$

where, x and y are in centimetres. The other details are as follows: Applied voltage = 15 V, Over potential = 0.67 V, f = 0.75 mm/min, work material = copper (A = 63.57, Z = 1), ρ = 8.96 g/cm^3, k =

0.2/Ω-cm.
Determine the equation of the required tool surface geometry and write the assumptions, if any.

(iii) Derive an equation for the maximum permissible feed rate during ECM. Also deduce the relationship for electrolyte temperature change for a given feed rate of tool.

(iv) Calculate the maximum permissible feed rate if boiling temperature of electrolyte = 98 °C, room temperature = 38 °C, ρ_e = 1.05 g/cm^3, η = 100%, electrolyte flow rate = 22 litres/s, electrolyte flow velocity = 20 m/s and specific heat = 1.07 cal/g- °C.
Derive the following equation:

$$(T_o - T_i) = \frac{1}{\alpha}\,[\exp(Ak_i\,\alpha\,x)\,(1-\alpha_v)^n - 1]$$

where, T_o and T_i are outlet and inlet temperatures respectively, k_i is inlet electrolyte conductivity, and A is given by

$$A = \frac{(V - \Delta V)^2}{U\,\rho_e C_e y^2}$$

where, x = distance along electrolyte flow direction, α = coefficient of change in electrolyte conductivity with temperature, α_v = void fraction, n = exponent, C_e = specific heat of electrolyte, V = applied voltage and y = IEG.

(v) Derive one single equation for computing IEG during both zero feed rate as well as finite feed rate. Write the assumptions clearly.

(vi) While using numerical method like FEM, FDM, or BEM for anode shape prediction, which of the two equations-implicit equation or equation derived in (v) above-will you recommend? Give the reasons.

Q2. (i) During ECM, 10 V DC supply is used. Initial conductivity of the electrolyte used is 0.2 Ω^{-1} cm^{-1}. Feed rate used is 1.0 mm/min. Calculate the IEG at 10, 20 and 30 mm distance from the inlet after 10s. Use the following given conditions (assume linear variation of α_v along *x* direction):
T = 30 °C, α_v = 0.0 at x = 0.00 mm,
T = 60 °C, α_v = 0.3 at x = 40.0 mm

Take n (exponent) = 1.75, atomic weight of anode = 55.9 g, density of anode = 7.86 g/cm^3, initial IEG = 0.4 mm, valency of dissolution = 2, overpotential = 1.5 V, and $\alpha = 0.002$. Also calculate the mean equilibrium gap.

(ii) Using Faraday's laws of electrolysis, derive and equation to compute maximum permissible feed rate. Modify this equation to take into consideration the effect of temperature and void fraction on the electrolyte conductivity.

(iii) The composition of nimonic alloy turbine blade is 18% cobalt, 62% Ni, and 20% chromium. Take I = 1500 A, and $\rho_a = 8.3$ g cm^{-3}. Use valency of dissolution for chromium, nickel and cobalt as 6, 2 and 2, respectively. Gram atomic weights are as follows: Ni = 58.7, Cr = 52.0 and Co = 59.9. Calculate the volumetric MRR, and state assumptions, if any.

(iv) Explain the 'self regulating (or adjusting) feature' of ECM.

NOMENCLATURE

A	Atomic mass, area of cross-section
C	Specific heat
E	Gram equivalent mass of material
f	Feed rate
F	Faraday's constant
H	Heat required
I	Current
J	Current density
k	Electrolyte conductivity
m	Mass of the material removed
n	Number of the constituent elements
q	Volumetric flow rate of electrolyte
R	Resistance
t	Time
T	Temperature
ΔT	Change in temperature
U	Electrolyte flow velocity
V	Applied voltage
$\overline{V}_e$	Volume of electrolyte flowing
ΔV	Over voltage

X_i Proportion by weight of ith element
x Coordinate direction
y Interelectrode gap, coordinate direction
Z Valency of dissolution
z Coordinate direction
α Temperature coefficient of specific conductance
ρ Density
η Machining efficiency

Subscripts

a Alloy, anode (workpiece)
b Boiling
e Equilibrium, electrolyte
g Grams
i Inlet
l Linear
m Maximum
o Condition at time t = 0, or initial condition, outlet
s Specific, second
t At any time 't'
v Volumetric

Acronyms

AC Alternating current
DC Direct current
ECM Electrochemical machining
GFRP Glass fiber reinforced plastics
IEG Interelectrode gap
m/c Machine
m/t Machine tool
MRR Material removal rate
SCRs Silicon controlled rectifiers
TWR Tool wear rate

CHAPTER 11
AT-A-GLANCE
ELECTROCHEMICAL MACHINING
(ECM)

- ECM PROCESS IS ALSO KNOWN AS : CONTACTLESS ELECTROCHEMICAL FORMING PROCESS

ECM PRINCIPLE

- FARADAY'S LAWS OF ELECTROLYSIS
- ELECTROCHEMICAL ENERGY DETACHES METAL FROM ANODE $\rightarrow$ ATOM BY ATOM
- POSITIVE IONS CONVERT AS HYDROXIDES
- ELECTROLYTES ——> | ———> ACIDIC | ———> ALKALINE | ———> NEUTRAL
- $2H_2O + 2e \rightarrow H_2\uparrow + 2\,OH^-$
- $2Fe - 2e \rightarrow 2Fe^{++}$
- $Fe^{++} + 2OH^{--} \rightarrow Fe(OH)_2\downarrow$
- $Fe(OH)_2 \rightarrow$ "INSOLUBLE IN WATER"
- SMALLER IEG $\rightarrow$ HIGHER MRR
- REACTION PRODUCTS $\rightarrow$ BARRIER TO THE FLOW OF ELECTROLYZING CUURRENT <- TO MINIMIZE THIS EFFECT $\rightarrow$ ELECTROLYTE FLOW VELOCITY 20-30 m/s (2 - 35 kgf/cm^2)
- FLOWING ELECTROLYTE ——>
  - |$\rightarrow$ LIMITS ION CONCENTRATION AT THE ELECTRODE SURFACE $\rightarrow$ HIGHER MRR
  - |$\rightarrow$ DILUTES REACTION PRODUCTS
  - |$\rightarrow$ DISSIPATES HEAT AT A FASTER RATE
- FEED RATE = MRR $\rightarrow$ CONSTANT IEG $\rightarrow$ CONSTANT MRR
- FACTORS GOVERNING ANODE PROFILE
  - ELECTROLYTE PROPERTIES, COMPOSITION, CONCENTRATION, TEMPERATURE, ETC. $\rightarrow$ CHANGE IN CURRENT DENSITY $\rightarrow$ MRR
  - FEED RATE
  - VOLTAGE
  - TOOL SHAPE
- ELECTROCHEMICAL DISSOLUTION BASED MACHINING OPERATIONS (FIG. 11.2)

ECM MACHINE TOOL

- POWER SOURCE
- ELECTROLYTE CLEANING AND SUPPLY SYSTEM
- TOOL & TOOL FEED SYSTEM
- WORK & WORK HOLDING SYSTEM

POWER SOURCE

- RECTIFIER CONVERTS & TRANSFORMS 220V/440V AC → LOW VOLTAGE → (5 - 20 V) HIGH CURRENT → (AS HIGH AS 40 kA) DC
- VOLTAGE REGULATION → ± 1 %
- SPARKING DETECTED (WITHIN 10 μs) → DEVICES LIKE SCR → PEEVENT DAMAGE TO TOOL & W/P
- TO MINIMIZE DAMAGE DUE TO A SPARK → USE OF SCR BANKS ACROSS DC INPUT

ELECTROLYTE SUPPLY AND CLEANING SYSTEM

- PUMP, FILTERS, PIPINGS, CONTROL VALVES, HEATING / COOLING COILS, PRESSURE GAUGE & RESERVOIR
- SUPPLY PORTS IN TOOL / WORK / FIXTURE ← MODE OF ELECTROLYTE FLOW
- TO MAINTAIN SMALL IEG (≤ 1 mm) → SMOOTH FLOW OF ELECTROLYTE AND NO BLOCKAGE OF GAP
- PROPER CLEANLINESS OF ELECTROLYTE USING FILTERS MADE OF ANTI-CORROSIVE MATERIALS
- PIPING SYSTEM : STAINLESS STEEL, GFRP, PLASTIC LINED M.S, OTHER ANTI–CORROSIVE MATERIALS
- : METALLIC PIPING ← EARTHED TO PREVENT ANODIC CORROSION
- TABLES, FIXTURES : EARTHED/CARRY CATHODIC POTENTIAL TO PREVENT CORROSION
- ELECTROLYTE TANK CAPACITY : 500 gal/1000 A
- ELECTROLYTE TEMPERATURE: ± 1°C → HEATIN/COOLING
- FILTERS —> |
  - |—> PERIODICALLY CLEANED FOR PROPER FUNCTIONING
  - |—> PLACED IN SUPPLY PIPE JUST PRIOR TO THE WORK ENCLOSURE
- SINGLE TANK SYSTEM (FIG 11.3)
- ECM : DIFFERENT METALS & ALLOYS, OPTIMUM MACHINING CONDITIONS, REQUIREMENTS OF ACCURACY & SURFACE FINISH
- USE OF SINGLE TANK → LOSS OF TIME & WASTAGE OF ELECTROLYTES RECOMMENDATION OF MORE THAN ONE TANK.

TOOL AND TOOL FEED SYSTEM

- TOOL MATERIAL ——> |-HIGH THERMAL & ELECTRICAL CONDUCTIVITY
  |-ANTI-CORROSIVE IN NATURE
  |-HIGH MACHINABILITY
- TYPES : ALUMINIUM, BRASS, COPPER, MONEL, BRONZE, ETC
- INSULATION OF TOOL / BIT TYPE OF TOOL
- ACCURACY & SURFACE FINISH OF TOOLS → AFFECT W/P ACCURACY
- FIXTURES ——> |—> NON-CORROSIVE & ELECTRICALLY NON-CONDUCTING MATERIALS
  |—> MINIMUM VIBRATION & DEFLECTION UNDER HYDRAULIC FORCES

WORKPIECE AND WORK HOLDING SYSTEM

- W/P MATERIAL : ELECTRICALLY CONDUCTIVE
- W/P HOLDING DEVICE :
  * ELECTRICALLY NON-CONDUCTING
  * GOOD THERMAL STABILITY
  * LOW MOISTURE ABSORPTION
  * GFRP, PLASTICS, PERSPEX

METAL REMOVAL IN ECM

- FARADAY'S LAWS OF ELECTROLYSIS
- IN ECM → W/P SHAPE & SIZE ARE CHANGED
- HIGH CURRENT DENSITY (MIN 8 A/mm^2)
- PREFERENTIAL REMOVAL OF METAL IS DESIRABLE
- $m = (E\ I\ t)/F$ (BASED ON SIMPLIFIED ASSUMPTIONS & DOES NOT ACCOUNT FOR CHANGE IN VALENCY, PROTECTIVE FILM, GAS EVOLUTION EFFECTS, OVER-POTENTIAL, ETC)
- PART OF CURRENT GOES AS A WASTE, $\eta \rightarrow 75\% - 100\%$
- $IEG \leq 1.0$ mm
- $f = MRR_l$ (MAINTAIN CONSTANT TOOL FEEDING)
- TOOL LIFE: THEORETICALLY INFINITE UNDER IDEAL CONDITIONS

ELECTROLYTE

- MEDIUM FOR CURRENT TO FLOW, TAKES AWAY HEAT GENERATED, REMOVES REACTION PRODUCTS, ETC
- NaCl COMMONLY USED
- RECIRCULATION
- VERY HIGH FLOW VELOCITY

MECHANICAL PROPERTIES OF ECM'D JOBS

- NO EVIDENCE ABOUT HYDROGEN EMBRITTLEMENT → HYDROGEN EVOLUTION AT THE CATHODE
- NO EFFECT ON DUCTILITY, YIELD STRENGTH, ULTIMATE STRENGTH, & MICRO-HARDNESS OF MACHINED COMPONENTS
- REDUCED FATIGUE STRENGTH AFTER ECM ← IMPROVE BY POST PROCESSING
- SURFACE FINISH AFFECTS FATIGUE STRENGTH
- FATIGUE CRACKS ORIGINATE FROM PITS
- DEPTH OF INTERGRANNULAR ATTACK ← 0.01 mm

ADVANTAGES

- MACHINES COMPLICATED SHAPES IN SINGLE PASS
- MACHINABILITY ← INDEPENDENT OF PHYSICAL & MECHANICAL PROPERTIES OF W/P
- STRESS & BURR-FREE
- GOOD FINISH (0.1-1.0 μm)
- GOOD ACCURACY
- LOW MACHINING TIME
- LOW SCRAP
- AUTOMATIC OPERATION

LIMITATIONS

- ONLY ELECTRICALLY CONDUCTIVE W/P
- ACCURACY → (TOOL DESIGN, DEGREE OF PROCESS CONTROL, SHAPE'S COMPLEXITY, ETC)
- DIFFICULTIES → HARD SPOTS, INCLUSIONS, SAND, ETC
- → SHARP CORNERS & EDGES

APPLICATIONS

- OPERATIONS : TURNING, TREPANNING, ETC
- INDUSTRIES : AERONAUTICS, NUCLEAR TECHNOLOGY, SPACE VEHICLES, AUTOMOBILES, ETC
- PRODUCTS : TURBINE BLADES, CURVILINEAR SLOTS, GEARS, INTEGRALLY BLADED NOZZLE RING, ETC
- MATERIALS : SPECIALLY DIFFICUTL-TO-MACHINE METALS AND ALLOYS
- FULL CAPABILITIES OF THE PROCESS HAVE NOT BEEN EXPLOITED

∇

PROBLEMS IN TOOL DESIGN

Chapter Twelve

ELECTROCHEMICAL GRINDING (ECG)

INTRODUCTION

Conventional grinding produces components with good surface finish and dimensional tolerances but such components are also associated with burrs, comparatively large heat affected zone (HAZ), and thermal residual stresses. These defects are not found in electrochemically ground workpieces (anodes). During electrochemical grinding (ECG), material is removed by mechanical abrasive action (about 10%) and by electrochemical dissolution (about 90%) of anodic workpiece. As in any other electrochemical dissolution based process, workpiece should be electrically conductive. Electrolyte is recirculated in ECG, hence, an effective and efficient electrolyte supply and filtration system is needed. The commonly used electrolytes are sodium chloride (NaCl) and sodium nitrate ($NaNO_3$).

In ECG, there is a **grinding wheel** (cathode) similar to a conventional grinding wheel except that the bonding material is electrically conductive. Electrolyte is supplied through inter electrode gap (IEG) between the wheel and the workpiece. The height of abrasive particles protruding outside bonding material of the wheel helps in maintaining a constant IEG because the abrasive particles act as spacers. *Life of the ECG wheel* is about ten times more than that of the conventional grinding wheel. Two factors are responsible for such a high wheel-life: only 10% contribution by abrasive action towards the total material removal, and very small length of arc of contact.

In ECG, the area in which machining is taking place can be divided into 3 *zones* (*named as, zone I, zone II, and zone III*) as discussed in the following. Fig. 12.1 shows a schematic diagram for ECG set-up.

In **zone I** (Fig. 12.2), material removal is purely due to *electrochemical dissolution* and it occurs at leading edge of the ECG wheel. Rotation of the ECG wheel helps in drawing electrolyte into the IEG. As a result of electrochemical reaction in zone I, reaction products (including gases) contaminate electrolyte resulting in lower conductivity. In fact, presence of sludge, to some extent, increases conductivity of the electrolyte [*Jain et al*, 1990], while that of gases decreases it. Net result is a decrease in the value of conductivity of the electrolyte. It yields a lower value of IEG. As a result, abrasive particles touch the workpiece surface and start removing material by abrasive action. Thus, a small part of material is removed in the form of chips. Further, electrolyte is trapped between the abrasive particles and workpiece surface, and it forms a tiny electrolytic cell as shown in Fig. 12.2. In each electrolytic cell, small amount of material from the workpiece is electrochemically dissolved.

The electrolyte is being forced into the IEG in **zone II** by rotational motion of the wheel. As a result, local electrolyte pressure increases in this part of the IEG (zone II). It suppresses formation of gas bubbles in the gap yielding higher MRR. Chemical or electrochemical reaction may result in the formation of passive layer on the workpiece-surface. In this zone II, abrasive grains remove material from the work surface in the form of chips and also remove *non-reactive* oxide layer. Most of the metal oxides formed are insoluble in water, and electrically nonconductive.

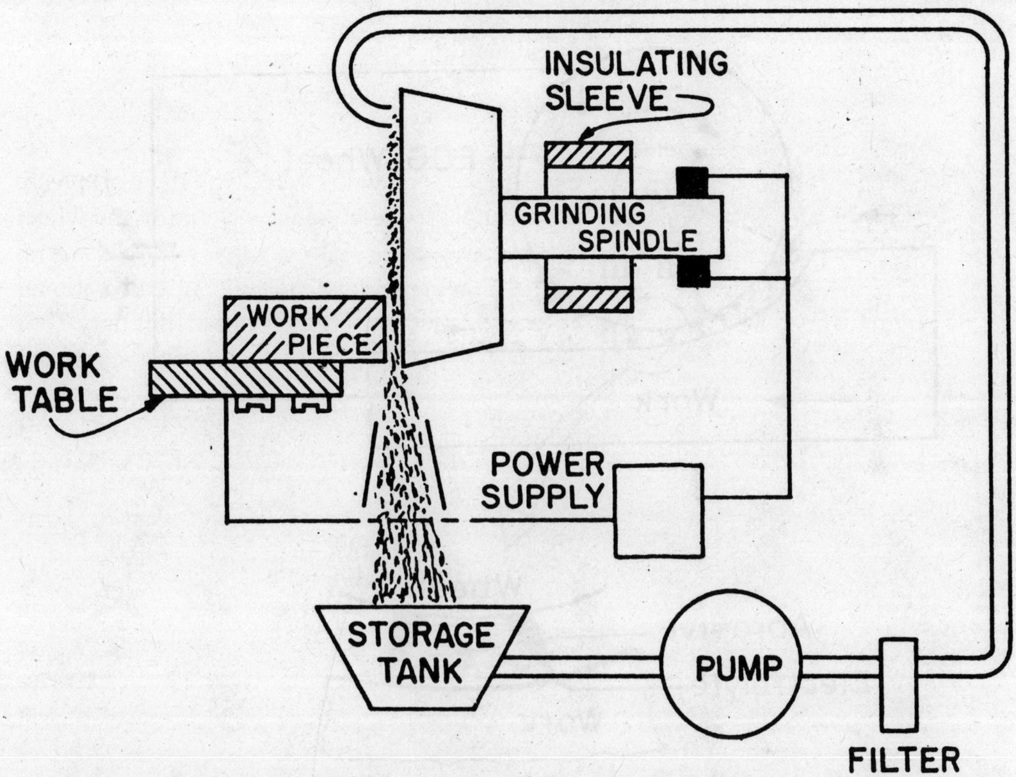

Fig. 12.1 Schematic diagram of electrochemical grinding set-up.

Removal of non-reactive oxide layer promotes electrolytic dissolution. It exposes fresh metal for further electrolytic action. Hence, it is also called [*Bhattacharyya, 1973*] as "mechanical assisted electrochemical grinding" process.

In **zone III**, material removal is totally by electrochemical dissolution. Zone III starts at the point where wheel lifts off the work-surface. In this zone, pressure is released slowly. This zone contributes to the removal of scratches or burrs that might have formed on the workpiece in zone II.

ECG MACHINE TOOL

The ECG systems are very similar to conventional grinders (Fig. 12.3). In ECG

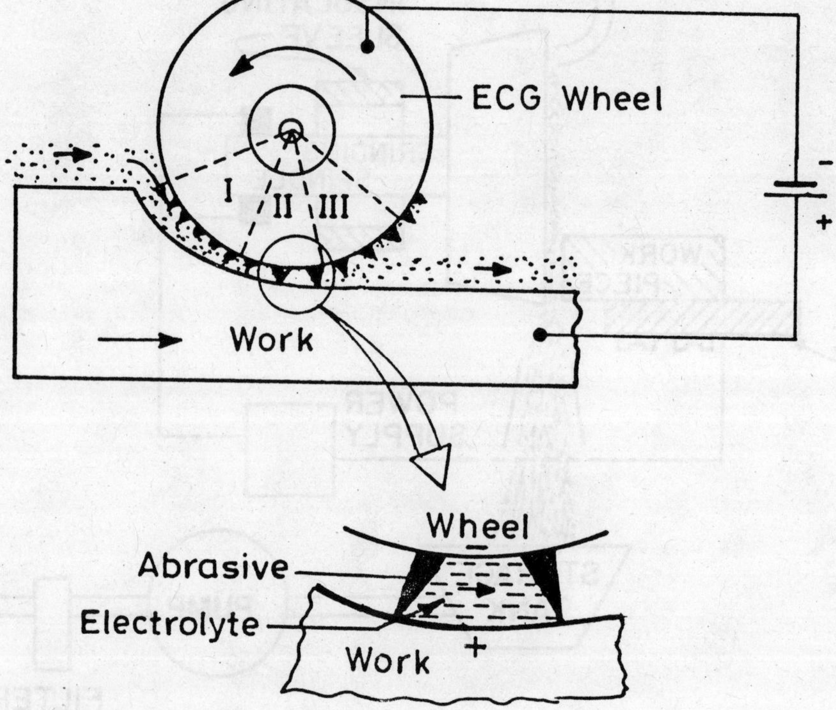

Fig. 12.2 Three machining zones and tiny electrochemical cell formation in ECG.

system, machining area is made up of non-corrosive materials. Power is supplied through spindle either with the help of brushes or mercury coupling. The latter can carry more current than the previous one. The probability of **short circuiting** during ECG is very low because of the presence of protruding abrasives which create a positive IEG. Hence, in this system, there is less need of having short circuit cut-off devices.

Five different **kinds of ECG operations** can be performed, viz electrochemical (EC) cylindrical grinding, electrochemical form grinding, EC surface grinding, EC face grinding and EC internal grinding. EC cylindrical grinding is the slowest process because of the limited area of contact between the

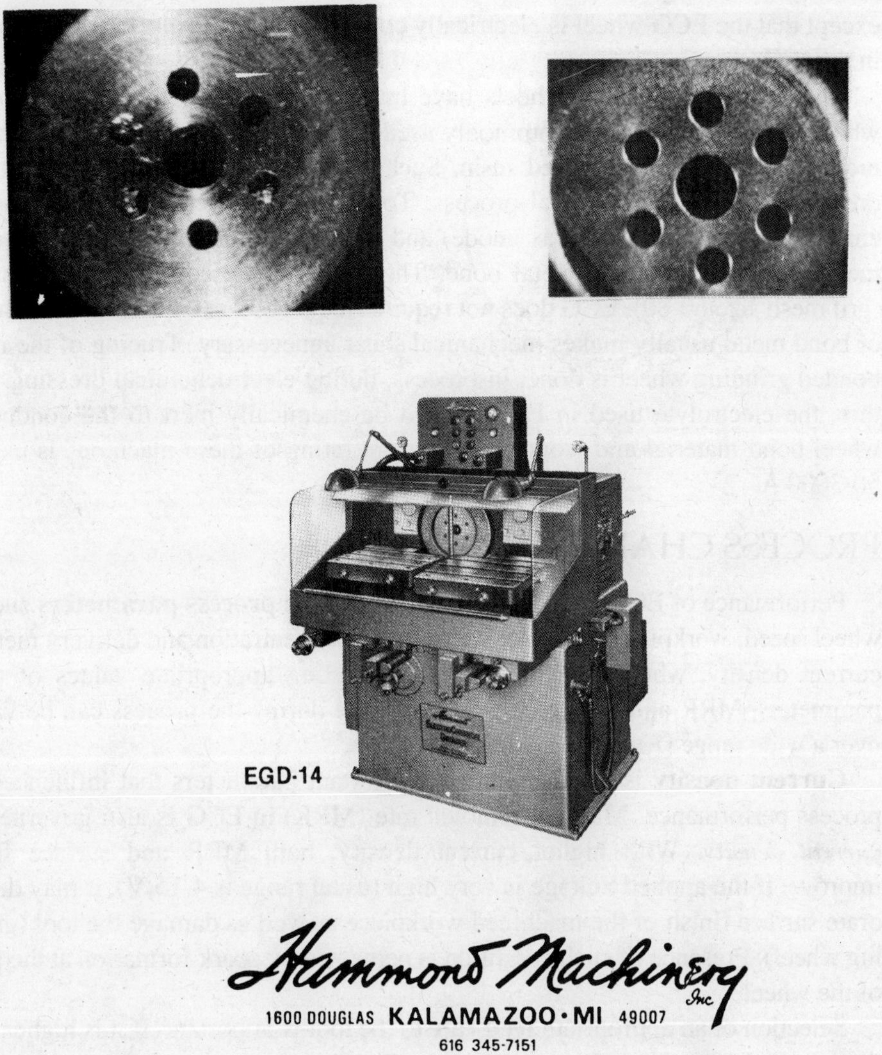

Fig. 12.3 Electrochemical grinder [*Hammond Machinery*, USA].

wheel and the workpiece. However, EC face grinding is the fastest process because of maximum area of contact between the anode and the cathode. Uneven wheel wear can be controlled by providing oscillating motion to the workpiece. In EC surface grinding, the workpiece reciprocates. EC internal grinding and EC form grinding are the same as conventional internal and form grinding operations except that the ECG wheel is electrically conducting and the electrolyte is present in the IEG.

Metal bonded grinding wheels have many advantages over resinoid bonded wheels. In ECG wheel, the commonly used **bonding materials** are copper, brass, nickel, or copper impregnated resin. Such metal bonded wheels are effectively dressed using electrochemical process. To prepare (or dress) them, reverse the current (or make the wheel as anode) and do the grinding on the scrap piece of metal. It will deplate the metal bond. The commonly used abrasive is alumina (grit mesh size 60-80). ECG does not require frequent wheel dressing. Dissolution of bond metal usually makes mechanical shear unnecessary. Trueing of the metal bonded grinding wheel is done, in-process, during electrochemical dressing. Further, the electrolyte used in ECG should be chemically inert to the conductive wheel bond material and workpiece. Current rating of these machines is usually 50-3000 A.

PROCESS CHARACTERISTICS

Performance of ECG process depends on various **process parameters** such as wheel speed, workpiece feed, electrolyte type, concentration and delivery method, current density, wheel pressure, etc. By selecting appropriate values of these parameters, MRR and surface finish obtainable during the process can be varied over a wide range [*Kuppuswamy, 1976*].

Current density is one of the most important parameters that influences the process performance. Material removal rate (MRR) in ECG is also governed by *current density*. With higher current density, both MRR and surface finish improve. If the applied voltage is very high (usual range is 4-15 V), it may deteriorate surface finish of the machined workpiece as well as damage the tool (grinding wheel). Presence of such condition is percepted by spark formation at the front of the wheel.

Selection of an appropriate **feed rate** to the tool is important. If it is higher than the required one, the abrasive particles will prematurely detach from the wheel, leading to excessive wheel wear. If it is lower than the required one, a large overcut (or poor tolerances) and poor surface finish will result. The IEG is usually a quarter of a millimetre while using a freshly dressed wheel. Surface speed of the wheel is in the range of 1200-1800 m/min. The depth of cut is usually below 2.5 mm and it is limited by the wheel contact arc length, which should never exceed

19 mm; otherwise electrolyte becomes ineffective because of higher concentration of H_2 gas bubbles and sludge [*Benedict, 1987*].

MRR achieved during ECG may be high as 10 times compared to conventional grinding on hard materials (hardness ≥ 65 HRC). But tolerances obtained in ECG are poorer (± 0.0025 mm). Minimum inside corner radius of 0.25 mm and outside corner radius of 0.025 mm can be produced by this process. Abrasive particles maintain electrical insulation between cathode and anode, and determine the effective gap between them (may be as low as 0.025 mm). Surface finish obtained by ECG ranges from 0.12 and 0.8 μm. The **surface produced** by ECG is free of grinding scratches and burrs. Surface finish produced on non-homogeneous materials during ECG is better than that produced during conventional grinding. In spite of more initial investment, the cost of EC grinding is lower than that of conventional grinding due to much higher MRR during ECG. Risk of thermal damage is also reduced.

Electrochemical grinding of WC-Co has been reported [*Levinger and Malkin, 1979*]. The initial specific etching rate of cobalt phase is higher than that of WC phase. In addition to the direct dissolution of material, the electrolysis process in ECG weakens the cermet material by selective removal of cobalt. When the in-feed rate during ECG is less than the initial specific etching rate of cobalt phase, selective etching of cobalt occurs that reduces mechanical power requirement for machining. At higher in-feed velocities, the reduction in mechanical power requirement is marginal. Study of surface roughness produced during ECG [*Geva et al.*, 1976] clearly indicates that there is selective electrochemical etching of the metal phase (Cobalt). It weakens the composite material and, thereby, reduces the forces required for mechanical grinding.

Kuppuswamy and Venkatesh [1979] conducted experimental study to investigate the **effect of magnetic field** on electrolytic grinding using diamond, SiC, and $Al_2\ O_3$ wheels. It has been found that the magnetic field in case of a diamond wheel, improves the process performance but the same is not true for $Al_2\ O_3$ wheel. In case of SiC wheel, the improvement obtained is marginal. The magnetic field interacts with the moving charged particles and may affect the rates of both mass transport and charge transfer processes [*Dash and King*, 1972]. The magneto-hydrodynamic force leads to stirring of electrolyte particularly in the neighbourhood of electrodes. It leads to enhanced mobility and hence an increased rate of electro-chemical reactions. *Gedam and Noble* [1971] concluded that fine grits and low concentration wheels show a tendency to draw more current and thereby achieve greater MRR than coarse grits and high concentration wheels.

ECG is a **cold process** (bulk temperature < 100°C), thus prevents structural damage and grinding cracks. However, electrochemically ground specimens show relatively poor fatigue strength, possibly due to stray current attack of the surface, which leaves a series of 'pits' that would act as sites for fatigue crack initiation [*DeBarr and Oliver*, 1968].

Declogging of the grinding wheel can be done by reversing its polarity (making the tool as anode) for a short period. It may, however, result in longer machining time and degeneration of the wheel-shape.

APPLICATIONS

Electrochemical grinding is economical for grinding carbide cutting tool inserts. Microscopic study of electrochemically ground surfaces of the cemented carbide do not reveal any damage to the microstructure, microcracks, or any other defects. This process is also used to reprofile worn locomotive traction motor gears. Usually wear marks from the gear tooth surfaces are removed by removing as much as 0.38 mm thick layer of material. ECG does not have any effect on gear hardness. ECG is also used for burr-free sharpening of hypodermic needles, grinding of superalloy turbine blades, and form grinding of fragile aerospace honeycomb metals.

BIBLIOGRAPHY

1. Benedict G.F. (1987), *Non-traditional Manufacturing Processes*, Marcel Dekker Inc., London.
2. Bhattacharyya A. (1973), *New Technology*, The Institution of Engineers (I), Calcutta.
3. Buttner A. and Lindenback D.A. (1969), Electrolytic dressing of diamond wheels for use in steel grinding, *Ind. Diamond Rev.*, pp. 450-454.
4. Dash J. and King W.W. (1972), Electrothinning and electrodeposition of metals in magnetic fields, *J. Electrochem Soc.*, Vol. 119, pp. 51-56.
5. DeBarr A.E. and Oliver D.A. (1968), *Electrochemical Machining*, MacDonald, London.
6. Gedam A. and Noble C.F. (1971), An assessment of the influence of some wheel variables in peripheral electrochemical grinding, *Int. J. Mach. Tool*

Des. Res, Vol. 11, pp. 1-12.

7. Kuppuswamy G. (1976), Electrochemical grinding: an investigation of the metal removal rate, *Proc.* 7th *AIMTDR Conf.* held at Coimbatore (India) pp. 337-340.
8. Kuppuswamy G. and Venkatesh V.C. (1979), Wheel parameters and effect of magnetic field on electrolytic grinding, *J. Inst. Engrs. (I), Part ME(I)*, Vol. 60, pp. 17-20.
9. Levinger R. and Malkin S. (1979), Electrochemical grinding of WC-Co cemented carbides, *ASME Paper No. 78-WA/PROD-26*, *WAM*, San Francisco, pp. 1-10.
10. McGeough J.A. (1988), *Advanced Methods of Machining*, Chapman and Hall, London.
11. Pandey P.C. and Shan H.S. (1980), *Modern Machining Processes*, Tata-McGrawHill, New Delhi.
12. Jain V.K., Tandon S. and Kumar Prashant (1990), Experimental investigation into electrochemical spark machining of composites, *Trans. ASME, J. Engg. Ind.*, Vol. 12, pp. 194-197.
13. Verma M.M. (1981), A study of process parameters in electrochemical grinding, *M. Tech. Thesis*, IIT Kanpur (India).

REVIEW QUESTIONS

(1) Why there is no need to have 'short circuit' protection device in ECG system?

(2) Explain the mechanism of material removal during ECG and how it is different from ECM.

(3) Why the life of the ECG wheel is much higher than conventional grinding wheel?

(4) Why some researchers prefer to call this process as "mechanically assisted ECG"?

(5) What are the factors responsible to improve ECG (with diamond wheel) process performance while using magnetic field?

(6) What do you understand by in-process dressing during ECG? Write its merits and demerits.

CHAPTER 12

AT-A-GLANCE

ELECTROCHEMICAL GRINDING

(ECG)

CONVENTIONAL GRINDING

- GOOD SURFACE FINISH & TOLERANCES
- PRESENCE OF BURRS, HAZ, & THERMAL RESIDUAL STRESSES
- 100 % REMOVAL BY ABRASIVE ACTION

ECG

- W/P → ELECTRICALLY CONDUCTIVE
- ELECTROLYTE RECIRCULATED
- GRINDING WHEEL BONDING MATERIAL → ELECTRICALLY CONDUCTIVE
- MECHANICAL ABRASIVE ACTION → 10 %
- ELECTROCHEMICAL DISSOLUTION → 90 %
- LONG WHEEL LIFE
- COMMON ELECTROLYTES : $NaCl$ & $NaNO_3$

3 ZONES OF IEG

ZONE 1

- PURE ELECTROCHEMICAL DISSOLUTION
- WHEEL ROTATION HELPS IN DRAWING THE ELECTROLYTE
- CONTAMINATION BY REACTION PRODUCTS & GASES
- CONDUCTIVITY CHANGES
- ELECTROLYTE TRAPPED ⇒ GRIT & W / P → ELECTROLYTIC CELL FORMATION
- SMALL AMOUNT OF MATERIAL REMOVED

ZONE 2

- HIGHER PRESSURE SUPPRESSES FORMATION OF GAS BUBBLES → HIGHER MRR
- ABRASIVE GRAINS → REMOVE W/P MATERIAL IN THE FORM OF CHIPS
- REMOVAL OF OXIDE LAYER HELPS IN EC DISSOLUTION

ZONE 3

- MATERIAL REMOVAL BY ELECTROCHEMICAL DISSOLUTION
- WHEEL STARTS LIFTING OFF THE WORK SURFACE
- REMOVES SCRATCHES & BURRS ← FORMED IN ZONE 2

- VERY HIGH VOLTAGE → SPARKING → DETERIORATES SURFACE FINISH & DAMAGES GW
- MATERIAL REMOVAL IS AS HIGH AS 10 TIMES OF CONVENTIONAL GRINDING ON HARD MATERIAL (> 65 HRC)

OPERATIONS

- CYLINDRICAL, FORM, SURFACE, FACE & INTERNAL ECG
  - CYLINDRICAL →
    - SLOWEST
    - LIMITED AREA OF CONTACT
  - FACE →
    - FASTEST
    - MAXIMUM AREA OF CONTACT

GRINDING WHEEL

- BONDING MATERIAL →
  - COPPER, BRASS, NICKEL
- DRESSING →
  - REVERSE THE CURRENT & DO ECG ON A SCRAP PIECE
  - DEPLATE THE METAL BOND EXPOSING ABRASIVE PARTICLES

APPLICATIONS

- ECONOMICAL IN GRINDING CARBIDE CUTTING TOOLS
- EC GROUND CEMENTED CARBIDE W/P → NO DAMAGE TO MICROSTRUCTURE & NO MICROCRACKS
- FOR RE–PROFILING WORN LOCOMOTIVE TRACTION MOTOR GEARS
- NO EFFECT ON GEAR HARDNESS
- BURR-FREE SHARPENING OF HYPODERMIC NEEDLES
- SUPERALLOY, TURBINE BLADES & HONEYCOMB METALS

Chapter Thirteen

ELECTROSTREAM DRILLING (ESD)

INTRODUCTION

Standard electrochemical drilling (ECD) process uses a hollow metal tube as cathode through which electrolyte flows at high velocity. This tube moves in the hole as it is drilled hence, the hole diameter is always bigger than the outside diameter of the cathode tube. There are certain practical limitations because of which the cathode tube diameter cannot be reduced below a certain value. To drill a hole of diameter smaller than this value, a process named **electrostream drilling** was invented [*Annon*, 1976]. This process was employed for drilling thousands of small cooling holes in nickel and cobalt superalloys [*Benedict*, 1987].

In electrostream drilling (ESD), electrically negatively charged, high velocity, acid electrolyte stream is passed through electrically non-conducting nozzle (Fig. 13.1). This stream strikes positively charged workpiece and removes material in the same way as in conventional ECM. This dissolved material (the sludge dissolved in the acid electrolyte) is flushed out from the machining zone in the form of metal ions in the solution. Since there is no sludge to restrict the flow of electrolyte, the limit on the minimum diameter of the hole that can be drilled is relaxed. This process can be used to drill very small holes at steep angles or curved holes (Fig. 13.2). By this method, it has been possible to drill holes of diameter as small as 0.127 mm to as large as 0.89 mm. But, the voltage used during ESD is very high (say, 150-850 V).

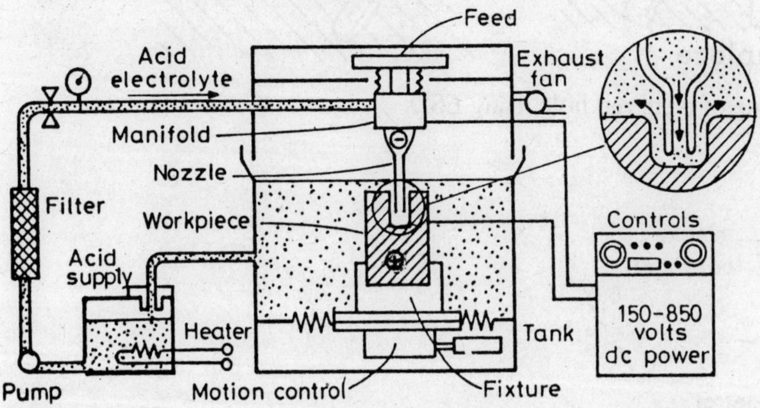

Fig. 13.1 Schematic diagram of electrostream drilling (ESD) set-up [*Benedict*, 1987].

In ESD, dilute solution of H_2SO_4 or HCl is used as **electrolyte**. Hydrochloric acid has proved to be better electrolyte for drilling materials like aluminium, titanium, etc. Sulfuric acid is the electrolyte preferred for drilling in carbon steel, cobalt alloys and stainless steel [*Bellow and Kohls*, 1982].

ESD can be performed in two ways, by giving no feed (or zero feed rate) and by providing finite feed to the nozzle. The first one is known as "**dwell drilling**" which is used when shallow, and less accurate holes are required. This technique is also employed under the circumstances when workpiece configuration or m/c capabilities do not permit movement of the nozzle. Nozzle tip is fixed at a predetermined distance from the work surface, and drilling is done by the electrolyte stream, but it limits the depth of the drilled hole and obtainable accuracy (Fig. 13.3).

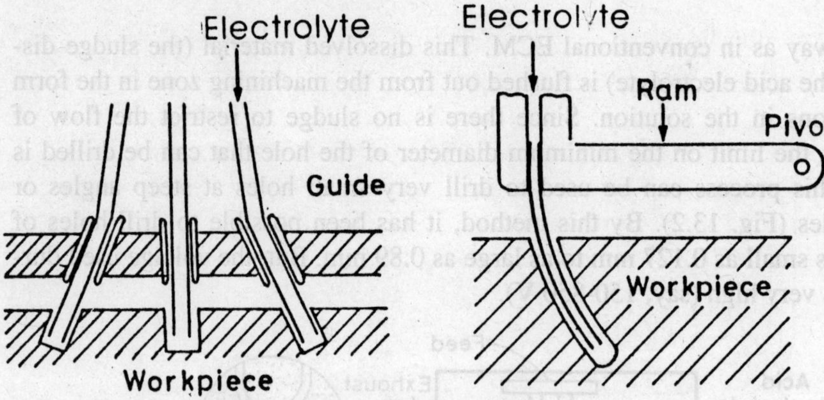

Fig. 13.2 Drilling of a curved hole using ESD.

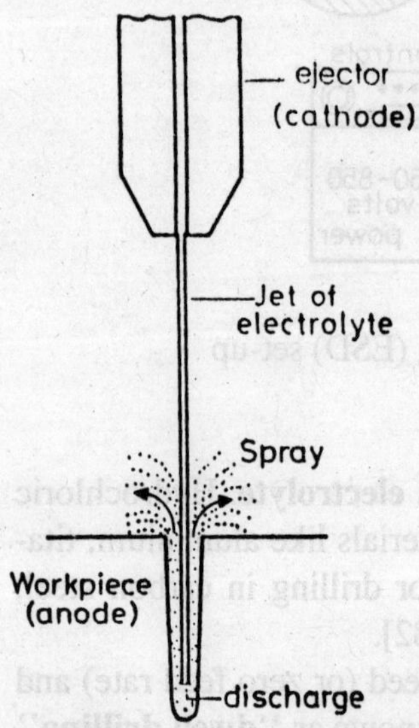

Fig. 13.3 Schematic of "Dwell drilling".

The second kind of ESD is known as "**penetration drilling**", and is used for deep and accurate hole drilling (Fig. 13.4). During ESD, the nozzle is fed towards the workpiece with a finite feed rate to maintain a constant inter electrode gap

(IEG). A gap sensing device is used to monitor the current being drawn, to slow down the feed, and trigger full power when the proper nozzle workpiece gap is detected.

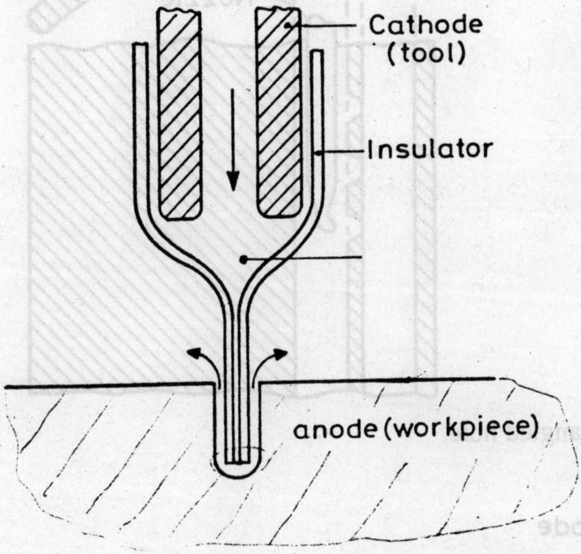

Fig. 13.4 Schematic of "Penetration drilling".

Components of **m/c and fixtures** must be made of acid resistant materials. Nozzles designed for drilling round holes are made of glass tubing that is drawn to a small diameter, thus forming a capillary at one end. If a hole in the cavity is to be drilled at such an angle that the line of sight access is not available, a tool with a right angle bend at the tip is used (Fig. 13.5).

The outside diameter of the nozzle tip should be such that it fits within the hole being drilled as well as allows room for the repeatable escape of the used electrolyte [*Bellow*, 1967]. The length of the small diameter part of the tube should always be slightly greater than the depth of the hole to be drilled.

Charging of electrolyte can be done in two ways, either by the use of a metallic sleeve or by a small titanium wire (Fig. 13.6) which is placed inside the large diameter section of the ES nozzle. Metallic sleeve or titanium wire is kept as close to the throat as possible. Multiple nozzle applications require the use of a junction manifold with individual wires running from the electrolyte manifold to each nozzle.

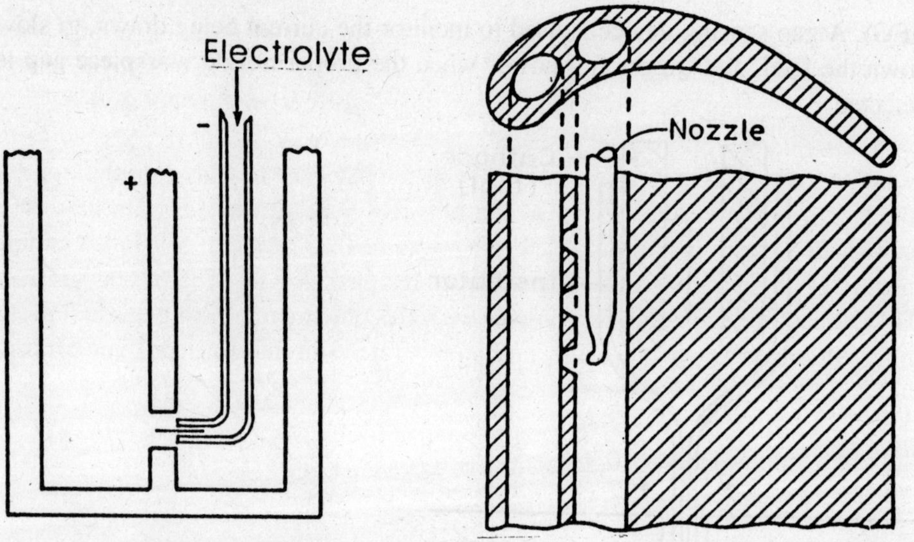

Fig. 13.5 Drilling of a right angled hole.

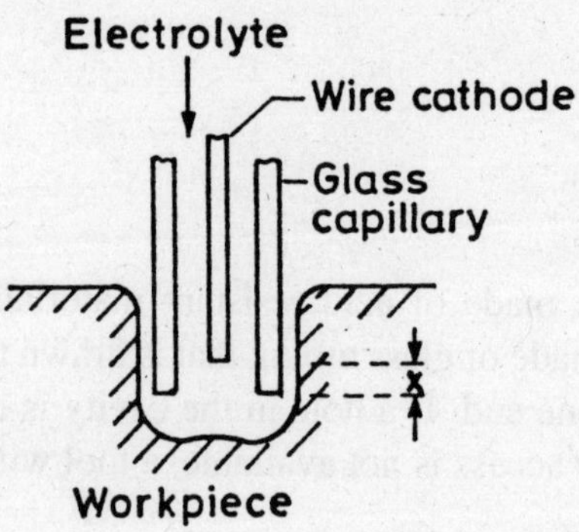

Fig. 13.6 Use of a small titanium wire to charge electrolyte.

PROCESS PERFORMANCE

The ESD process is being used for drilling holes having large depth and small diameter, ie high aspect ratio (in penetration drilling 40:1 and in dwell drilling 10:1), in any electrically conductive material without affecting its properties [*Anon, 1976*]. The maximum depth to which holes can be drilled in dwell drilling and penetration drilling are 5 mm and 19 mm, respectively. Machining rate that

that has been achieved in single nozzle or multiple nozzles (as many as 100 holes) is 1.5 mm min^{-1}. In a specific case, as many as 96 holes have been simultaneously drilled in a turbine blade in a period of just 1.5 min [*Bellow et al.*, 1982]. By penetration drilling, shallow holes up to an angle of 75° from the normal to the work surface, tolerance up to ± 5% of the diameter of the hole (but no less than 0.025 mm), and surface finish in the range of 0.25 μm to 1.67 μm have been achieved. Taper can be controlled to a value of 0.03 mm/cm. Machined component has burr free surfaces with no induced residual stresses. There is insignificant HAZ (heat affected zone) and no tool wear. Electrostream drilling has bell mouth hole entrance. To increase the production rate, simultaneous drilling of small holes is done.

BIBLIOGRAPHY

1. Annon (1967), Liquid stream drill holes in tough metals steel, *Steel*, April 24, pp. 49-62.
2. Benedict G.F., (1987), *Non-traditional Manufacturing Processes*, Marcel Dekker Inc., London.
3. Bellow G. (1967), New Non-conventional machining processes, *Am. Soc. Tool Manuf. Engg.*, Paper No. MR, pp. 67-141.
4. Bellow G. and Kohls J.B. (1982), Drilling without drills, *Am. Machinist*, pp. 173-188.

REVIEW QUESTIONS

1. Explain the working principle of ESD. In what respects it is different from conventional ECM?
2. Write a major difference between dwell drilling and penetration drilling.
3. In your opinion, what is the most interesting application of ESD?
4. What do you understand by charging of electrolyte.

CHAPTER 13
AT-A-GLANCE
ELECTROSTREAM DRILLING (ESD)

- IN CONVENTIONAL ELECTROCHEMICAL DRILLING ⇒ HOLLOW METAL TUBE AS CATHODE
  - ELECTROLYTE FLOWS AT HIGH VELOCITY
  - CATHODE PENETRATES IN HOLE
  - HOLE DIAM ≥ CATHODE OUTSIDE DIAM
- TO DRILL A HOLE OF SMALL DIAM. → ESD
  - SMALL COOLING HOLES IN SUPERALLOYS
  - HIGH VELOCITY NEGATIVELY CHARGED ACID ELECTROLYTE STREAM
  - 0.127 mm <HOLE DIAM < 0.890 mm
  - VOLTAGE → 150 - 850 V
  - ESD ⇒ ZERO FEED RATE ⇒ DWELL DRILLING ⇒ SHALLOW & LESS ACCURATE HOLES

↓

- LIMITED DEPTH OF HOLE
- W/P CONFIGURATION, OR M/C CAPABILITIES DO NOT ALLOW NOZZLE MOVEMENT
- MAXIMUM DEPTH → 5 mm

↓

- FINITE FEED RATE (PENETRATION DRILLING)

↓

- DEEP & ACCURATE DRILLING
- CONSTANT GAP USING FINITE FEED RATE
- A GAP SENSING DEVICE TO MAINTAIN CONSTANT GAP
- MAXIMUM DEPTH ⇒ 19 mm
- MACHINING RATE → 1.5 mm/min
- NO BURR
- NO RESIDUAL STRESSES
- INSIGNIFICANT HAZ
- NO TOOL WEAR
- BELL MOUTH HOLE ENTRANCE
- MULTIPLE HOLE DRILLING

- M/C & FIXTURES MADE OF → ACID RESISTANT MATERIAL
- NOZZLE AND GLASS TUBING
- A TOOL WITH RIGHT ANGLE BEND ALSO
- LENGTH OF TUBE > DEPTH OF THE HOLE
- CHARGING OF ELECTROLYTE
  ~ METALLIC SLEEVE
  OR
  ~ TITANIUM WIRE PLACED INSIDE TUBE.
- MULTIPLE NOZZLE APPLICATION

↓

JUNCTION MANIFOLD
|
|
——> INDIVIDUAL WIRES RUNNING TO EACH NOZZLE

- H_2SO_4 / HCl → ELECTROLYTE
  |
  |
  ——> Al, Ti, ETC.
  * CARBON STEEL
  * COBALT ALLOYS
  * STAINLESS STEEL

Chapter Fourteen

ELECTROCHEMICAL DEBURRING (ECDe)

INTRODUCTION

A designer during the design phase of a component usually considers the aspects like material, form, dimensional accuracy, surface texture and heat treatment, but not the surface integrity and edge quality [*K. Takazawa*, 1988]. However, the last two factors are very important from the point of view of performance and the life of the product. Fig. 14.1 shows the concepts of surface technology. In this chapter, some discussion on how to achieve the desired edge quality (shape, dimension, tolerance and surface roughness) through electrochemical deburring is presented.

When a component is processed by a conventional machining method, usually, it is left with burrs specifically along the two intersecting surfaces. Such burrs are undesirable from the viewpoint of performance of a component as well as safety of an operator, or for that matter whosoever works with this component. Such burrs can be removed by one of the deburring processes. Deburring is an important phase for manufacturing quality products, especially in large scale industries. The problems of burrs are still persisting and unsolved in many industries. Attempts are made to reduce burr level by various means, viz. by fixing speed, feed rates, and tool life, but this could not be achieved for certain reasons. Hence, the quality of the products is affected.

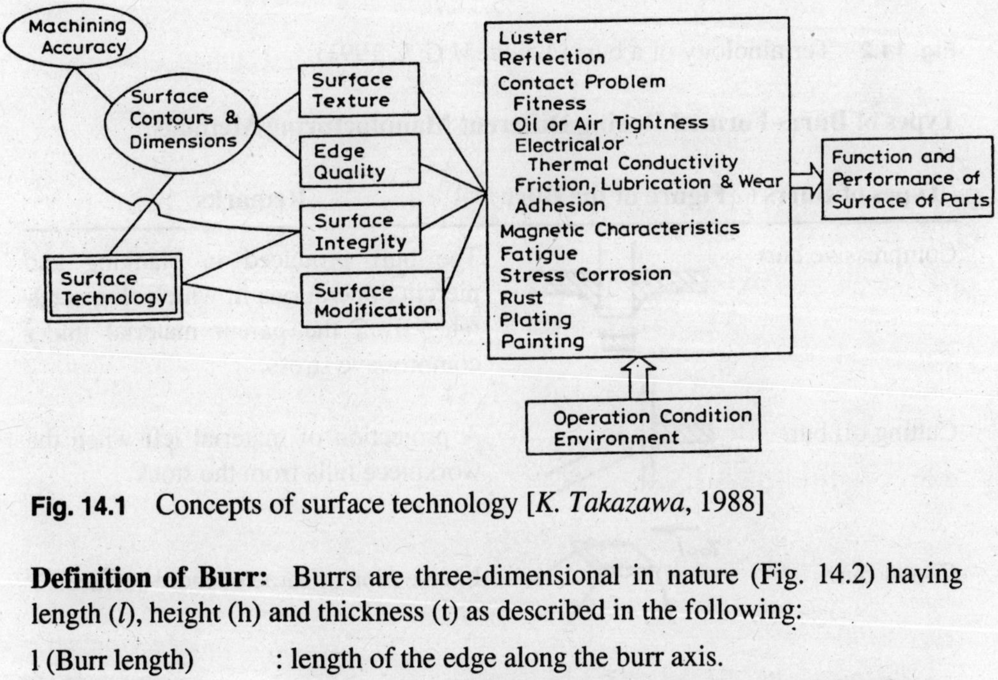

Fig. 14.1 Concepts of surface technology [*K. Takazawa*, 1988]

Definition of Burr: Burrs are three-dimensional in nature (Fig. 14.2) having length (*l*), height (h) and thickness (t) as described in the following:

l (Burr length) : length of the edge along the burr axis.

h (Burr height) : distance of the burr projected above the parental surface.

t (Burr thickness) : thickness or width of the burr where it joins the parental surface.

Burr Hardness : Hardness of the burr in the vicinity of the base of the burr.

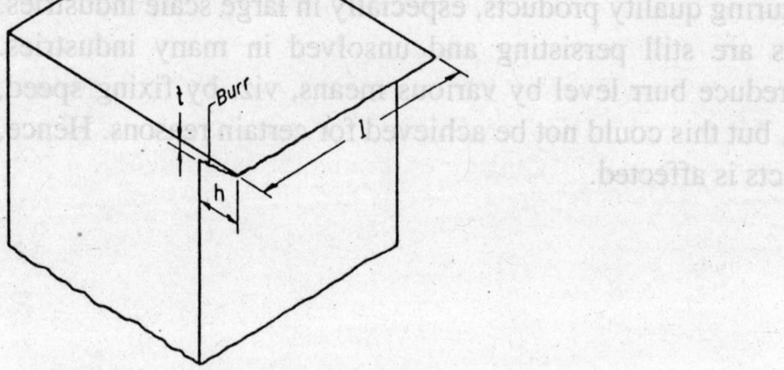

Fig. 14.2 Terminology of a burr [*Naidu, M.G.J.*, 1991].

Types of Burrs Formed During Different Manufacturing Methods:

Types of Burrs	**Figure of the Burr**	**Remarks**
Compressive burr		The burr produced in blanking and piercing operations in which slug separates from the parent material under compressive stress.
Cutting off burr	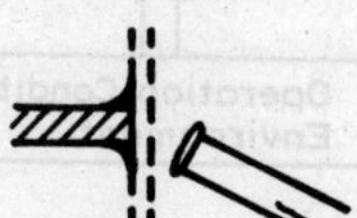	A projection of material left when the workpiece falls from the stock.
Corner burr	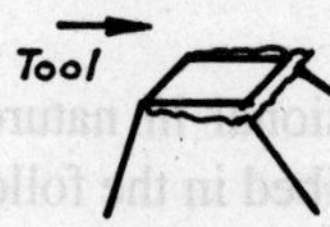	Intersection of three or more surfaces.
Edge burr		Intersection of two surfaces.

Types of Burrs	Figure of the Burr	Remarks
Entrance burr		Cutting tool enters in the workpiece.
Exit burr		Cutting tool exits the workpiece.
Feather burr		Fine or thin burr.
Flash burr		Portion of flash remaining on the part after trimming.
Hanging burr		Loose burr not firmly attached to the workpiece.
Roll over burr	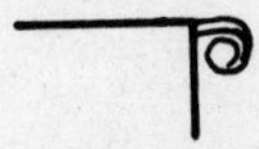	Burr formed when it exits over a surface and allows the chips to be rolled away.
Tear burr		Formed from the sides of the tool as the tool tears the edge.

Basic Approach on Deburring

In the modern industrial technology, the deburring process has attained great importance because of rigid quality standards.

In analyzing the specific situations, one has to know 'why deburring is required?' Burrs, to some extent, can be reduced by controlling cutting conditions, but cannot be eliminated completely. Hence, deburring becomes inevitable. However, deburring cost should not be very high. This requires careful considerations regarding effects of the presence of burrs on functional, physiological and aesthetic requirements [*Naidu, 1991*], as outlined in the following chart.

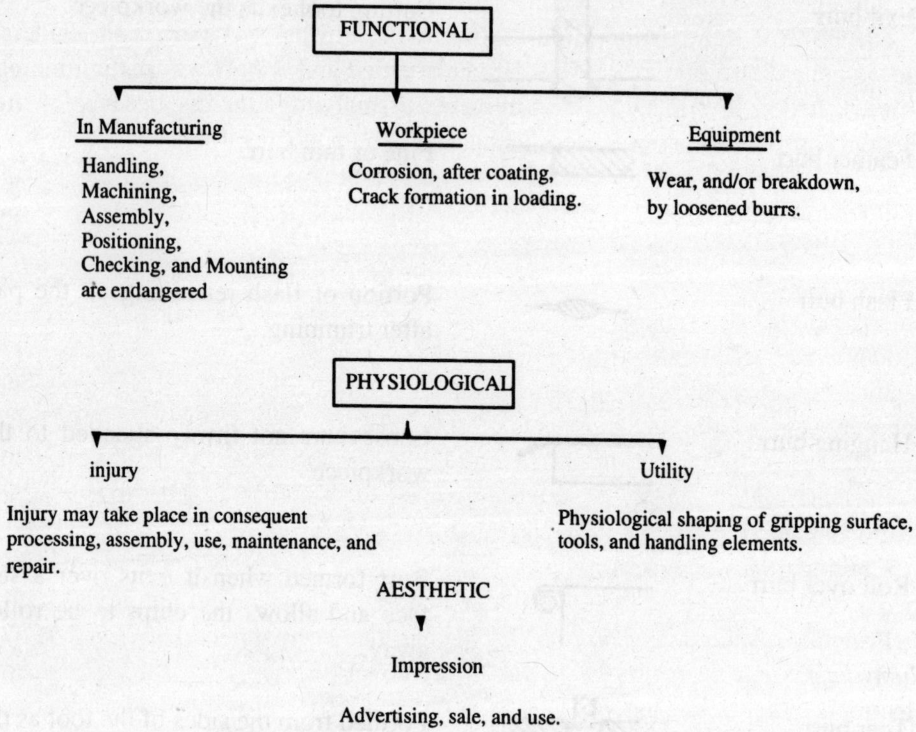

CLASSIFICATION OF DEBURRING PROCESSES

Deburring processes can be classified (Fig. 14.3) as:

1. Mechanical deburring,
2. Abrasive deburring,
3. Chemical and Electro-chemical deburring, and
4. Thermal deburring.

Mechanical deburring, using cutting tools, brushes, scrapers, belt sanders, etc, is generally unreliable as it is labour oriented and is only a burr-minimizing process. It does not meet the requirements of high edge quality because of the existence of fine burrs even after mechanical deburring operation.

In **abrasive deburring** like tumbling, barrel finishing, vibratory deburring, liquid abrasive flow, sand blasting,etc, the selection of abrasive medium, its shape and quantity play an important role. Deburring by this method generally affects other areas on the component where deburring is not required and has limitations on edge quality. The reliability, uniformity and MRR of these processes are low, and tend to charge the workpiece with grits.

In **thermal deburring** (TDe), components to be deburred are placed in the deburring chamber. The chamber is closed and filled with combustible gas mixture of oxygen and hydrogen. After ignition by an electric spark, the gas burns in few milliseconds and the temperature attained is over 3500°C. Due to this short heat wave, burrs and sharp edges on the component burn away.

Chemical deburring is the process where the burrs are dissolved in chemical media. It also may affect the areas where material removal is not required.

Principle of *anodic dissolution* (ECM) has also been applied [*Benedict*, 1987; *Rumyantsev and Davydev*, 1989] for the removal of burrs. However, in case of electrochemical deburring (ECDe) as compared to ECM, magnitude of current, electrolyte flow rate, and electrolyte pressure are all low. Secondly, tool is stationary. Fig. 14.4 shows an ECDe system.

Electrochemical deburring is generally employed for far away located as well as inaccessible places where other deburring processes are not effective. This process involves the use of flowing electrolyte for conducting electric current for the electrochemical reaction to take place. The current rating and duration of the current flow to suit a particular component are determined after extensive trials for each type of the component. The electrolyte commonly used is either sodium chloride or sodium nitrate. Because of the corrosive nature of the electrolyte

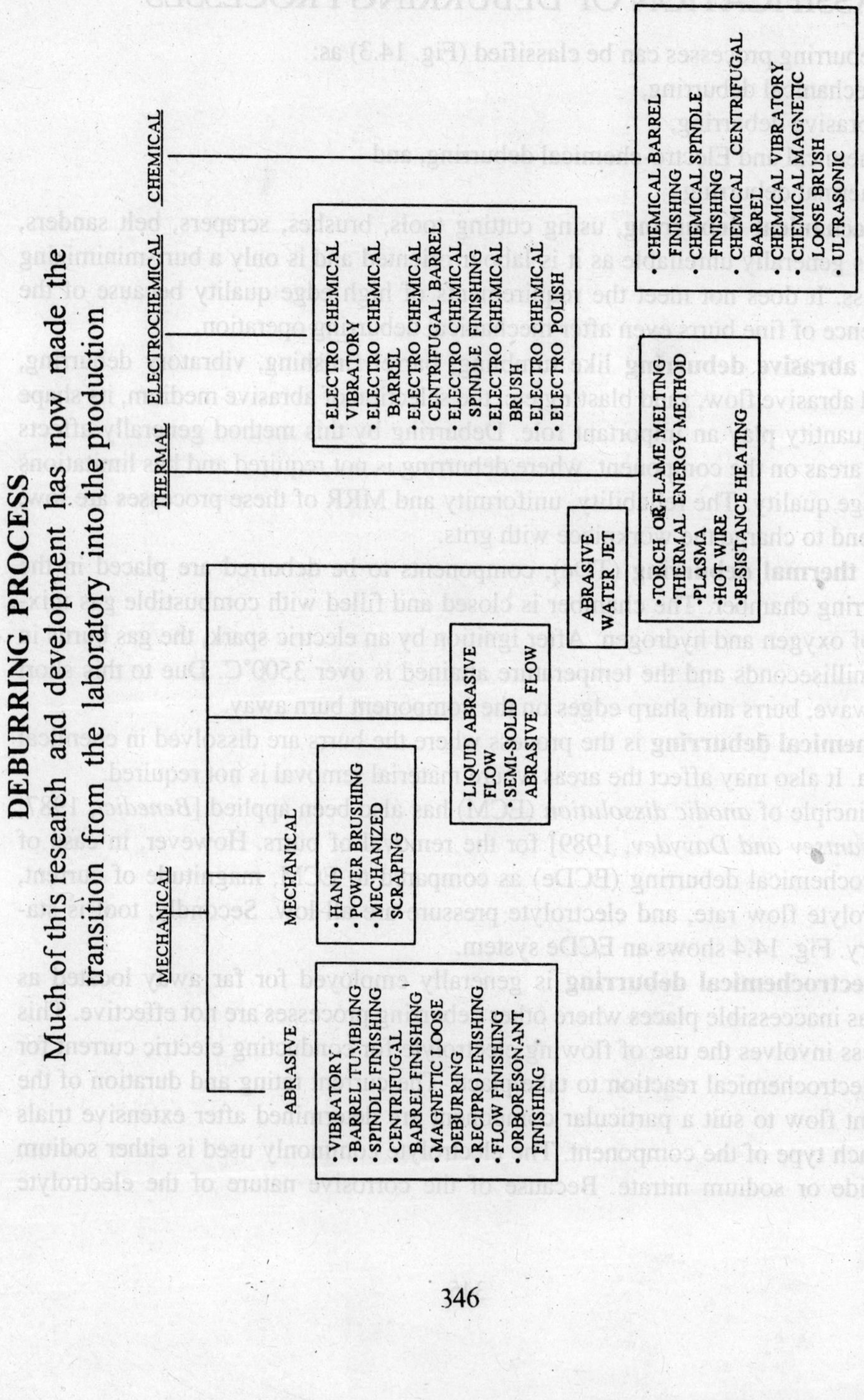

Fig. 14.3 Classification of deburring processes.

and ferrous hydroxide released by the process, machines are built with non-corrosive materials.

Electro polishing deburring (EPDe) is different from EC deburring. The metal removal and polishing simultaneously take place on all the surfaces in EPDe, whereas in ECDe, the metal removal is localised. In this process, a component which has roughness and some burrs, is subjected to selective electrochemical attack. As a result, microprofile is smoothened and levelled.

Selection of a process depends on the edge quality and other requirements of the component. Most of the modern industries are switching over to modern technologies like ECDe, EPDe and TDe due to obvious advantages including cost savings.

ELECTROCHEMICAL DEBURRING (ECDe)

This process has been tried out successfully on contours where the conventional deburring tools can not be used. The performance of the process is improved with higher current intensities. It requires specially engineered equipment for its use as manufacturing unit. It consists of:

1. Electrolyte system which provides high velocity to the electrolyte flow,
2. Electrical power system which supplies the electrolyzing current,
3. Mechanical structure which locates and provides movement/mounting to the electrodes, and
4. Separator which separates the sludge.

Principle of working

When a voltage is applied between two metal electrodes immersed in an electrolyte, current flows through the electrolyte from one electrode to the other. Unlike the conduction of electric current in the metals in which only the electrons move through the structure of the material, 'ions' (electrically charged groups of atoms) physically migrate through the electrolyte. The transfer of electrons between the ions and electrodes completes the electrical circuit and also brings about the phenomenon of metal dissolution at the positive electrode or anode (workpiece). Metal detached atom by atom from the anode surface appears in the main body of the electrolyte as positive ions, or as precipitated semi-solid of the metal hydroxide, which is more common in electrochemical deburring process.

The **tool** is usually insulated on all surfaces except a part which is adjacent to the burr(s). Instead of insulated tool, a **bit type of tool** [*Jain and Pandey*, 1982] can also be used. The electrolyte is made to flow through inter electrode gap. However, setting of dimensions of the bare part of the tool, time of machining, and other machining conditions are all decided by '*trial and error*' method. The inter electrode gap (IEG) is usually kept in the range of 0.1-0.3 mm. The deburring tool-tip should normally overlap with the area to be worked by 1.5 to 2.0 mm.

ECDe machine tool (M/T) is usually designed with multiple work stations served from a single power supply. The electrolyte is properly filtered out before its re-circulation. Criteria for selection of tool material used is the same as for general ECM.

It is considered as a high-tech method when compared to the conventional methods of deburring. Before applying ECDe method for a particular type of job, one should know about thickness, shape, and repeatability of burrs on the job in hand. Almost identical shape and size of the burrs should be on the job otherwise efficient burr removal may not take place. Further, the part of the tool supposed to remove burrs should be shaped as a replica of the contour of the work. Fig. 14.5 shows a tracing from the micrograph of a sample deburred electrochemically.

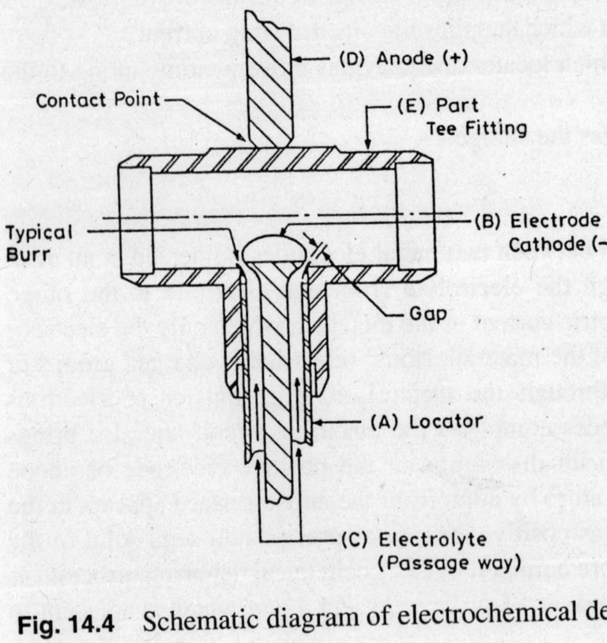

Fig. 14.4 Schematic diagram of electrochemical deburring.

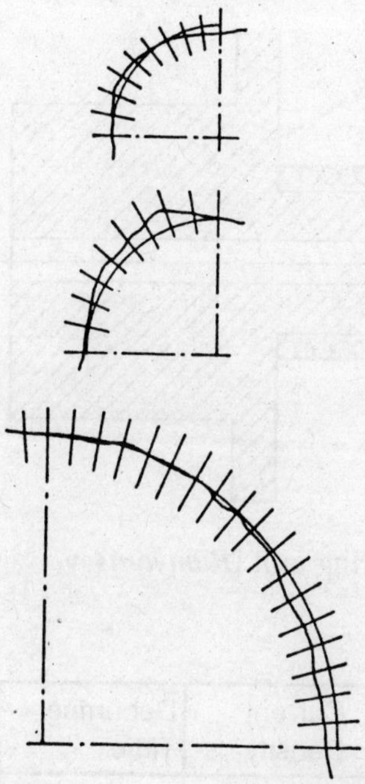

Fig. 14.5 Tracing from the micrograph of an electrochemically deburred samples [Rumyantsev and Davydev, 1989].

In some cases, deburring can also be done with the help of a **movable ECD$_e$** unit [*Rumyanstev and Davydov*, 1989] consisting of deburring gun, electrolyte supply tank, and power supply unit (≈ 50 A max. current). The deburring gun is supplied with electrodes of different diameter and length so that it can be used to deburr hard-to-get places. However, in some cases, tool of varying length may be needed. For this purpose, a *flexible tool* is used which consists of wire coiled into a closely wound spring. The details of the tool are shown in Fig. 14.6 (1-insulating spherical tip, 2-copper tip with a hole for electrolyte supply, 3-spring, 4-PVC sleeve, 5-copper shank press fitted to the spring. The flexible electrode is attached to the gun by means of shank 5). The deburring gun is also used to deburr edges of sheet metals. Deburring speed may be as 400-500 mm/min.

The data in the following table may be useful for practical deburring purposes:

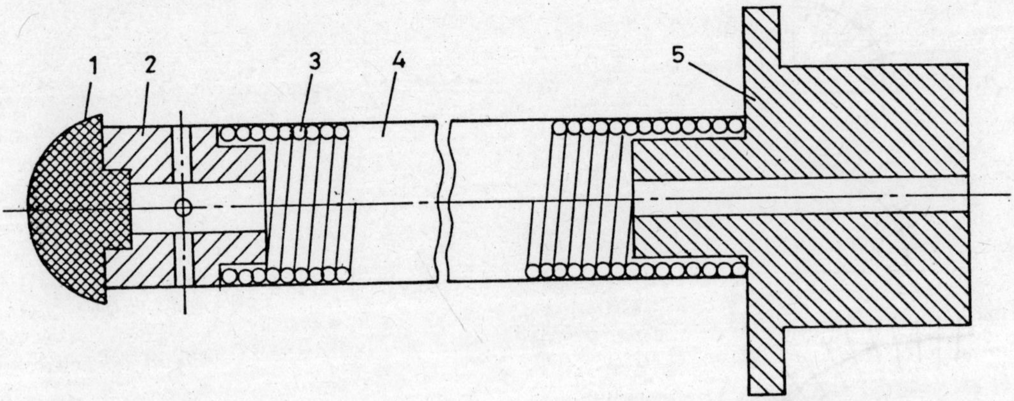

Fig. 14.6 A tool for a movable electrochemical deburring unit [*Rumyantsev and Davydev*, 1989].

Material	Electrolyte	Applied Voltage	Current Density	Deburring Time
Carbon and low carbon steel	5–15% $NaNO_3$ + 2–5% $NaNO_2$			
Copper alloys	5–15% $NaNO_3$	12–24V	5–10 A/cm^2	5–100 s
Aluminium alloys	15–20% $NaNO_3$			
Stainless steels	5% $NaNO_3$+NaCl			

ECDe is found suitable for removing burrs from tubes and pipes widely varying in configuration, length and cross-sectional area.

Functions of Electrolyte and its Importance

Depending upon the requirements of the process, sodium chloride (NaCl) and sodium nitrate ($NaNO_3$) are generally used as electrolytes. The other electrolytes like hydrochloric acid, potassium chlorate, etc. have certain disadvantages in the process application. Sodium nitrate and sodium chloride have certain variations in usage [*Naidu*, 1991] which are as follows:

Details	Sodium Nitrate	Sodium Chloride
1. Voltage requirements	High	Low
2. Reaction	Normal	Aggressive slow
3. Increase of pH value	Fast	Slow
4. Machining efficiency		
• Low carbon steel	60–80%	80–100%
• High carbon steel	Good	Poor
• Aluminium	Good	Poor
5. Machined surface		
• Roughness	About 5 μm	About 1 μm
• Colour	Dark grey	Bright grey
• Edges	Sharp & distinct	Blunt & dull
• Dimensional accuracy	Accurate	Not accurate
• Reach	Up to 1 mm	Up to 5 mm
• Effect on surrounding surface	Limiting stray machining and pitting of adjacent areas	Removes material from surrounding areas and pitting damages adjacent areas.
6. Cost	Costly	Not very expensive.

The deburred component shows a **localized deposit** (dark grey) which is a reaction product of the process. The composition of the reaction product while deburring ferrous component is Fe_3O_4. It is a magnetic oxide of iron and its thickness is less than 1 μm. The adhesion of the deposit is very strong and it can be removed only mechanically. In specific cases, this deposit can be removed from unhardened components (unalloyed or low alloyed steels) as follows:

1. Anneal under air flow for more than 40-60 min at 430°C,
2. Pickle in hydrochloric acid for 1/2 min,
3. Anodically clean in alkaline bath for about 2 min.

In most of the cases, the deposit is electrolytically conductive and does not come in the way of galvanic deposits. This deposit disappears during heat treatment of the components.

APPLICATIONS

ECDe has applications in *industries* like consumer appliances, biomedical, aerospace, automobile, etc. ECDe is used for the components like gears, splines, drilled holes and milled components, fuel supply and hydraulic system components, etc. It is very successful even in the situations where two holes cross each other like crank shaft. Apart from economics of the process, it gives higher reliability, reduced operation time, and more uniformity. This process can be automated in an easier way than other processes. Fig. 14.7 shows a gear before and after deburring.

SPECIFIC FEATURES OF ECDe M/C

Specific features of an ECDe M/C are summarized as follows:

1. *Application*: Removal of thin and thick burrs at inaccessible and irregular areas within the restricted zone, and wherever edge rounding is required.
2. *Principle*: High velocity electrolyte (sodium chloride/sodium nitrate) is passed through the IEG between the tool-cathode and the component-anode connected to the electric potential. The undesired superfluous projections on the component are dissolved electrochemically within a preset cycle time of a few seconds. The dissolved burrs in the form of hydroxides settle down and the electrolyte is regenerated. The hydroxide is disposed through outlet drain.
3. *Equipment*: Consists of specially designed DC-power pack, working station, storage tank, and necessary controls. All the controls and safety requirements are interlocked in a logic system.
4. *Capacity*: It is decided *based on the area of deburring* and production quantity. Approximately 200 mm diameter gear with the tooth profile and spline slots are deburred within 45-60 s cycle time on a 500 A capacity machine.
5. *Consumables*: A 500 A capacity machine requires approximately 20 kg of electrolyte salt per week, working in two shifts. The cathode electrode wears out at the insulating areas where the component has heavy burrs, hence the life of the electrode is approximately 3000-25000 *components per electrode*.

The cost of such electrode varies from Rs. 1000 to 5000.

6. *Adoption*: It can be adopted to any material through the selection of an appropriate electrolyte. Generally sodium nitrate ($NaNO_3$) is used on steels, cast iron and aluminium for a precise deburring, and sodium chloride (NaCl) is used on steels for an aggressive deburring.

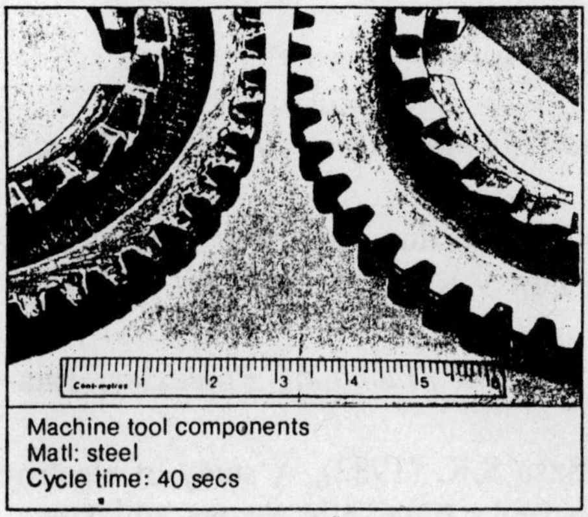

Fig. 14.7 A gear before and after deburring [*Courtesy*, *EleChem Technik*, Bangalore].

7. *Pollution*: Electrolyte is a domestic salt and solution is free from health hazard. **The hydroxide removed from the drain valve is extensively being used as a raw material for the lapping paste.**
8. *Quality*: The component having the burr root thickness less than 0.2 mm and the height less than or equal to 1 mm, can be precisely deburred by maintaining the edge rounding from 0.2–0.5 mm radius, and the cycle time for approximately 20–30s. If the burr size is larger than the above stated one, the edge quality after deburring would be of approximately 0.5–1.5 mm radius and cycle time required will be 60-90s depending upon the burr level. There will be discolouration at the adjacent area but no material removal; however, as the cycle time increases more than 30s, there will be slight material removal (approximately 0.05 mm) on a width of 1-2 mm adjacent to the deburred area. It can be controlled through tooling, if desired.
9. *Components Cleaning*: Before deburring, the components should be free from loose burrs which damage the electrodes, and from grease/oil which

contaminates the electrolyte. Hence, it is preferred to thoroughly wash the components just before deburring. After deburring, it should be dipped immediately in running water followed by dewatering fluid, thus protecting against corrosion.

Deburring is advised to be done usually before any surface treatment. There will be no hydrogen embrittlement.

BIBLIOGRAPHY

1. Benedict G.F. (1987), *Non-traditional Manufacturing Processes*, Marcel Dekker Inc., New York.
2. Naidu M.G.J. (1991), Electrochemical deburring for quality production, *Proc. Winter School on AMT* (Eds.: Jain V.K. and Choudhury S.K.), IIT Kanpur (India).
3. Pandey P.C. and Shan H.S. (1980), *Modern Machining Processes*, Tata-McGraw Hill, New Delhi.
4. Pramanik D.K., Dasgupta R.K. and Basu S.K. (1982), A study of electrochemical deburring using a moving electrode, *Wear*, Vol. 82, pp. 309-316.
5. Rumyantsev E. and Davydev A. (1989), *Electrochemical Machining of Metals*, Mir Publishers, Moscow.
6. Takazawa K. (1988), The challenge of burr technology and its worldwide trends, *Bull. JSPE*, Vol. 22, No. 3, pp. 165-170.
7. *Tool Design Manual for Electrochemical Deburring*, Chemtoal Inc., Minnesota, U.S.A.

REVIEW QUESTIONS

Q1. (i) From which viewpoints it is important to have a deburred component?
(ii) Which type of burr is expected in the following type of machining operations: drilling, milling, turning and shaping?
(iii) How the presence of burrs at the intersecting holes will affect the functional requirements of a given component?
(iv) How can you protect the areas where dissolution of the work material during ECDe, is not desirable?
(Hint: See chemical machining).

Q2. Write names of the two types of components from each of the five different industries, in which cases ECDe can be applied.

Q3. What is the usual feed rate in case of ECDe?

ACRONYMS

ECDe	Electrochemical deburring
ECM	Electrochemical machining.
EPD_e	Electropolish deburring
IEG	Interelectrode gap
M/C	Machine
M/T	Machine Tool
TDe	Thermal Deburring

CHAPTER 14
AT-A-GLANCE
ELECTROCHEMICAL DEBURRING
(ECDe)

- MACHINING OF MATERIALS BY CONVENTIONAL METHODS → BURRS SPECIFICALLY AT INTERSECTING SURFACES

⇓

UNDESIRABLE
- ⟶ PERFORMANCE OF COMPONENTS
- ⟶ SAFETY OF OPERATORS / USERS

- REMOVAL OF BURRS
  - MECHANICAL → TUMBLING, BLASTING, ETC → NON-UNIFORMITY, LOW MRR, CHARGED WITH GRITS, ETC
  - THERMAL
  - ELECTROCHEMICAL → BASED ON ECM PRINCIPLE
- ECDe TOOL
  - INSULATED EXCEPT A PART ADJACENT TO BURRS
  - DIMENSIONS OF BARE PART OF TOOL, MACHINING TIME & MACHINING CONDITIONS
- ECDe M/C TOOL
  - MULTIPLE WORK STATIONS
  - SINGLE POWER SUPPLY
  - ELECTROLYTE IS RECIRCULATED AFTER FILTRATION
  - FOR PROPER DEBURRING → BURR CHARACTERISTICS SHOULD BE KNOWN

APPLICATIONS

- BIOMEDICAL, CONSUMER APPLIANCES, AEROSPACE, AUTOMOBILE, ETC
- COMPONENTS → GEARS, SPLINES, DRILLED HOLES, ETC
- HIGHER RELIABILITY, REDUCED OPERATION TIME & MORE UNIFORMITY

Chapter Fifteen

SHAPED-TUBE ELECTROLYTIC MACHINING (STEM)

INTRODUCTION

For drilling smaller diameter but deep holes in electrically conductive materials, shaped-tube electrolytic machining (STEM) process is used [*Benedict*, 1987]. Shaped-tube electrodrilling is quite commonly done in aerospace industries and difficult-to-machine superalloys. It was developed primarily for drilling high aspect ratio (up to 300:1) round and shaped holes in turbine engine airfoils. It uses acid based electrolyte which ensures that the reaction products formed during electrolytic deplating are dissolved and carried away as metal ions. It eliminates clogging of the electrolyte flow path around the electrode. STEM uses an electrode as a hollow, shaped-tube covered with insulating coating on all exterior surfaces except at the tip [*Bellow and Kohls*, 1982].

Fig. 15.1 shows different **elements** *of a STEM system*. STEM is a low-voltage process (5-15 V DC). At higher voltages, drilling rates can be increased but boiling of electrolyte, plating of electrodes, and damage to the electrode coating may lead to serious problems. Feed rate depends on the machining parameters; however, it can be as high as 5 mm/min against a normal feed rate of 1.5 mm/min. For clarity, magnified view of overcut is shown in Fig. 15.1. Value of the overcut can be changed by varying the applied voltage and electrolyte flow velocity with the help of computer control. During **STEM**, taper obtained is 0.015 mm/cm and surface finish achieved is 0.8 to 3.1 μm. Surface finish depends upon work material and machining parameters.

Electrolyte normally used is 10% concentration of sulphuric acid or hydrochloric acid in water. Temperature of electrolyte is maintained between 37-40°C. To maintain a constant gap between tool and workpiece, a servo system is used. After drilling is over, the workpiece should be thoroughly washed by water to neutralize and remove residual acid. Tube material is pure titanium to resist acidic action of the electrolyte. Coating on the tube serves dual purpose, eg it saves from the attack of the electrolyte as well as eliminates stray cutting at the walls of the hole. Best results are obtained if the tip is dressed at an angle of 10°. Pinhole scratches and delaminations in the coating would result in stray cutting (i e non-uniform cutting). Multiple electrodes even with varying shapes and sizes, would be able to produce as many as 100 holes simultaneously. The components of the M/C which are likely to come in contact with electrolyte, should be made of corrosion resistant materials like titanium, stainless steel, plastics, and ceramics. Proper ventilation of machining chamber must be provided to extract corrosive electrolyte mist and hydrogen gas byproducts of the process.

Material removal is performed basically by the foot (or tip) of the tube, hence by changing its shape, the holes of different shapes can be obtained without changing the shape of the full tube. To increase productivity and to make the process more versatile, these tips may be made of different shapes and sizes even in multiple hole machining. An alternative method is to make these tips as independent, separate, **bit type of tube** of the desired shape and size, and then fasten (by threading or otherwise) them to the bit holder (or electrode) [*Jain and Pandey*, 1982]. Electrolyte mist and hydrogen gas byproducts of the process should be properly removed from the machining chamber.

BIBLIOGRAPHY

1. Benedict G.F. (1987), *Non-traditional Manufacturing Processes*, Marcel Dekker Inc., New York.
2. Bellow G and Kohls J.B. (1982), Drilling without drills, *Am. Machinist*, pp. 173-188.
3. Jain V.K. and Pandey P.C. (1982), Investigations into the use of bit as a cathode in ECM, *Int. J. Mach. Tools Des. & Res.*, Vol. 22(4), pp. 341-352.

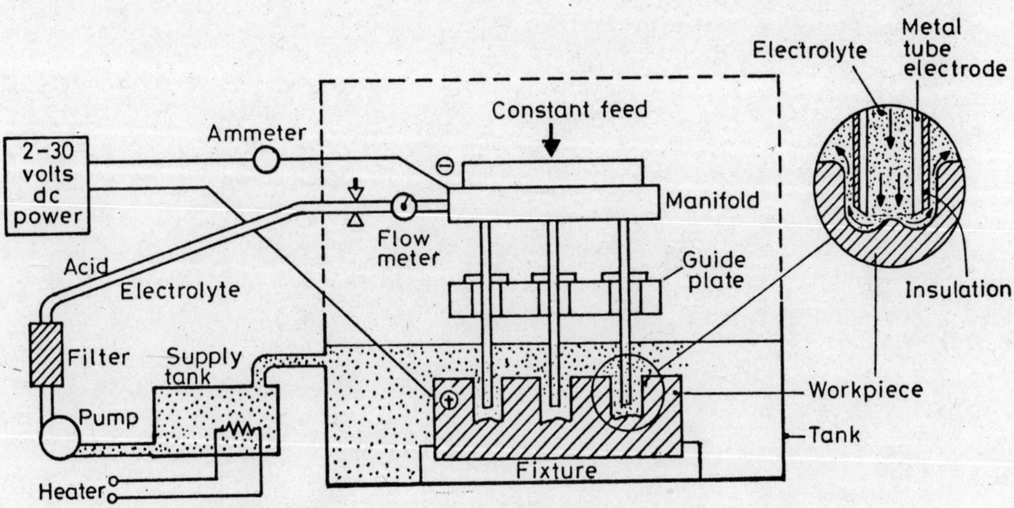

Fig. 15.1 Schematic diagram of a shaped-tube electrolytic machining system.

ACRONYMS

DC	Direct current
M/C	Machine
STEM	Shaped-tube electrolytic machining

CHAPTER 15

AT-A-GLANCE

SHAPED - TUBE ELECTROLYTIC MACHINING

(STEM)

- SMALL DIAMETER DEEP HOLES IN ELECTRICALLY CONDUCTIVE MATERIALS
- HIGH ASPECT RATIO ⇒ ROUND & SHAPED HOLES
- SUPERALLOY MACHINING
- ACID-BASED ELECTROLYTE
- REACTION PRODUCTS ⇒ DISSOLVED
- SHAPED TUBE COATED ON EXTERIOR SURFACE EXCEPT TIP
- LOW VOLTAGE ⇒ 5 - 15 V
- AT HIGHER VOLTAGE ⇒ HIGHER MRR, DAMAGE TO ELECTRODE COATING, AND BOILING OF THE ELECTROLYTE.
- OVERCUT CONTROL ⇒ APPLIED VOLTAGE & ELECTROLYTE FLOW VELOCITY
- TAPER OBTAINED → 0.015 mm/cm AND SURFACE FINISH ⇒ 0.8 - 3.1 μm

↓

WORKPIECE & MACHINING PARAMETERS

- ELECTROLYTE ⇒ 10% H_2SO_4 + WATER.

TEMPERATURE : 37 - 49 °C.

- COATED TUBE (TOOL) ⇒ ELIMINATES STRAY CUTTING

↑

PIN HOLES, SCRATCHES & DELAMINATION OF COATING

- MULTIPLE ELECTRODES ⇒ 100 HOLES SIMULTANEOUSLY
- PROPER EXHAUST SYSTEM ⇒ EXTRACT CORROSIVE ELECTROLYTE MIST & H_2 GAS
- TIP REMOVES MATERIAL ⇒ CHANGE ITS SHAPE RATHER THAN FULL TUBE

Chapter Sixteen

CHEMICAL MACHINING (ChM)

INTRODUCTION

In ancient days, this process was used by artists for engraving the metals. However, they were unaware of how the process worked. Chemical machining (ChM) is a process used to remove material by dissolution in a controlled manner, from the workpiece by application of acidic or alkaline solution (i e *etchant*). *Maskants* (chemically resistant coatings) are used to cover the surfaces which are not to be machined. Maskants are the materials which do not allow etchants to penetrate through to reach the work material to dissolve. This technique is quite useful for producing complex configurations in delicate parts that otherwise would get damaged by the application of forces in case of the conventional machining processes. This process is used in many industries, viz aviation industries for making aircraft wing panels, printed circuit boards (PCB), jewelery, etc.

It is also used for the manufacture of very thin laminations without burrs. *Other* **applications** include large turbine engine containment rings [*Benedict*, 1987], milling of spun aluminium pressure vessel bulkheads, and also shuttle components. It is also an useful process for finishing of parts to close tolerances and to achieve precision fits. Some sheet metal components are manufactured by ChM in place of welding or riveting.

Chemical milling and chemical blanking are the two most common versions of ChM. During *chemical milling*, the material is removed to produce ''blind'' details (pockets, channels, etc) or to reduce the weight. During *chemical blanking*, the details that usually penetrate the material entirely (holes, slots etc) are produced. This is also the process of blanking complete parts from the sheet by chemically etching the periphery of the desired shape. Fig. 16.1 shows a set-up used in chemical machining which involves various steps listed in the following.

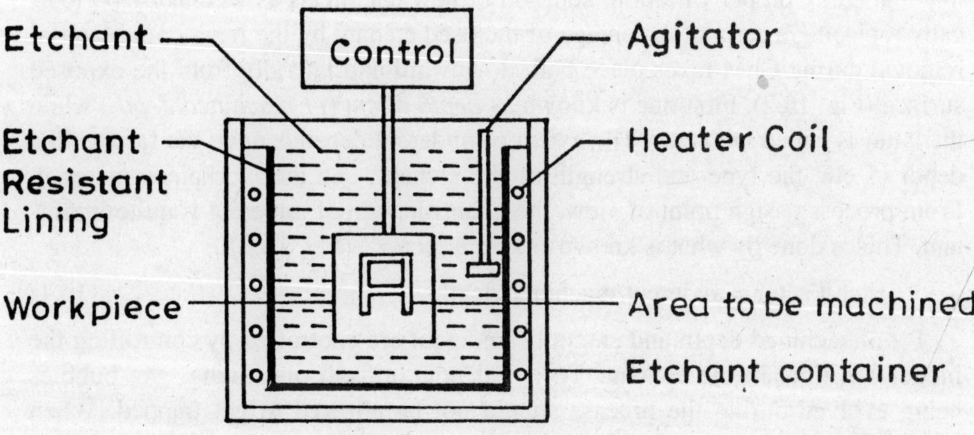

Fig. 16.1 Schematic of chemical milling/machining.

- Clean the workpiece with alkaline solution and then wash with fresh water.
- Apply a thin coating of maskant by 'cut and peel' method, screen method, or by photoresist method (details are given elsewhere).
- Dip the workpiece in the etchant bath and let it be there for the desired duration of time.
- Take the workpiece out from the etchant bath and remove the maskant layer, if any.
- Wash the workpiece thoroughly under fresh water.

Chemical **cleaning** is necessary to ensure proper adhesion of the masking material to workpiece. In case of debonding of maskant, stray etching would occur. The type of the cleaning (vapour degreasing, alkaline etching, etc) to be used depends upon the kind of maskant, kind of work material, and required machined depth. However, *cleaning of porous workpiece* material is rather difficult.

To avoid uneven removal of material from the workpiece, the fresh etchant is either continuously sprayed or the component is submerged in the etchant tank. To increase material removal rate, the etchant is agitated and, if necessary, heated as well.

The **strength of the etchant** contaminated with the reaction products can be maintained by proper filtration, addition of new chemicals as well as by periodically replacing a certain percentage of the used etchant by the fresh one. Material removal during ChM takes place both downward and laterally from the exposed surface (Fig. 16.2). First one is known as *depth of cut (or machined depth)* while the latter is called *undercut*. The extent of undercut depends upon the factors like depth of cut, the type and strength of the etchant, and the workpiece material. From process design point of view, the quantification of undercut is quite important. This is done by what is known as ''*etch factor*'' (Fig. 16.3).

Etch Factor = undercut/machined depth ...(16.1)

Total machined depth and extent of undercut are controlled by controlling the immersion time in the etchant. To avoid non-uniform machining, gas bubbles being evolved during the process should not be allowed to get trapped. When etching is over, the parts are cleaned either mechanically (in case of thicker and more durable mask), or chemically (more appropriate for thinner mask and sophisticated parts).

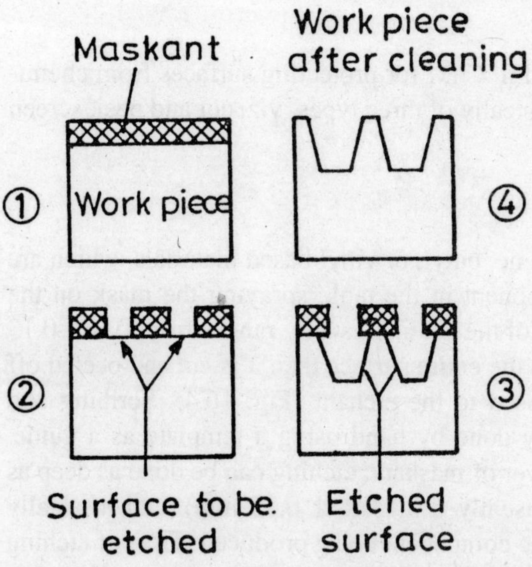

Fig. 16.2 Downward and lateral material removal

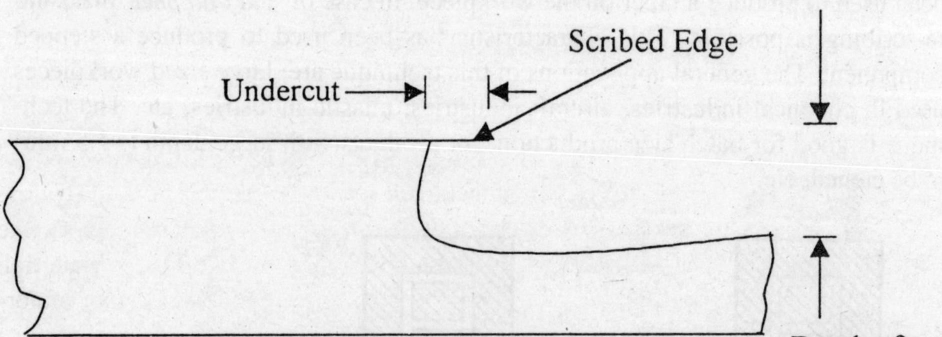

Fig. 16.3 Chemical machining: undercut and machined depth during ChM.

Performance of ChM process is influenced by many **process parameters**, viz type of etchant, temperature of etchant, type of maskant, method of applying the maskant, and method of circulating the etchant. The surface finish produced during ChM is about 0.75-3.75 μm.

MASKANTS

Maskants are the substances used in ChM for protecting surfaces from chemical etching by etchants. They are basically of three types, viz cut and peel, screen print, and photoresist.

Cut and Peel

Cut and peel maskants are neoprene, butyl, or vinyl based materials, which are applied by either dipping the component in the tank, spraying the mask on the part, or flow coating. The thickness of the coating usually ranges from 0.025-0.13 mm. This maskant is first applied to the entire surface then it is cut and peeled off from the selected areas to be exposed to the etchant (Fig. 16.4). Scribing and peeling off of the maskant is easily done by hand using a template as a guide. Because of large thickness of the layer of maskant, etching can be done as deep as 13 mm but accuracy is not high (usually ± 0.13 to ± 0.75 mm), and it usually depends upon the size and type of the component being produced. Time of etching controls the depth to which material removal takes place. This characteristic has been used to produce a taper on the workpiece. In case of '*cut and peel*' maskant, re-scribing is possible. This characteristic has been used to produce a stepped component. The general applications of this technique are: large sized workpieces used in chemical industries, aircraft industries, missile industries, etc. The technique is good for batch size production, for products with large depth (>1.5 mm) to be etched, etc.

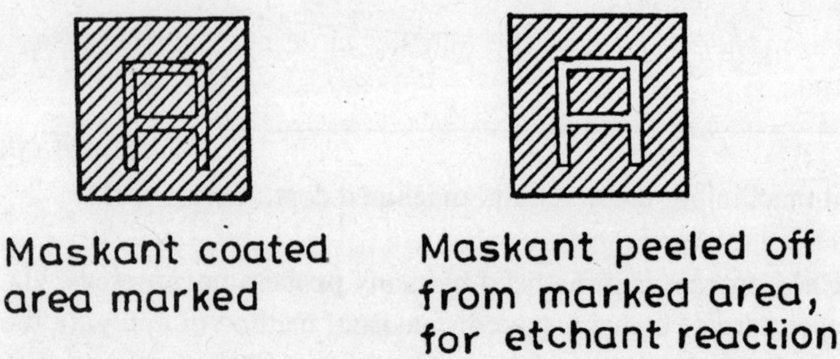

Fig. 16.4 "Cut and peel off" method of chemical machining.

Screen Printing

This technique is good for high volume production with low accuracy, comparatively low etching depth (< 1.5 mm), and for parts sized normally not more than 1.2 m × 1.2 m. In this method, a screen (usually stainless steel) is used. The screen has the areas blocked off that are to be selectively etched. Press the screen against the surface of the part and roll up the maskant. Now, remove the screen and dry the part by baking. The exposed (unmasked) area is the one from which material is to be etched.

Photoresist Maskant

This technique has become very common and is called ''**photochemical machining (PCM)**''. This method is used to produce complicated but accurate shapes. In this method, the following successive operations are carried out:

(i) Prepare the enlarged drawing of the part accounting for the effects of undercut.

(ii) Photograph the artwork and produce the reduced, highly accurate, photographic master transparency. An alternative to this procedure is to use a laser pattern-generator which receives design data from a CAD system and images (or produces) the features directly onto the photo master film.

(iii) The part to be chemically machined is thoroughly cleaned to remove the dirt and oxides.

(iv) Now, apply a thin coating of photoresist (light activated etchant resistant material) and dry by baking.

(v) Hold the photo master and workpiece together in intimate contact. Throw the strong ultraviolet light through the photo master on the workpiece so that after development, the coating is removed from all areas where etching is to take place.

(vi) The parts are now chemically etched, stripped off residual photo resist, and inspected. Accuracies of 0.025 mm and repeatability of 0.0005 mm are possible. The choice of maskant depends upon many factors, viz it should be resistant to the etchant, it should be easily removable after machining is over, it should not have any chemical effect on the workpiece, and it should be stable at the high temperature of the etchant bath.

ETCHANT

Etchant dissolves metal by changing it into a metallic salt that goes into solution. Some of the etchants are $FeCl_3$ (for Al, Cu, Ni & their alloys), $FeNO_3$ (for Ag), HF (for Ti), and HNO_3 (for tool steel). Other etchants used are chromic acid and ammonium persulphate. While selecting an etchant, the factors that should be considered are surface finish, material removal rate, depth of penetration, type of workpiece, type of maskant, damage to the workpiece by etchant, availability, and cost.

ADVANTAGES AND LIMITATIONS

ChM can remove material uniformly. It is also possible to make tapered sheets and structural members with close tolerances and good surface finish. Also, highly skilled operators are not required. The process has certain limitations. Only a few metals can be machined by this process because it is not possible to find appropriate etchant and maskant for other materials. Evolved gases collect under the maskant and result in uneven etching of the work. Such gases when they come out of the system, attack the surrounding equipment. Material removal rate obtained is very low. Chemicals used are corrosive in nature and may be toxic as well. As a result, it may be harmful. In case of chemical machining of alloys, differential machining rates result in poor surface finish. Further, material removal rate goes down because the etchant gets contaminated with the reaction products.

BIBLIOGRAPHY

1. Allen D. M. (1986), *The Principles and Practice of Photochemical Machining and Photo-etching*, Adam Hilger, Bristol.
2. Benedict G.F. (1987), *Nontraditional Manufacturing Processes*, Marcel Dekker Inc., New York.
3. Bhattacharyya A. (1971), *New Technology*, The Institution of Engineers (I), Calcutta.
4. DeBarr A.E. and Oliver D.A. (1975), *Electrochemical Machining*, Macdonald and Co. Ltd., London.
5. Gunter D. (1984), Photochemical machining, *Engineering* (London), 224(2), I–IV.

6. Pandey P.C. and Shan H.S., (1980), *Modern Machining Processes*, Tata-McGraw-Hill, New Delhi.
7. Snoeys R., Staelens F., and Dekeyser W. (1986). Current trends in non-conventional material removal processes, *Annls CIRP*, Vol. 35(2), p. 467.
8. Taniguchi N. (1983), Current trends in and future trends of ultra-precision machining and ultra-fine materials processing, *Annls CIRP*, Vol. 32(2), p. 18.

REVIEW PROBLEMS

(1) What do you understand by ''etch factor''?

(2) Name various parameters that influence the performance of ChM process.

(3) Write in table form, the basic differences between ChM and ECM.

(4) Why surface finish obtained in case of ChM of alloys is poor? Explain, in brief.

ACRONYMS

ChM	Chemical machining
MR	Material removal
PCM	Photochemical machining
PCB	Printed circuit board.

CHAPTER 16

AT-A-GLANCE

CHEMICAL MACHINING

(ChM)

INTRODUCTION

- ANCIENT PROCESS ⇒ ENGRAVING OF METALS
- REMOVES MATERIAL IN A CONTROLLED MANNER
- APPLICATION OF MASKANT AND ETCHANT

MASKANT ↓	ETCHANT ↓
DOES NOT ALLOW ETCHANT TO REACH AND REACT WITH WORKPIECE TO DISSOLVE IT	DISSOLVES WORKPIECE MATERIAL BY CHEMICAL ACTION

- APPLICATIONS
  - AVIATION INDUSTRIES
  - PRINTED CIRCUIT BOARDS
  - JEWELLERY
  - TURBINE ENGINES
  - PRESSURE VESSEL BULKHEADS
- CHEMICAL MILLING ⇒ PRODUCTION OF BLIND HOLES ⇐ POCKETS, CHANNELS, ETC
- CHEMICAL BLANKING ⇒ PRODUCTION OF THROUGH CAVITIES (THROUGH HOLES, SLOTS, ETC)
- STEPS TO BE FOLLOWED ⇒

(1) CLEAN THE WORKPIECE ⇒ ALKALINE SOLUTION AND THEN WITH FRESH WATER

(2) APPLY THE MASKANT COAT ⇒
- ⊗ CUT & PEEL METHOD
- ⊗ SCREEN METHOD
- ⊗ PHOTORESIST METHOD

(3) IF CUT & PEEL METHOD ⇒ CUT AND PEEL OFF THE MASKANT

(4) DIP THE WORKPICE IN THE ETCHANT BATH AND LET IT BE THERE FOR THE DESIRED TIME

(5) TAKE THE WORKPIECE OUT FROM ETCHANT BATH. REMOVE THE MASKANT LAYER IF ANY

(6) WASH UNDER FRESH WATER

(Contd...)

CLEANING

TO ENSURE PROPER ADHESION OF MASKANT ON WORKPIECE

↓

IF DEBONDING ⇒ STRAY ETCHING

- TYPE OF CLEANING → ϕ (MASKANT, WORKPIECE MATERIAL, & MACHINED DEPTH)
- CLEANING OF POROUS MATERIAL ⇒ DIFFICULT
- HIGHER MRR ⇒ AGITATE & HEAT THE ETCHANT
- HOW TO MAINTAIN HIGH MRR ⇒
  - FILTRATION OF ETCHANT
  - PERIODICAL ADDITION OF NEW ETCHANT
  - DISCARDING SOME PERCENT OF USED ETCHANT

* MR IN TWO DIRECTIONS DOWNWARD → DEPTH OF CUT; LATERAL → UNDERCUT

ETCH FACTOR = UNDERCUT/DEPTH OF CUT

* FOR UNIFORM MACHINING DON'T ALLOW GAS TO BE TRAPPED

MASKANT ——> CUT & PEEL
——> SCREEN/PRINTING
——> PHOTORESIST MASKANT

↓

* PROTECTS THE WORKPIECE SURFACE FROM CHEMICAL ATTACK

- CUT AND PEEL
  - NEOPRENE, BUTYL, OR VINYL BASED MATERIAL ← MASKANTS
  - DIPPING, SRAYING, OR FLOW COATING
  - COATING THICKNESS → 0.025-0.13 MM
  - APPLY TO ENTIRE SURFACE
  - "CUT AND PEEL OFF" FROM THE AREA TO BE EXPOSED TO ETCHANT
  - ACCURACY USUALLY OBTAINED ± 0.3 TO ± 0.75 MM
  - MACHINED DETH → ETCHING TIME → TAPERED WORKPIECE
  - RE-SCRIBING → STEPPED COMPONENT
  - GOOD FOR LARGE SIZED PRODUCTS, LARGE DEPTHS
- SCREEN PRINTING
  - HIGH VOLUME PRODUCTION, LOW ACCURACY, LOW ETCHING DEPTH, & PART SIZE < 1.2 m × 1.2 m
  - SCREEN BLOCKS OFF AREAS TO BE MACHINED
  - ROLL UP THE MASKANT
  - REMOVE SCREEN & DRY BY BAKING
- PHOTORESIST MASKANT
  - FOR COMPLICATED & ACCURATE SURFACES
  - ENLARGE DRAWING ACCOUNTING FOR UNDERCUT → PRODUCED REDUCED & ACCURATE MASTER TRANSPARENCY OR PHOTOMASTER FILM (LASER)
  - CLEAN THOROUGHLY THE WORKPIECE
  - APPLY A THIN COATING OF PHOTORESIST & DRY BY BAKING
  - HOLD PHOTOMASTER & WORKPIECE TOGETHER

- THROW ULTRAVIOLET LIGHT THROUGH PHOTOMASTER ON WORK-PIECE → AFTER DEVELOPEMENT → REMOVE COATING FROM AREAS WHERE ETCHING IS TO TAKE PLACE
- ETCH. STRIP OFF RESIDUAL PHOTORESIST & INSPECT

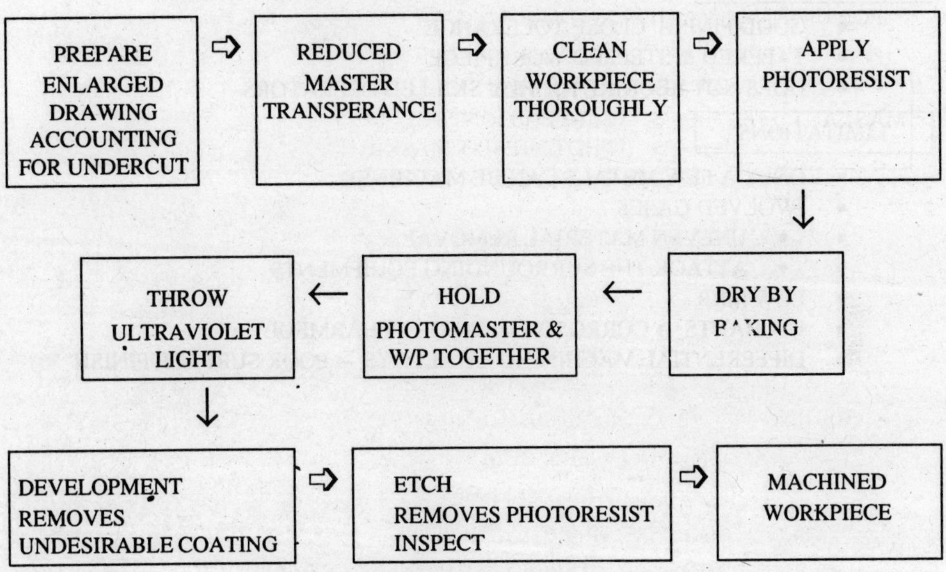

CHOICE OF MASKANT
- ——> TYPE OF ETCHANT
- ——> TYPE OF WROKPIECE MATERIAL
- ——> STABLE AT HIGH TEMPERATURE
- ——> NO CHEMICAL EFFECT ON WORKPIECE

ETCHANT

- * CHANGES METAL INTO METALLIC SALT
- * $FeCl_3$; (Al, Cu, Ni), $FeNO_3$ (Ag), HF (Ti), HNO_3 (TOOL STEEL)

❑ FACTORS TO BE CONSIDERED DURING SELECTION OF ETCHANT
- SURFACE FINISH
- DEPTH OF PENETRATION
- TYPE OF WORKPIECE

- TYPE OF MASKANT
- DAMAGE TO WORKPIECE
- AVAILABILITY & COST

ADVANTAGES

- GOOD FINISH, CLOSE TOLERANCE
- TAPERED & STEPPED WORKPIECE
- DOES NOT REQUIRE HIGHLY SKILLED OPERATORS

LIMITATIONS

- ONLY A FEW METALS CAN BE MACHINED
- EVOLVED GASES
  - UNEVEN MATERIAL REMOVAL
  - ATTACK THE SURROUNDING EQUIPMENTS
- LOW MRR
- ETCHANTS → CORROSIVE & TOXIC → HARMFUL
- DIFFERENTIAL MACHINING OF ALLOYS → POOR SURFACE FINISH

Chapter Seventeen

ANODE SHAPE PREDICTION AND TOOL DESIGN FOR ECM PROCESSES

INTRODUCTION

As described in earlier chapters, ECM is the process of controlled anodic dissolution and has got wide **applications** as a manufacturing process, viz. turning [*Bergsma*, 1968; *Hofstede et al.* 1970; *Dietz et al.*, 1979], grinding [*Kuppuswamy et al* 1978], fine hole drilling [*Bannard, 1978*], deburring [*Rolsten et al*, 1969], etc. This process is extensively used in aeronautics, nuclear [*Sorkhel et al*, 1972], bio-medical, space and similar other industries. Many research data have been published about ECM but the potential of this production process has not been fully exploited due to complex nature of the process, the practical difficulties faced in the design of tools and the prediction of anode shapes.

Tool design in *ECM* deals basically with the computation of tool shape which under specified machining conditions would produce a workpiece having the prescribed shape and accuracy [*Tipton*, 1964; *Tsuei et al* , 1977; *Dietz et al*, 1973]. Since the actual tool design problems are still difficult to solve, little progress has been made in this direction. Conversely, procedures, empirical or otherwise, have been well established for the *prediction of anode (or work) shape* obtainable from a tool while operating under the specified conditions of machining. For **anode shape prediction**, different models have been developed ranging from the one based on simple 'cos θ' principle to those based on approximate numerical techniques (Fig. 17.1) like finite difference technique (FDT) [*Tipton*, 1971; *Nanayakkara*, 1976], finite element technique (FET) [*Jain*, 1980; *Jain et al* , 1987; *Jain and Murugan*, 1987], and boundary element technique (BET) [*Narayanan et al*, 1986].

Anode shape prediction in case of ECM with parallel electrodes is simpler than the one of electrochemical drilling of blind holes. However, different models are used to predict work profile in different zones of electrochemically drilled blind hole, viz., side, transition, front, and stagnation zones as shown in Fig. 17.2.

The classical **theory of** ECM is valid only under ideal conditions. '*Trial and error*' methods for tool profile correction have been extensively used. An alternative method for tool profile correction is to calculate the working gap and account for it during tool manufacture. With the availability of high speed computers, it has been possible to determine, by successive approximations, the tool shape that would yield a workpiece of specified geometry. Since the computing time is only a small fraction of the time required for carrying out the experiments, application of tool correction technique has proved to be time saving and economical.

A number of models for anode shape prediction and tooling design in ECM have been developed. Ability to predict variations in IEG for any given operating conditions is a prerequisite for proper design of ECM tools and anode shape prediction. Hence, the various models in practice, therefore, have been discussed in terms of the equilibrium gap.

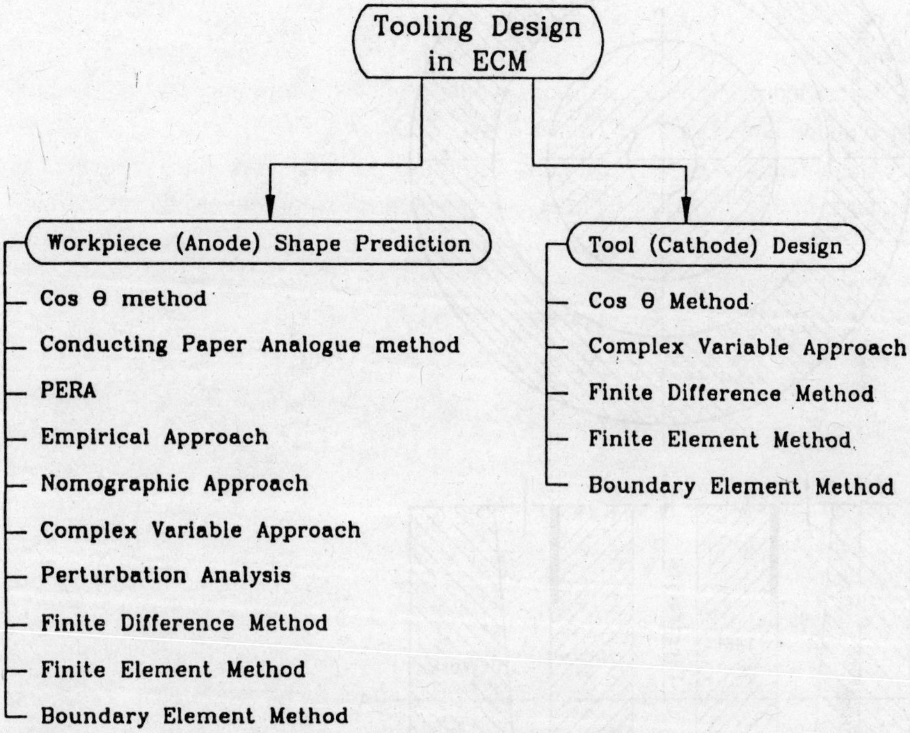

Fig. 17.1 Various models proposed for tooling design in ECM

ANODE SHAPE PREDICTION

Present day techniques require the use of high speed computers for anode shape prediction in ECM. Step by step procedure followed for anode shape prediction is as follows.

- Decide whether the analysis to be made is one-, two-, or three-dimensional.
- Collect all the initial data in the proper units and, if necessary, preliminary computation (for example, flow velocity from the volumetric flow rate) is also done.

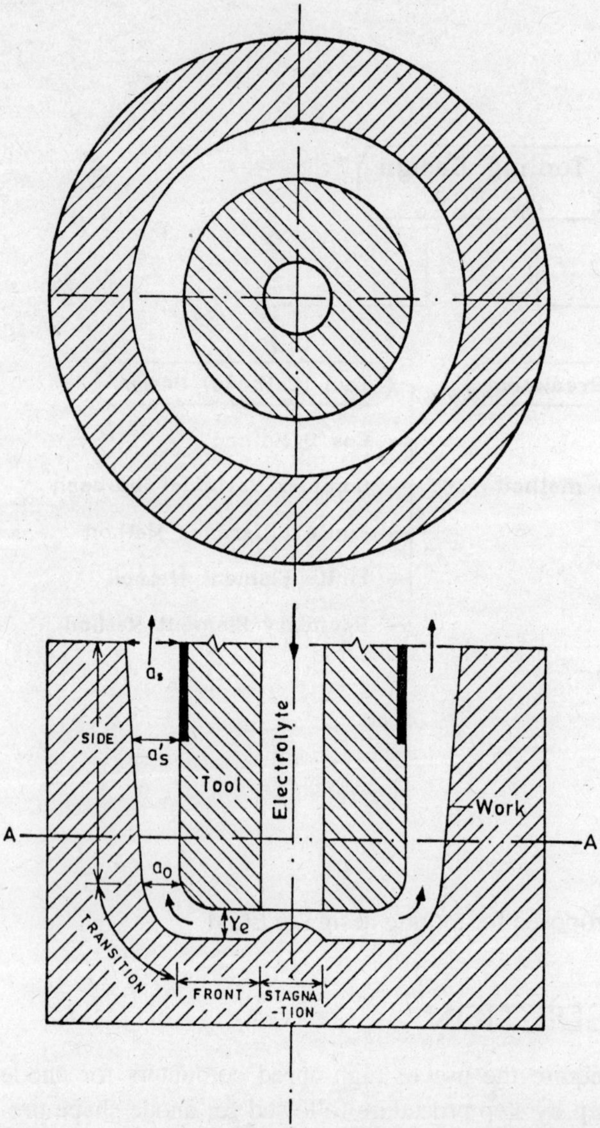

Fig. 17.2 Schematic diagram of electrochemical drilling using partially coated tool with outward mode of electrolyte flow. Zone 1 (Stagnation), Zone 2 (front), Zone 3 (transition), and Zone 4 (side).

- Now, decide whether the problem is to be solved as a temperature distribution, or electric field potential distribution problem.

There are mainly two alternatives (A) and (B), as described in what follows, for solving basic anode shape prediction problem. For this purpose, (A) either evaluate temperature distribution, or (B) evaluate the electric field potential distribution in the domain of interest. After this evaluation, intermediate computations are performed to finally arrive at the IEG at different nodal points as discussed in the following.

(A) (i) Use appropriate differential equation to evaluate **temperature distribution** along the electrolyte flow direction. This equation can be solved by one of the approximate numerical methods (FDM, FET, or BET).

(ii) From the above temperature distribution, calculate electrolyte conductivity (Ch. 11) at different points.

(iii) Find out effective feed rate at different points.

(iv) Assuming that Ohm's law is followed, compute current density (J) (Ch. 11).

(v) Using the value of J, calculate IEG at different points along the electrolyte flow direction using a generalized equation for IEG (Ch. 11) which is applicable for both zero and finite feed rates.

(vi) Once the IEG at different points is known the anode shape can be plotted.

However, in the above discussed approach the assumption that the Ohm's law is obeyed is not very true. Lines of electric potential are not always straight and normal to the electrode surface specially when a curved zone is analyzed. To take care of this, another approach (B) may be followed:

(B) (i) The **electric field potential** at different points is calculated by solving Laplace equation in one-, two-, or three-dimensions as the case may be. This may be done by one of the above stated approximate numerical methods (FDM, FEM, BEM).

(ii) Using the above electric field distribution data, current density is calculated.

(iii) With the help of Eq. (11.22), temperature at different points is found and is used to modify the electrolyte conductivity.

(iv) Now, IEG is computed at different points in the domain of interest so that the expected anode shape can be predicted.

However, there are certain approaches for anode shape prediction in which a few of the above steps may not be needed and some new ones may have to be added. A concise description of different approaches used for anode shape prediction in ECM will be given now.

Cos θ Method

According to the theory proposed by *Tipton* [*1964*], if a tool is having its face normal to the feed direction, the equilibrium gap is given by Eq. (11.4). However, if the feed direction is inclined at an angle θ to the normal to the tool face or vice versa, the equilibrium gap would be equal to $y_e/\cos\theta$. Thus, in this method, equilibrium work shape is computed corresponding to the tool whose profile has to

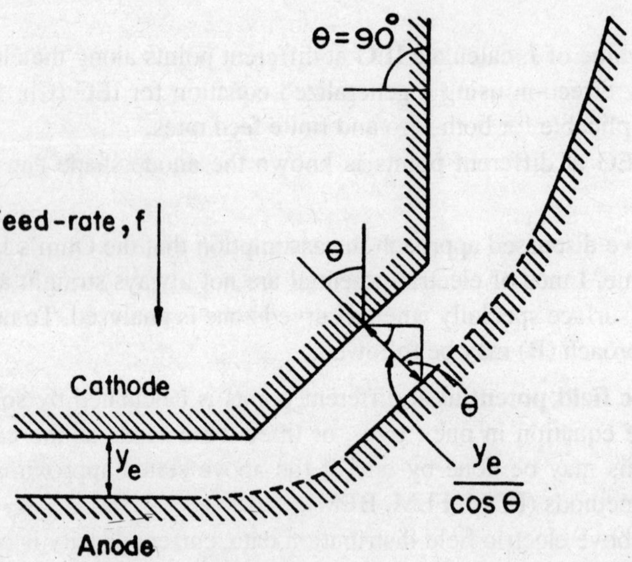

Fig. 17.3 Schematic diagram to explain the principle of tool design by 'cos θ' method [after *Tipton*, 1964].

be approximated by a large number of planar sections inclined at different angles (Fig. 17.3). Thus, this method is based on the computation of equilibrium gap for the given conditions but excludes the considerations of the mode of electrolyte flow, overvoltage, variation in electrolyte conductivity, heat transferred to the environment, etc. However, the scope of this theory is limited due to the following reasons:

- Regions with sharp corners cannot be analyzed.
- Generally applicable only if $\theta < 45°$.
- It is not possible to account for the effects of the mode of electrolyte flow, void fraction, change in electrolyte temperature and conductivity, overvoltage, heat conducted away to the tool and workpiece, etc.

There are conflicting opinions about the reference surface for the measurement of angle θ. *Tsuei et al* [1977] suggested that θ should be taken as an angle between tool feed direction and normal to the anode surface,and not the cathode surface as suggested by others [*Tipton, 1968*]. However, in view of the above stated approximations involved, use of cos θ method is not recommended, specially when complex shaped workpieces are to be analyzed [*Jain and Pandey, 1981*].

Empirical Approach

The exact path of the electric current flow lines within the IEG is difficult to determine analytically. Therefore, normally, the chordal distance between the two stations is taken as the length of current flow line. This approximation in majority of the cases is an important factor responsible for the discrepancy between the analytical and experimental results. It is easier to determine equilibrium gap in the front zone but difficult to evaluate the same in transition zone. It has also been found that the conformity of the surface radii of the tool and anode cavity decreases as the angle θ increases. Therefore, attempts have been made to derive empirical (based solely on the experimental observations rather than theory) equations for the evaluation of IEG in the transition and side zones.

Based on the analysis of experimental results, *Konig et al*, [1977] have suggested the following equation to evaluate interelectrode gap (a_o) at the start of transition zone (Fig. 17.2).

$$a_o = (r_c^{0.35})0.35\,[(10e^*)\,y_e]^{0.5} \quad \text{for } 0.15 \le y_e \le 0.6 \text{ and } 0.5 \le r_c \le 5\text{mm}$$

where, e* is Euler's number, r_c is tool corner radius, y_e is equilibrium gap and a_o is IEG in the transition zone.

For the case of tools partially or fully coated on side wall and having $1 \le r_c \le 5$ mm, following equations have been suggested.

$$a_o = (0.1 + y_e)(0.314r_c + 1.17) \quad \text{for} \quad b_b = 0 \quad \text{and} \quad 0.1 < y_e < 0.7\text{mm}.$$

$$a_o = 2y_e + 0.1\,[6.283(r_c - 1)]^{0.5} \quad \text{for} \quad b_b \ge 1\text{ mm},$$

where, b_b is length of the bare part of the tool in side zone (Fig. 17.2).
The above equations are reported to yield erroneous results at low feed rates ($f \le 0.006$ mm s^{-1}) and large equilibrium gap ($y_e \ge 1$ mm). Regression analysis of experimental data [*Jain, Jain and Pandey*, 1984] has yielded the following simple equation for the evaluation of a_o,

$$a_o = R_{c_1} \cdot y_e + R_{c_2}. \qquad \text{...(17.1)}$$

The values of R_{c_1} and R_{c_2} (constants) depend upon many factors like tool and work material combination, electrolyte, etc.
Following equations also have been suggested to evaluate IEG (a_s and a_s') in the side zones under coated and bare part of the tool (Fig. 17.2).

$$a_s = [2b_b y_e + r_c^{0.7}\, 0.123(10e^*)y_e]^{0.5} + 0.65y_e \qquad \text{...(17.2)}$$

$$a'_s = (2b_b \cdot y_e + a_0^2)^{0.5} \qquad \text{...(17.3)}$$

where, a_s is side gap in the coated part of the tool, and a'_s is side gap in the bare part of the tool (Fig. 17.2).

Test results show that the overcut in ECM is a function of machining parameters but does not depend on the tool dimensions. Further, it should be noted that above Eqs. (17.2) & (17.3) are valid only for the case of non-passivating electrolytes, specified tool-work combination, and specified machining conditions. The discrepancy between experimental and analytical results increases with an increase in the value of kEV/f.

From above discussion, it is evident that for the evaluation of side and front gaps, y_e can be used as a basic parameter. In some of the cases, side gap has been demonstrated to be independent of the machining time which, however, does not seem true. It is to be noted that in majority of the cases the effect of electrolyte flow mode (inward, outward, and side) has been neglected. Further, empirical equations are normally valid only under specified working conditions which further limit their use.

Nomographic Approach

Nomograph is a set of scales for the variables in a problem which are so distorted and placed that a straight line connecting the known values on some scales will provide the unknown values at its intersections with other scales.

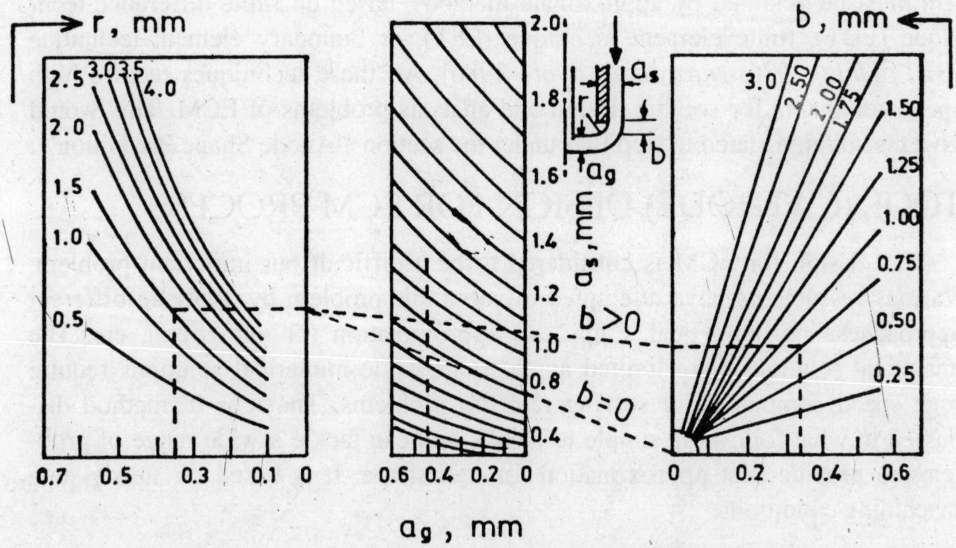

Fig. 17.4 A nomogram proposed by *Konig & Pahl* [1970] for the evaluation of side gap.

For the evaluation of equilibrium anode shape, a nomographic approach has also been used. *Konig and Pahl* [1970] have prepared a nomogram for the evaluation of side gap for the known equilibrium gap, tool radius and bare tool length. Fig. 17.4 shows one such nomograph for $b_b = 0$ and $b_b > 0$ for a given tool. *Heitman* [1968] has used nomograms for evaluation of electrolyte temperature rise (ΔT) while working under specified conditions. Nomograms have proved to be useful in planning an ECM operation. However, the nomograms are very often prepared based on a number of simplified assumptions.

Numerical Methods

Conducting paper analogue method represents physical analogue of the ECM process whereas computer method is based on a mathematical analogue of the process. It is necessary to solve Laplace's equation over the domain of interest to determine the potential at a large number of points. Laplace's equation can be analytically solved only for simple shapes but in majority of the cases one comes across complicated shapes in ECM. Hence, in general, solution of the field problem must be obtained by approximate methods, based on finite difference technique (FDT), finite element technique (FET), or boundary element technique (BET) [*Jain*, 1980; *Narayanan et al*. 1986]. All these techniques require high speed computers for solving design and analysis problems of ECM. This would give the solution stated in step B(i) under the section ''Anode Shape Prediction''.

TOOL (CATHODE) DESIGN FOR ECM PROCESS

Tool design for ECM is considered to be a difficult but important problem. Various researchers have attempted to solve this problem by applying different approaches, viz. analytical, empirical, approximation (or numerical), etc. The analytical solutions have limited applications while numerical solutions require high speed computers for solving real life problems. The 'cos θ' method discussed in what follows is simple to handle and can tackle a wide range of problems to provide first approximation for tool shape. It is based on steady-state machining conditions.

'Cos θ' Method

Fig. 17.5 shows a tool having angle θ between feed direction and normal to the tool surface. For a general case ($0^\circ < \theta < 90^\circ$), the value of equilibrium gap is given by $y_e/\cos\theta$. Note that the application of this method is based on the assumption that the electric field is normal to the surfaces of the electrodes. Further, this method is applicable only in those areas where the local radii of curvture of the anode and cathode surfaces are larger compared with the equilibrium gap. The assumption is that the surfaces are smooth with gentle variations so that the electric current flow lines are straight and parallel to each other. However, if this assumption is not valid (say, in case of complex shaped workpieces) then the numerical solution to the electric field is desirable.

Let the workpiece surface be expressed as

$$y = f(x). \qquad \text{...(17.4)}$$

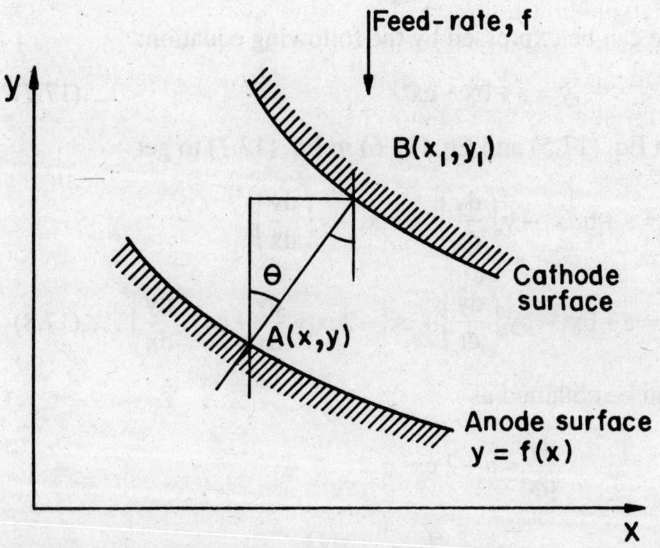

Fig. 17.5 Cathode design by 'cos θ' method.

Let a point A (x, y) on the workpiece surface correspond to an equivalent point B (x_1, y_1) on the tool surface. From Fig. 17.5,

$$y - y_1 = AB \cdot \cos\theta$$

$$= \frac{y_e}{\cos\theta} \cdot \cos\theta \qquad \text{...(17.5)}$$

or,

$$y_1 + y_e = y.$$

Also,

$$x_1 - x = AB \cdot \sin\theta = \frac{y_e}{\cos\theta} \cdot \sin\theta$$

$$= y_e \tan\theta$$

$$= y_e \frac{dy}{dx}$$

or, $$x = x_1 - y_e\left(\frac{dy}{dx}\right). \qquad \text{...(17.6)}$$

Suppose the anode surface can be expressed by the following equation:

$$y = a + bx + cx^2. \qquad \text{...(17.7)}$$

Substitute the values from Eq. (17.5) and Eq. (17.6) in Eq. (17.7) to get

$$y_1 + y_e = a + b\left\{x_1 - y_e\left(\frac{dy}{dx}\right)\right\} + c\left\{x_1 - y_e\left(\frac{dy}{dx}\right)\right\}^2$$

$$= a + bx_1 - by_e\left(\frac{dy}{dt}\right) + cx_1^2 - 2cx_1y_e\frac{dy}{dx} + cy_e^2\left(\frac{dy}{dx}\right)^2 \qquad \text{....(17.8)}$$

From Eq. (17.7), dy/dx can be obtained as

$$\frac{dy}{dx} = b + 2\,cx$$

$$= b + 2c\left(x_1 - y_e\frac{dy}{dx}\right).$$

Simplifying it to get

$$\frac{dy}{dx} = \frac{b + 2cx_1}{1 + 2cy_e}. \qquad \text{(17.9)}$$

Now, substitute the value from Eq. (17.9) in Eq. (17.8), simplify it, and ignore the term $cy_e^2\,(dy/dx)^2$ assuming it to be insignificant. Then, we get

$$y_1 = \underline{a + bx_1 + cx_1^2 - y_e} \; - \; \underline{y_e\left[\frac{(b + 2cx_1)^2}{1 + 1cy_e}\right]}. \qquad \text{...(17.10)}$$

In the above Eq. (17.10), the first four terms indicate anode surface displaced by a distance equal to the equilibrium gap, y_e. The last term indicates the correction for the curvature of the anode.

Above analysis can be repeated for a surface y = f (x, z) taking into account the changes in the z-direction.

$$y = a + bx + cx^2 + dz + ez^2 + gxz. \qquad \ldots(17.11)$$

Following the same procedure as in case of 2-D geometry, the tool shape can be represented as:

$$y_1 = a + bx_1 + cx_1^2 + dz_1 + ez_1^2 + gx_1z_1 - y_e - y_e\left[\frac{(b + 2cx_1 + gz_1)^2(d + 2ez_1 + gx_1)^2}{1 + 2(c + e)y_e}\right] \qquad \ldots(17.12)$$

Correction Factor Method

Recently, a correction factor method using the finite element technique has been proposed [*Reddy et al*, 1988; *Jain et al*, 1991] for one- and two-dimensional tool design in ECM. The correction factor concept has been applied to modify, in different stages, tool shape assumed in the first design cycle. The modification process continues until the difference between the predicted work shape and the desired work shape is within the specified tolerance. Tool design procedure involves the following five major steps:

1. Using the anode shape prediction model based on FEM (or any other method), the anode profile is estimated for an assumed (or modified) tool shape and specified machining conditions. The initial tool shape is assumed to be approximately complementary to the workpiece shape.
2. The computed and the desired work shapes are compared at every node and the difference between the two is evaluated as an error at that node.
3. If the error is more than the specified tolerance value the correction factor is calculated.
4. The tool shape is modified by applying the correction factor.
5. Steps 1-4 are repeated until the desired tool shape is achieved. This concept has been applied for designing one-and two-dimensional tools for plane par-

electrodes machining and electrochemical drilling. This concept of correction factor can also be applied to other tool design models employing finite difference or boundary element techniques for anode shape prediction.

Fig. 17.6 shows a comparison of an experimental bare tool and the bare tool designed by using the correction factor model in conjunction with FET. The difference between the experimental and the computed work shapes at the top of a drilled hole is depicted in some examples [*Reddy et al.*, 1988]. It is due to the fact that the effect of stray current attack has not been incorporated in the present model. The disagreement between the designed tool shape and the experimental tool shape has been attributed to the inaccuracy inherited in the anode shape prediction model and some built-in inaccuracy in the measurement technique used for the measurement of overcut in the transition zone.

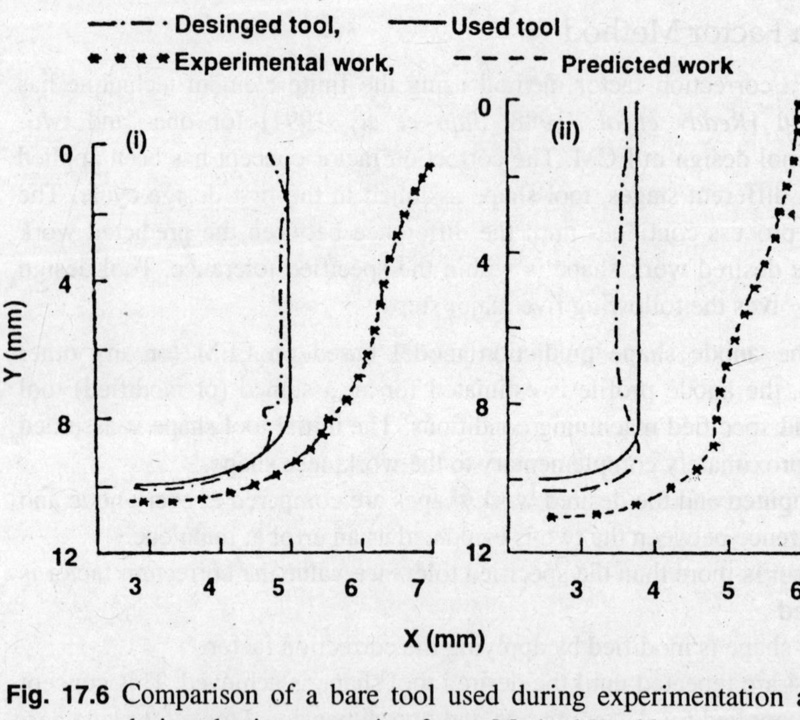

Fig. 17.6 Comparison of a bare tool used during experimentation and the one designed using correction factor Method [*Reddy et al* 1988]. (i) and (ii) indicate two different test coditions.

BIBLIOGRAPHY

1. Bannard J. (1978), Fine hole drilling using electrochemical machining, *Proc. 19th Int. MTDR Conf.*, pp. 503-510.
2. Bergsma F. (1968), Electrochemical machining of metals, *Annls CIRP*, pp. 93-99.
3. Dietz H., Gunther K.G. and Otto K. (1973), Reproduction accuracy with electrochemical machining: Determination of side gap, *Annls CIRP*, Vol. 22/1, pp. 61-62.
4. Dietz H., Gunther K.G., Otto K. and Stark G. (1979), Electrochemical turning considerations on machining rates which can be attained, *Annls CIRP*, Vol. 28/1, pp. 93-97.
5. Heitman H. (1968), Surface roughness formation and determination of optimum working conditions in electrochemical diesinking, *Proc. 2nd AIMTDR Conf.*, pp. 41-47.
6. Hofstede A. and Berkel Van den. (1970), Some remarks on electrochemical turning, *Annls CIRP*, Vol. 18, pp. 93-106.
7. Jain V.K. (1980), *An Analysis of ECM Process for Anode Shape Prediction*, Ph.D. thesis, University of Roorkee, Roorkee, India.
8. Jain V.K. and Murugan S. (1987), Workshape prediction in electrochemical drilling and electrochemical boring operations, *Microtechnic*, pp. 42-47.
9. Jain V.K. and Pandey P.C. (1981), Tooling design for ECM: A finite element approach, *Trans. ASME, J. Engg. for Ind.*, Vol. 103, pp. 183-190.
10. Jain V.K., Raju K. Ravi, Lal G.K. and Rajurkar K.P. (1991), Investigations into tool design for electrochemical drilling, *Process. Adv. Mater.*, Vol. 1, pp. 105-121.
11. Jain V.K., Jain V.K., and Pandey P.C. (1984), On the reproduction of anode corner profile during electrochemical drilling of blind holes, *Trans. ASME, J. Engg. Ind.*, Vol. 106, No. 1, pp. 55-61.
12. Jain V.K., Yogindra P.G. and Murugan S. (1987), Prediction of anode profile in ECBD and ECD, *Int. J. Mach. Tools Manuf.*, Vol. 27(1), pp. 113-134.
13. Konig W. and Pohl D., (1970), Accuracy and optimal working conditions in ECM, *Annls CIRP*, Vol. 18(2), pp. 223-230.
14. Konig W. and Humb H.J. (1977), Mathematical model for the calculation of the contour of the anode in electrochemical machining, *Annls CIRP*, Vol. 25, pp. 83-87.

15. Kuppuswamy G. (1978), Effect of Magnetic field on electrolytic grinding, *Proc. 8th AIMTDR Conf.*, pp. 546-549.
16. Nanayakkarara M.B. and Larsson C.N. (1976), Computation and verification of workpiece shape in electrochemical machining, 20th *Int. Mach. Tool Des. Res. Conf.*, pp. 617-624.
17. Narayanan O.H., Hinduja S. and Noble C.F. (1986), The prediction of workpiece shape during electrochemical machining by the boundary element method, *Int. J. Mach. Tool Des. & Res.*, Vol. 26(3), pp. 323- 338.
18. Reddy M.S., Jain V.K. and Lal G.K. (1988), Tool design for ECM: Correction factor method, *Trans. ASME, J. Engg. for Ind.*, pp. 111-118.
19. Rolsten R.F. and Manning E.J. (1969), Electrochemical deburring, *Tool & Manufg. Engr.*, pp. 14-16.
20. Sorkhel S.K, *et al.* (1972), Electrochemical machining, *Proc. 5th AIMTDR Conf.*, pp. 651-656.
21. Tipton H. (1964), The dynamics of electrochemical machining, *Proc. 5th Int. Mach. Tool Des. & Res. Conf.*, pp. 509-522.
22. Tiption H. (1968), The determination of tool shape for electrochemical machining, *Machinery & Prod. Engg.*, pp. 325-328.
23. Tipton H. (1971), Calculation of tool shape for ECM, in *Fundamentals of Electrochemical Machining*, (Ed.) C.L. Faust, Electrochemical Society, New Jersy pp. 87-102.
24. Tsuei Y.G. and Nilson R.H. (1977), Effects of variable electrolyte conductivity on side gap geometry in ECM, *Int. J. Mach. Tool Des. Res.*, Vol. 17, pp. 169-178.

QUESTIONS

1. (a) Derive an equation to determine tool geometry when the work geometry is given by the parabolic equation.

   (b) Using the above derived equation find the equation for the tool shape for the following work shape:

   $Y = 10 + 0.3\,X - 0.05\,X^2$ (X and Y are in cm)

   Given: applied voltage = 15 V, over potential = 0.67 V, feed velocity = 0.75 mm/min, atomic weight = 63.57, Z = 1, density = 8.96 g/cm^3, k = 0.2 S/cm.
2. Write the tool equation if the workpiece surface is approximated by a straight line given as:

   $Y = a + bX.$

NOMENCLATURE

A, b, c, d, e, g,	Constants
a_o	Inter-electrode gap at the start of transition zone
a_s	Side gap in the coated part of the tool (Fig. 17.2)
a_s^1	Side gap in the bare part of the tool (Fig. 17.2)
b, b_b	Length of the bare part of the tool
e*	Euler's constant
E	Electrochemical equivalent
f	Feed rate
J	Current density
k	Electrolyte's electrical conductivity
r, r_c	Tool corner radius
Rc_1, Rc_2	Constants
ΔT	Rise in electrolyte temperature
V	Applied voltage
x	Distance along x-coordinate
y	Interelectrode gap
y_e	Equilibrium gap
θ	Angle between feed direction and a normal to the tool face under consideration.

ACRONYMS

BET	Boundary element technique
ECD	Electrochemical drilling
ECM	Electrochemical machining
FDM	Finite difference method
FEM	Finite element method
IEG	Inter-electrode gap
PERA	Production engineering research association
W/P	Workpiece

CHAPTER 17

AT-A-GLANCE

ANODE SHAPE PREDICTION

&

TOOL DESIGN FOR ECM PROCESSES

- ECM PROCESS CAPABILITIES ARE NOT FULLY EXPLOITED →
    - COMPLEX NATURE OF THE PROCESS
    - TOOL DESIGN PROBLEMS
    - ANODE SHAPE PREDICTION PROBLEMS
- TOOL DESIGN IN ECM → COMPUTES TOOL SHAPE WHICH UNDER SPECIFIED MACHINING CONDITIONS, PRODUCES DESIRED ANODE (W/P) SHAPE, A DIFFICULT PROBLEM
- ANODE SHAPE PREDICTION IN ECM → PREDICTION OF OBTAINABLE W/P SHAPE & SIZE FROM GIVEN TOOL AND SPECIFIED MACHINING CONDITIONS
- MANY METHODS AVAILABLE FOR TOOL DESIGN AND ANODE SHAPE PREDICTION → COS θ, NOMOGRAPHIC, EMPIRICAL, ETC
- TRIAL & ERROR METHOD OF TOOL PROFILE CORRECTION → STILL WIDELY USED
- APPLICATION OF HIGH SPEED COMPUTERS → SAVES TIME

ANODE SHAPE PREDICTION

* <u>GENERAL PROCEDURE</u>

- DECIDE ABOUT THE TYPE OF ANALYSIS (1-D, 2-D, OR 3-D)
- DATA COLLECTION & PRELIMINARY CALCULATIONS
- TYPE OF PROBLEM TO BE SOLVED (TEMPERATURE OR FIELD POTENTIAL CALCULATION)
- SOLVE TEMPERATURE DISTRIBUTION PROBLEM USING NUMERICAL / APPROXIMATE METHOD, CALCULATE k, f_{eff}, AND J. EVALUATE 'IEG' TO PLOT ANODE SHAPE

<u>OR</u>

- SOLVED LAPLACE EQ BY NUMERICAL METHOD AND CALCULATE 'J' AND USE IT TO FIND 'T' TO MODIFY k COMPUTE 'IEG' TO PREDICT ANODE PROFILE

<u>COS θ METHOD</u>

- THIS METHOD IS NOT APPLICABLE FOR THE REGIONS WITH SHARP CORNERS
- COMPUTE EQUILIBRIUM GAPS AT DIFFERENT PLANAR SECTIONS INCLINED AT DIFFERENT ANGLES
- PLOT ANODE PROFILE USING COMPUTED EQUILIBRIUM GAPS

EMPIRICAL APPROACH

DISCREPANCY BETWEEN THE ANALYTICAL AND EXPERIMENTAL RESULTS → EMPIRICAL EQS. DERIVED TO EVALUATE IEG IN THE TRANSITION & SIDE ZONES

NOMOGRAPHIC APPROACH

WITHOUT CALCULATIONS, IT PROVIDES DESIRED UNKNOWN VALUES OF PARAMETERS LIKE SIDE GAP, ELECTROLYTE TEMPERATURE, ETC, USEFUL IN PLANNING AN ECM OPERATION

NUMERICAL METHODS

USED TO SOLVE LAPLACE EQN. FOR DETERMINING POTENTIAL OR TEMPERATURE DISTRIBUTION IN IEG. THEN EQUILIBRIUM SHAPE IS DETERMINED AS DISCUSSED EARLIER

CATHODE DESIGN FOR ECM PROCESS

COS 'θ' METHOD

- θ ANGLE BETWEEN FEED DIRN. AND NORMAL TO THE TOOL SURFACE
- NOT FOR SHARP CORNERS
- DETERMINE TOOL SHAPE FOR THE GIVEN WORKPIECE SHAPE

CORRECTION FACTOR METHOD

MODIFY ASSUMED TOOL SHAPE IN DIFFERENT STAGES UNTIL THE DIFFERENCE BETWEEN THE PREDICTED W/P SHAPE AND DESIRED W/P SHAPE IS WITHIN SPECIFIED TOLERANCES

Chapter Eighteen

ELECTROCHEMICAL SPARK MACHINING (ECSM)*

INTRODUCTION

Electro Chemical Machining (ECM) and Electro Discharge Machining (EDM) are the processes which are being more commonly used in the industries for several years as compared to other Advanced Machining Processes (AMPs). However, each process has its own merits and demerits, or limitations. ECM and EDM have many merits but one of the main limitations of both these processes is that they are applicable only for electrically conducting materials. Hence, attempts have been made by different researchers to combine these two processes so that a new process is developed which can be employed to machine both electrically conducting as well as non-conducting materials. This new ECM and EDM combined process (or, *a hybrid process*) was named as *Electrochemical Spark Machining (ECSM)* process. Thus, ECSM has evolved as a hybrid of thermoelectric (EDM) and electrochemical (ECM) processes. As can be seen later in this chapter, ECSM process is a complex hybrid process [*Jain et al*., 1991, 1996, 2008]. Several names have been given for

* Author thanks Dr. Anjali Kulkarni of IIT Kanpur for providing appropriate papers and other information which helped in writing this chapter.

this process namely, Electrochemical Discharge Machining (ECDM), Spark Assisted Chemical Engraving (SACE) and Electrochemical Spark Machining. The diversity of the names assigned to this process illustrates the complexity of different functional mechanisms associated with it. Sparks are the only desirable form of discharges for material removal, and it is associated with hardly any damage to the workpiece hence this process has been named as *Electrochemical Spark Machining* (ECSM). In this process, thermal erosion is more productive mechanism of material removal than others.

In ECSM process, electrolyte is used as in ECM but the mechanism of material removal is melting and vaporization as in case of EDM. In ECSM process, sparking takes place between the electrode (usually, cathode)-electrolyte interface (more accurately, between electrode and gas bubble) in the proximity of workpiece, while in case of EDM the sparking takes place between the two electrodes that is anode and cathode. In EDM, material removal takes place from both the electrodes while in case of ECSM, it is from cathode and workpiece (normally, electrically non-conducting). (Please note that material removal can take place from the electrically conducting workpiece also without making it as an electrode. However, the process is basically developed for machining electrically non-conducting materials).

Figure 18.1(a) shows a schematic diagram and Figure 18.1(b) shows a photographic view of the main set-up of the ECSM process [*Jain and Priyadarshani,* 2014]. The ECSM cell contains anode and cathode as electrodes, and workpiece as a separate entity to be machined. Anode is made up of non-consuming material while cathode (tool) is usually made of copper or high electrically conducting material. NaOH is commonly used as electrolyte. Other electrolytes or a mixture of electrolytes can also be used depending upon the requirements of the process and type of workpiece material. As in ECM, when a voltage is applied across the electrodes, electrolysis takes place and hydrogen gas is released at the cathode and oxygen is released at the anode. When the applied voltage is increased beyond a threshold value, hydrogen gas bubbles grow in size, and sparking takes place across the bubbles between the electrode-electrolyte interface. Sparking always takes place at the smaller electrode-electrolyte interface. If anode is smaller, then the sparking takes place

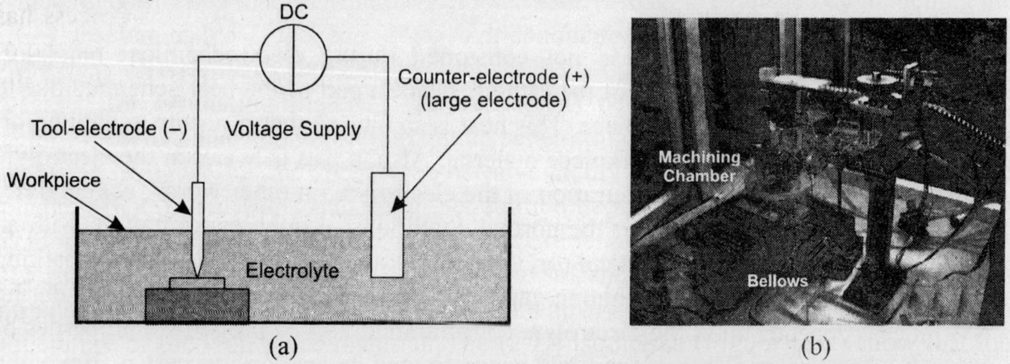

Fig. 18.1: (a) Schematic diagram of ECSM process (not to the scale), (b) Photographic view of the ECSM Set-up [*Jain and Priyadarshani*, 2014]

across oxygen bubbles (on the anode) in place of hydrogen bubbles (at the cathode) [*Jain and Adhikary*, 2008]. Machining takes place on the workpiece surface kept near the cathode tip (or, anode tip) wherever the sparking is taking place. The ECSM process is a standalone process unlike other processes utilized for machining. This process does not demand intermediate processing steps such as: masking, pattern transfer, sample passivation, and sample preparation.

During ECM, electrochemical reactions at the electrodes (say, gas evolution, plating, electrode dissolution, etc.) and chemical reactions in the bulk of the electrolyte (the complex compound formation) take place. The Electrochemical reactions occur at the electrode-electrolyte interface and result in the transfer of ions in the electrolytic solution by diffusion, migration in an electrical field and convection in the flowing medium. When a DC power is applied between the tool (cathode) and the auxiliary electrode (or anode), different reactions at cathode and anode take place during ECSM process [*Kulkarni and Jain*, 2015; *Jiang et al.*, 2014]. The same are reproduced as follows.

ELECTROCHEMICAL REACTIONS

At Cathode

In ECSM process, the anode material used is non-consuming type (graphite) hence there is hardly any possibility of electroplating of cathode. Hydrogen gas bubbles are evolved at the cathode, which result in sparking at the cathode-electrolyte interface due to high potential gradient across the bubbles. With copper as a cathode material and NaOH as electrolyte, following reactions take place.

$$Cu^{2+} + 2e^{-} \rightarrow Cu \qquad \text{... (18.1)}$$

$$2H^{+} + 2e^{-} \rightarrow H_2\uparrow \qquad \text{... (18.2)}$$

$$Na^{+} + e^{-} \rightarrow Na \qquad \text{... (18.3)}$$

$$2Na + 2H_2O \rightarrow 2NaOH + H_2\uparrow \qquad \text{... (18.4)}$$

$$2H_2O + 2e^{-} \rightarrow H_2\uparrow + 2OH^{-} \qquad \text{... (18.5)}$$

It is to be noted that NaOH is not consumed during electrochemical reactions. Workpiece is placed in the vicinity of the cathode hence a part of the heat generated due to sparking is transferred to the workpiece. This heat is so intense that it results in melting or, melting and vaporization of the workpiece material. Also, a part of water in the electrolyte is evaporated. It changes the concentration of the electrolyte, in other words, conductivity of the electrolyte is changed. Under the normal conditions of bubbles formation; with an increase of the supply voltage, a critical (or, threshold) voltage is reached wherein sparking starts across the bubbles in the machining gap. The voltage at which sparking starts during ECSM process depends upon the electrolyte concentration and its properties, properties of gas (say, H_2) bubbles as dielectric, and tool geometry and its material properties. Beyond a certain voltage, violent sparking starts taking place which may damage the workpiece also.

At Anode

In ECSM process, two types of anodic reactions take place: dissolution of anode as metal ions in the electrolytic solution and evolution of oxygen gas at the anode surface. However, in ECSM, the dissolution of anodic material does not take place as it is made up of non-dissolving material, say, graphite. But, oxygen evolution as a reaction in NaOH electrolyte, takes place as follows.

$$4(OH)^{-} \rightarrow 2H_2O + O_2 \uparrow + 4e^{-} \qquad \ldots (18.6)$$

As the spacing between the cathode (tool) and anode is large, the reactions taking place at the anode-electrolyte interface are not affecting the machining process.

In Bulk Electrolyte

Reduction of electrolyte takes place as follows.

$$NaOH \rightarrow Na^{+} + OH^{-} \qquad \ldots (18.7)$$

As seen above, the electrochemical reactions continuously supply electrons from cathode to the solution and from solution to the anode. This ionic and electronic current of the order of few milliamperes, contributes towards the average process current. At the onset of sparking, there is a momentary, transient high current due to the electrons. How this spark energy results in material removal from the workpiece has been explained else where?

MECHANISM OF SPARKING IN ECSM PROCESS

The electrochemical spark has been modelled as a phenomenon similar to that which occurs in 'arc discharge valves' [*Jain et al.*, 1996]. This phenomenon is used to explain various experimental results, on the basis of circuit and arc discharge valve characteristics. The spark energy and the approximate order of hydrogen gas bubble diameter are computed by the proposed valve theory. Material Removal Rate (MRR) is evaluated by modelling the ECSM problem as a *3-D unsteady state heat conduction problem*. The problem is solved by the *finite element method* (FEM) to compute the temperature distribution which is post-processed for estimating Material Removal (MR) per spark, overcut obtained in the machined cavity, and attainable maximum penetration depth. Based on this study, the conclusion is drawn that the application of valve theory to the ECSM process seems to be realistic. MRR, overcut and maximum penetration depth computationally obtained using this model show a good agreement with the experimental observations.

During computational ECSM, volume of the softened material (defining the crater) is computed by generating (3-D) isotherms for the temperature equal to and above the melting temperature of the workpiece material. The (2-D) isotherms (Figure 18.2) show that the crater can be well approximated by semi-ellipsoid [*Jain et al.*, 1996]. It is interesting to note that these simulated isotherms are very close to the experimentally found signatures of the isotherms [*Kulkarni et al.*, 2002, 2011], Figure 18.3. The smaller circular zones are the result of low energy sparks striking the workpiece surface. The geometry of a single spark zone is circular in nature with an average diameter of 300 μm. The ring patterns of isotherms of graded temperature regions have been formed by melting and solidification of workpiece material due to the quenching effect of the electrolyte.

Following sections give a brief discussion on ***specific applications*** of this process in the macro and micro domains.

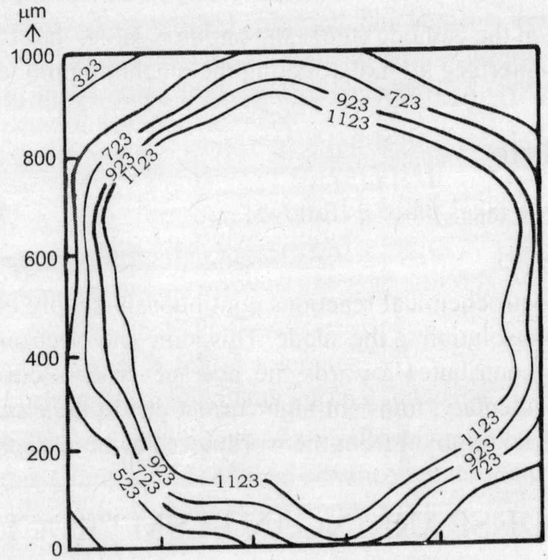

Fig. 18.2: Simulated isotherms obtained while machining soda lime glass [*Jain et al.*, 1996]

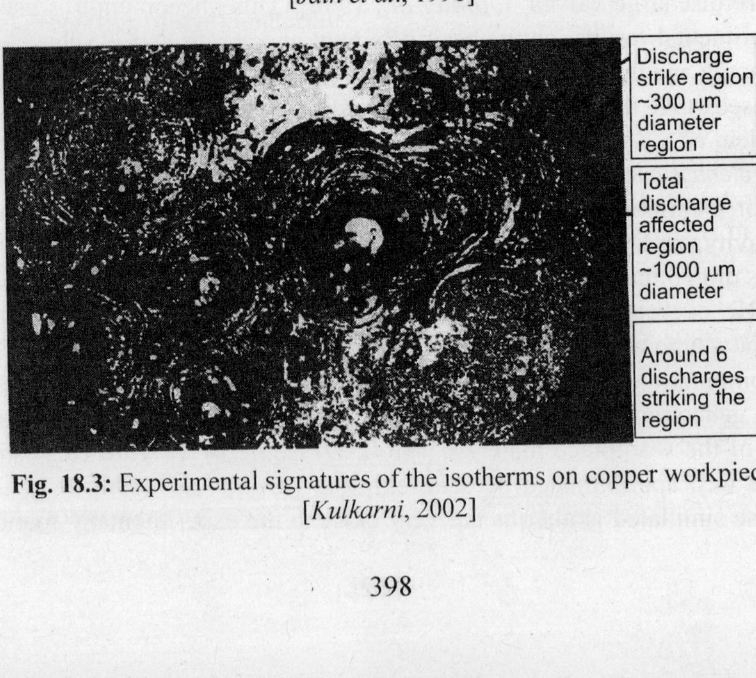

Fig. 18.3: Experimental signatures of the isotherms on copper workpiece [*Kulkarni*, 2002]

Machining of Kevlar-Epoxy Composites

Nesarikar et al. [1995] used a travelling wire ECSM process for slicing of thick Kevlar-epoxy composites. Various input parameters including specimen thickness were taken as independent variables while MRR and average diametric overcut were taken as responses. The flow of electrolyte helped to minimize the temperature variation of the electrolyte, to remove the graphite particles (due to anode) and the debris of the composite (workpiece material). It was concluded that with an increase in the specimen thickness, the spark zone shifted to the top surface of the specimen. It leads to the non-uniform material removal along the depth of cut. It also results in lower MRR for almost the same energy input because the energy generated on the wire away from the workpiece is wasted in heating the electrolyte. Secondly, the wire is in the close vicinity of the workpiece, the bubbles concentration on the wire in the groove reduces hence the energy input in the groove for removing the material is also reduced [*Kulkarni and Jain*, 2018].

The dependence of overcut on voltage at different values of electrolyte concentration shows an increase in overcut with an increase in the voltage. This can be explained by the fact that the spark energy increases with an increase in the voltage hence the electrolysis process also gets accelerated resulting in an increase in the rate of H_2 gas bubbles generation. These two factors together cause an increase in the overcut. *Jain et al.* [1991] correlate this phenomenon with an increase in the current density (due to increase in voltage up to a critical value) resulting in the production of large number of H_2 bubbles and hence increase in sparking rate, that is yielding wider cuts. Generation of large number of bubbles eventually reduces the effective area of wire tool for current flow and it increases current density. MRR and overcut are found to increase with increase in voltage, and electrolyte concentration up to 20%. Beyond around 20% concentration, MRR starts decreasing because at higher concentration ($\geq$ 20%) the specific conductivity of NaOH electrolyte starts decreasing.

Machining of Alumina (Al_2O_3) and Quartz

Jain et al. [2002] used ordinary and abrasive particles embedded tools (Figure 18.4) for machining of alumina and glass to enhance the process performance. Experiments were performed using DC voltage and NaOH as the electrolyte. MRR improved with the use of abrasive particles embedded tools. It was observed that, beyond a certain value of electrolyte temperature, the process performance deteriorated. Figure 18.4 also shows active and inactive abrasive particles (grains), bubbles formed at the interface, sparks, molten material, and the chips formed during machining. Active abrasive particles are those which take part in removing material from the workpiece by shearing action. During the ECSM process, sparks emanate from bonding material (which is electrically conducting) across the bubbles in the gap. If the gap height is large enough, then the sparking will not be very effective in removing material from the workpiece as well as in softening the workpiece material, because most of the heat will be convected away in the electrolyte, and partly it is conducted away in the tool. Sparks remove material from the workpiece by melting, or both by melting and evaporation. Sparks soften/melt the workpiece material, and the abrasive particles throw the molten material out of the cavity in addition to removing softened material by shearing action. It is concluded from the experimental results that the machined depth is higher for abrasive particles embedded cutting tool as compared to the bare cutting tool. Some micro cracks and grinding marks are clearly visible on the machined surface of alumina. In the case of cutting tool having no abrasive particles embedded in it, there are no scratches/grinding marks on the machined surface. It is also concluded that ECSM with cutting tool having embedded abrasive particles has shown the improved performance in terms of MRR and machined depth specially while machining electrically non-conducting materials.

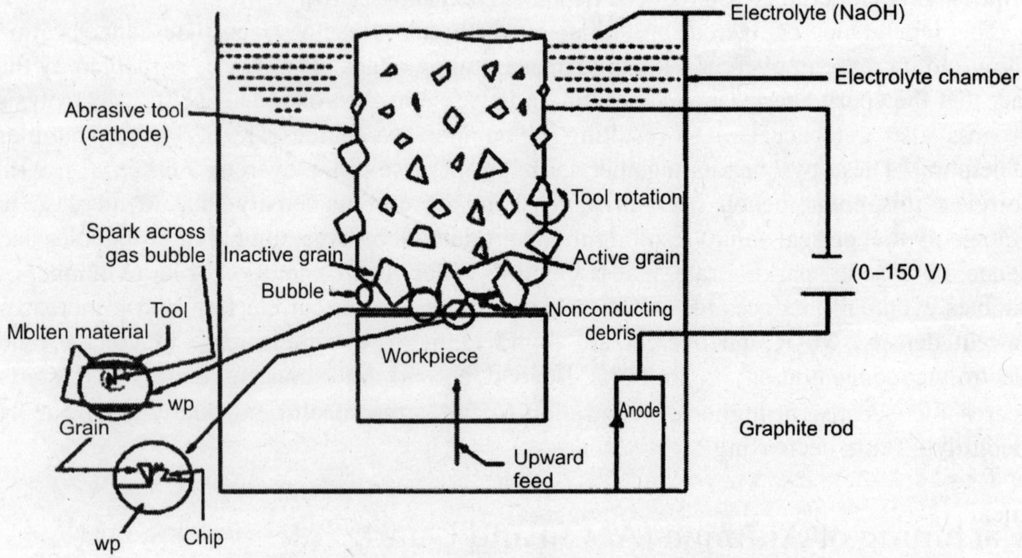

Fig. 18.4: Machining piece with abrasive cutting tool
(wp stands for workpiece) [*Jain et al.,* 2002]

Electrically non-conducting, hard, brittle, and high temperature resistant materials such as alumina and quartz have been machined employing ECSM process with orbital rotation of the tool [*Gautam and Jain*, 1998; *Chak and Rao*, 2007]. ***Trepanning operation (with orbital motion of tool)*** has been used efficiently and economically for drilling large size through holes (15 × 15 × 7 mm^3 size) by comparatively smaller electrode (or wire) [Jain and Chak, 2000]. However, at greater machined depth, the conventional electrode configurations and machining parameters show that machining performance gradually deteriorates with increase in machined depth and finally causes micro cracks on the machined surface due to thermal shock at high voltage. To minimize this problem and to enhance the machining performance during trepanning operation of Al_2O_3, a spring fed cylindrical abrasive electrode of 1.5 mm diameter has been used to study the effects of three influential parameters, namely, pulsed DC supply voltage, duty factor and electrolyte conductivity. This study revealed that the pulsed DC reduced the probability of cracking of workpiece at high supply voltage as compared to smooth DC. Further, the machining ability of the abrasive electrode was better than the bare copper electrode due to the presence of abrasive particles. In this process, the tool was continuously interacting with the workpiece surface with nominal force to ensure a constant gap between the tool electrode and workpiece through the active abrasive particles. For trepanning operation, eccentric rotation to the tool was provided by a slider attachment for adjusting the offset between the tool axis and the spindle axis with the help of a screw. Experiments were also conducted with the mixed NaOH and KOH electrolyte apart from either NaOH or KOH alone [*Chak and Rao*, 2007, Gautam and Jain, 1998].

POLARITY OF TOOL IN ECSM PROCESS

Experiments of ECSM in both polarities (*i.e.* reverse polarity (cathode or tool is negatively polarized) and direct polarity (tool is positively polarized)) have been successfully conducted, and the comparative study has been performed under the same experimental conditions by *Jain and Adhikari* [2008]. Smaller electrode (as a tool) is made cathode or anode by changing its polarity. Sparking occurs between the tool and electrolyte interface across the gas bubbles due to the breakdown of the dielectric (insulating gas bubbles) due to high potential gradient across them. MRR, Penetration Rate (PR), overcut and Tool Wear Rate (TWR) are higher in reverse polarity as compared to the direct polarity, Table 18.1. Therefore, it seems that material removal in reverse polarity is not only by melting and vaporization of workpiece material by the heat generated due to sparking. Deep crater formed in anode at anode-workpiece interface leads to the conclusion that some chemical reaction is also responsible for material removal from the workpiece.

Table 18.1: Comparison between reverse polarity and direct polarity performances (Machining parameters: 72 V, NaOH Electrolyte concentration 10% by weight) [*Jain and Adhikari,* 2008]

Tool	*Machining Time (min)*	*Workpiece (mg)*		*Tool (mg)*		*Overcut (mm)*	*Machined Depth (mm)*	*MRR (mg/min)*	*TWR (mg/min)*	*PR (mm/min)*
		Initial Weight	*Final Weight*	*Initial Weight*	*Final Weight*					
Positive ECSMWRP	23.724	0.456	0.421	0.429	0.416	0.175	6.66	1.47	0.548	> 0.339
Negative ECSMWDP	59.310	0.909	0.876	0.411	0.408	0.108	7.175	0.343	0.050	~0.120

ELECTROCHEMICAL SPARK MICROMACHINING (ECSMM)

The experimental images of spark affected zones show the ring patterns having diameter in the range of 200–500 μm depending upon the machining conditions and tool-workpiece material properties [*Shankar et al.*, 1997]. This feature size of the craters indicates its potential use in micron regime machining [*Kulkarni et al.*, 2011]. Figure 18.5 shows different elements of a setup for ECSM process. A spark is visible at the tool tip. In the ECSM cell, the electrolyte level is usually maintained at 1 mm above the surface of the workpiece. Commonly used electrolyte is NaOH having concentration in the range of 15–20% by weight. Different parts of the ECSM cell have been made of either Perspex or glass (electrically non-conductive) material to avoid any loss due to corrosion and electrical contacts.

ECSM process has been extensively studied by many researchers to make it a viable choice for industrial applications. As a result, this process has been employed in different configurations such as different tools material, tool geometries, abrasive particles embedded tool, rotating tool, travelling wire as tool, stagnant or flowing electrolyte, different types of electrolytes such as NaOH, NaCl, HCl, etc., different workpiece materials such as metals, non-metals, quartz, composites, alumina, glass, silicon [*Kulkarni et al.*, 2011], etc.

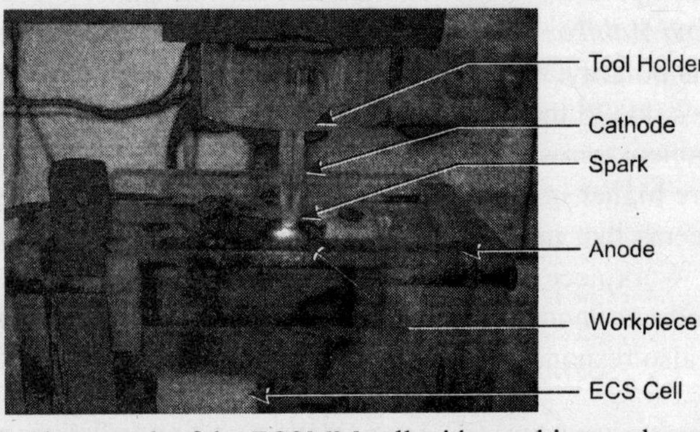

Fig. 18.5: Photograph of the ECSMM cell with graphite anode, work piece and spark at the micro tool tip and electrolyte interface [*Kulkarni et al.*, 2015]

CUTTING OF MICROCHANNELS

Fabrication of microchannels in quartz using ECSMM has been demonstrated by *Jain and Priyadarshani* [2014] for the prospective applications in micromixing of fluids. The optimal value of voltage and workpiece feed rate are found during the pilot experiments using copper as the tool material. The voltage at which sparking starts, depends upon the tool geometry and conductivity of the electrolyte. The smaller the diameter of the tool, the smaller is the spark. Use of Perspex as a tool holder material prevents lateral flow of current. It is highly corrosion resistant against NaOH. Use of Bakelite shaft to mount the workpiece prevents unwanted chemical reactions with the electrolyte. Use of drip system (normally used in hospitals) for electrolyte supply maintains the electrolyte flow. NaOH as electrolyte with 20% concentration by weight (approximately at maximum conductivity) is found to be suitable for micromachining of quartz. Experimentally, the voltage range obtained is 35–60 V because below 35 V there is no sparking and above 60 V the workpiece cracks and then breaks. Figure 18.6 shows an experimentally machined microchannel. Microchannels of different shapes are also generated. Figure 18.7 shows deep cut showing the fractured quartz piece.

Material removal takes place not only by melting (as evidenced by the presence of globules, Figure 18.7) and evaporation, but also by mechanical means as evidenced by the fractured piece in the figure. It may be due to repeated heating by sparks and cooling by the presence of electrolyte that would lead to microcrack formation and the growth of the microcracks is assisted by very high pressure. Collapse of plasma channel and release of pressure would lead to separation of the fractured quartz pieces (Figure 18.7).

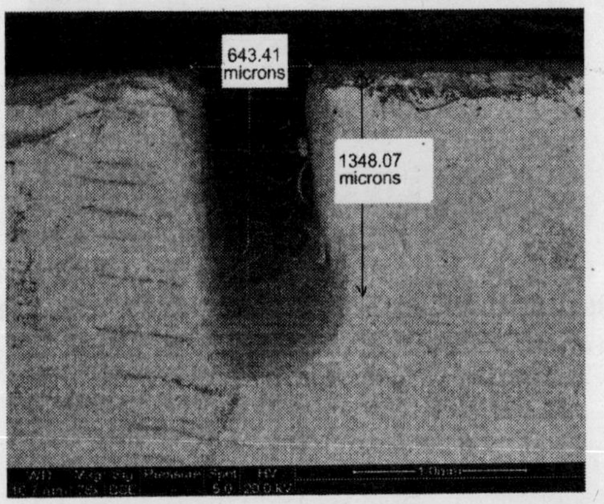

Fig. 18.6: Photograph of a microchannel on quartz material, with depth = 1348.07 μm and width = 643.41 μm, and aspect ratio = 2 : 1 [*Jain and Priyadarshani*, 2014]

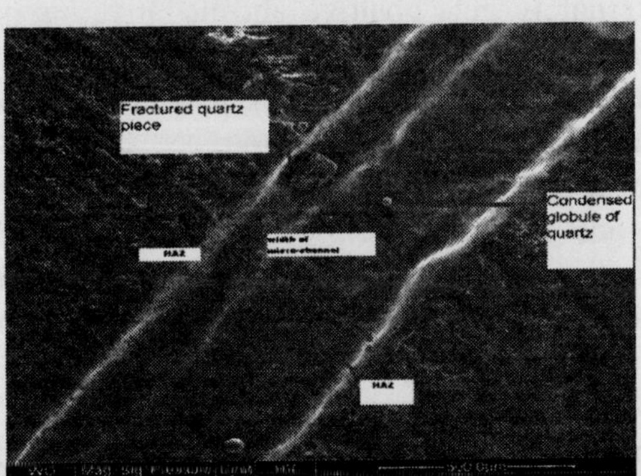

Fig. 18.7: Magnified view (160×) of 1 mm deep cut showing the fractured quartz piece and condensed globules of quartz [*Jain and Priyadarshani*, 2014]

ECSM process has been explored for microchannel fabrication on silicon using DC (60 V), AC (60 Vrms), and Half Wave Rectified (HWR) DC (60 Vrms) power supply [*Kulkarni et al.*, 2011, 2013]. Experiments are performed using tool made of copper of 200 μm wire diameter, electrolyte (NaOH) concentration of 14% by volume and X-Y table speed at 25 μm/s. Circular microchannels are fabricated using DC power supply on silicon sample of square size (50 mm × 50 mm) used as the workpiece while having machining gap between the tool tip and workpiece around 50 μm. The average width of the micro-channel is 82.5 μm, and average depth as 40 μm. Some microchannels have been machined in the shape of '∞', Figure 18.8. As expected, the depth increases as the table speed decreases.

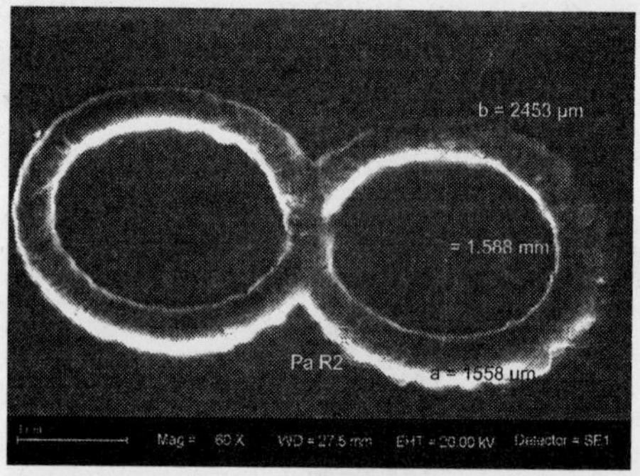

Fig. 18.8: SEM image at 60× of a glass sample (50 V DC, 16% electrolyte concentration) showing inner and outer diameters a and b respectively of the circular microchannel. Inner diameter = 1588 μm and outer diameter = 2453 μm [Kulkani *et al.*, 2011; Kulkarni, 2013].

REMARKS

As envisaged, it is found that micromachining, copper deposition and surface treatment, all are taking place simultaneously with AC supply source. The active electrode (say, cathode in this case) alternately gets positive and negative supply for 10 ms duration. Micromachining takes place while tool (cathode) is getting negative supply. In the positive half cycle, the tool material (copper in this case) is melting (in some cases, it may get vaporized also) and getting deposited on the workpiece surface. Over the one complete cycle of the AC supply, heat treatment of the workpiece surface may take place due to the repeated heating (sparks occurring at the tip of the smaller electrode that is tool in this case, i.e. cathode) and cooling. Sparking takes place across the bubbles at the electrode-electrolyte interface. Heat treatment aids in sticking the copper ions from copper cathode on to silicon surface during deposition cycle.

Traces of copper are found in different percentage (from 17.12% to 1.40%) distributed across the microchannels during the use of AC power supply in ECSM. It is zero outside the microchannel region. Thus, the deposition of copper material takes place within the microchannel and its magnitude can be controlled by varying the input parameters. However, no traces of copper material are found for DC and HWR DC cases.

BIBLIOGRAPHY

[1] Basak, I. and Ghosh, A. (1997). Mechanism of Material Removal in Electrochemical Discharge Machining a Theoretic Model and Experimental Verification, *Journal of Materials Processing Technology,* 71, 350–359.

[2] Bhattacharyya, B., Doloi, B.N. and Sorkhel, S.K. (1999). Experimental Investigations into Electrochemical Discharge Machining (ECDM) of Non-conductive Ceramic Materials, *Journal of Material Process Technology,* 95, 145–54.

[3] Bhuyan, B.K. and Yadava, V. (2014). Experimental Modelling and Multi-Response Optimization of Travelling Wire Electrochemical Spark Machining of Pyrex glass. *Proc. Inst. Mech. Eng. B: J. Eng.* 228(8): 902–16.

[4] Cao, X.D., Kim, B.H. and Chu, C.N. (2009). Micro Structuring of Glass with Features Less than100 μm by Electrochemical Discharge Machining. *Precis Eng.*, 33: 459–65.

[5] Cao, X.D., Kim, B.H. and Chu, C.N. (2012). Hybrid Micromachining of Glass Using ECDM and Micro Grinding. *Int J. Precis Eng. Manuf.,* 14(1): 5–10.

[6] Chak, S.K. and Rao, V.P. (2007). Trepanning of Al_2O_3 by Electro-Chemical Discharge Machining (ECDM) Process Using Abrasive Electrode with Pulsed DC Supply. *Int. J. Mach Tool Manuf.,* 47(14): 2061–70.

[7] Cheng, C.P., Wu, K.L., Mai, C.C., Yang, C.K., Hsu, Y.S. and Yan, B.H. (2010). Study of Gas Film Quality in Electrochemical Discharge Machining. *Int. J. Mach Tool Manuf.,* 50, 689–97.

[8] Cheng, C.P., Wu, K.L., Mai, C.C., Hsu, Y.S. and Yan, B.H. (2010). Magnetic Field Assisted Electrochemical Discharge Machining. *Journal of Micromechanics and Microengineering,* 20, 1–7.

[9] Crichton, I.M. and McGough, J.A. (1985). Studies of the Discharge Mechanisms in Electrochemical Arc Machining, *Journal of Applied Electrochemistry,* 115, 113–119.

[10] Fascio, V., Wüthrich, R. and Bleuler, H. (2004). Spark Assisted Chemical Engraving in the Light of Electrochemistry, *Electrochimca Acta,* 49, 3997–4003.

[11] Gautam, N. and Jain, V.K. (1998). Experimental Investigations into ECSD Process using Various Tool Kinematics, *Int. J. Mach. Tool Manuf.,* 39, 15–27.

[12] Gupta, P.K., Dvivedi, A. and Kumar, P. (2014). Developments on Electrochemical Discharge Machining: A Review of Experimental Investigations on Tool Electrode Process Parameters. *Proc. Inst. Mech. Eng., Sr. B: J. Eng.* 1–11.

[13] Han, M.S., Min, B.K. and Lee, S.J. (2011). Micro-electrochemical Discharge Cutting of Glass Using a Surface-Textured Tool. *CIRP, J. Manuf. Sci. Technol.,* 4: 362–9.

[14] Han, M.S., Min, B.K. and Lee, S.J. (2008). Modeling Gas Film Formation in Electrochemical Discharge Machining Processes Using a Side Insulated Electrode, *Journal of Micromechanics and Microengineering,* 18, 1–8.

[15] Huang, S.F., Liu, Y., Li, J., Hu, H.X. and Sun, L.Y. (2014). Electrochemical Discharge Machining Micro-Hole in Stainless Steel with Tool Electrode High-Speed Rotating. *Mater Manuf. Process,* 29(5): 634–7.

[16] Jain, V.K. and Chak, S.K. (2000). Electrochemical Spark Trepanning of Alumina and Quarts, *Machining Science and Technology,* Vol. 4, pp. 277–290.

[17] Jain, V.K., Dixit, P.M. and Pandey, P.M. (1996). On the Analysis of Electrochemical Spark Machining Process. *Int. J. Mach Tool Manuf.,* 39(1), 165–186.

[18] Jain, V.K. and Adhikary, S. (2008). On the Mechanism of Material Removal in Electrochemical Spark Machining of Quartz under Different Polarity Conditions. *Journal of Materials Processing Technology*, 200(1–3), 460–70.

[19] Jain, V.K., Rao, S., Chowdhury, S.K. and Rajurkar, K.P. (1991). Experimental Investigations into Traveling Wire Electrochemical Spark Machining (TWECSM) of Composites, *Trans. ASME, Eng. Ind*., 113, 75–84.

[20] Jain, V.K., Choudhury, S.K. and Ramesh, K.M. (2002). "On the Machining of Alumina and Glass." *Int. J. Mach. Tool Manuf.*, 42, 1269–1276.

[21] Jain, V.K. and Priyadarshini, D. (2014). Fabrication of Micro Channels in Ceramics (quartz) Using Electrochemical Spark Micro Machining (ECSM). *Journal of Advanced Manufacturing Systems*, 13(1), 5–16.

[22] Jalali, M., Maillard, P. and Wüuthrich, R. (2009). Toward a Better Understanding of Glass Gravity-Feed Micro-Hole Drilling with Electrochemical Discharges, *Journal of Micromechanics and Microengineering*, 19, 1–7.

[23] Jiang, B., Lan, S., Ni, J. and Zhang, Z. (2014). Experimental Investigation of Spark Generation in Electrochemical Discharge Machining of Non-Conducting Materials. *Journal of Materials Processing Technology,* 214(4), 892–898.

[24] Jui, S.K., Kamaraj, A.B. and Sundaram, M.M. (2013). High Aspect Ratio Micromachining of Glass by Electrochemical Discharge Machining (ECDM). *J. Manuf. Process;* 15(4), 460–6.

[25] Kim, D.J., Ahn, Y., Lee, S.H. and Kim, Y.K. (2006). Voltage Pulse Frequency and Duty Ratio Effects in an Electrochemical Discharge Micro Drilling Process of Pyrex Glass, *International Journal of Machine Tools and Manufacture,* 46(10), 1064–1067.

[26] Kulkarni, A.V., Sharan, R. and Lal, G.K. (2002). An Experimental Study of Discharge Mechanism in Electrochemical Discharge Machining, *Int. J. Mach. Tool Manuf.,* 42(10), 1121–1127.

[27] Kulkarni, A.V. (2013). Performance Investigation into Microfabrication using Electrochemical Spark, Ph.D. Thesis, *G.B. Technical University,* Lucknow (India).

[28] Kulkarni, A.V., Vijay Kumar, Jain and Krishna Avtar, Misra (2011). Electrochemical Spark Micromachining: Present Scenario, Position paper, *IJAT,* 5(1), 52–59.

[29] Kulkarni, A.V. and Jain, V.K. (2015). Design and Development of an Electrochemical Spark Micro Manufacturing Equipment, *International Journal of Mechanical Engineering and Robotics Research*, 4(4), 368–372. DOI: 10.18178/ijmerr.4.4.368–372.

[30] Kulkarni, A.V., Jain, V.K. and Misra, K.A. (2011). Electrochemical Spark Micromachining (Microchannels and microholes) of Metals and Non-Metals, *International Journal of Manufacturing Technology and Management*, 22(2), 107–123.

[31] Kulkarni, A.V., Jain, V.K. and Misra, K.A. (2013). Study of Microstructure and Transient, Instantaneous Current in Different Excitations During Silicon Micromachining, *J. Inst. Eng.*

India Ser. B, DO - 10.1007/s40031-012-0022-z, http://dx.doi.org/10.1007/s40031-012-0022-z, Springer-Verlag.

[32] Kurafuji, H. and Suda, K. (1968). Electrical Discharge Drilling of Glass, *Ann. CIRP*, 16, 415–419.

[33] Laio, Y.S., Wu, L.C. and Peng, W.Y. (2013). A Study to Improve Drilling Quality of Electrochemical Discharge Machining (ECDM) Process. *Procedia. CIRP;* 6, 609–14.

[34] Liu, J.W., Yue, T.M. and Guo, Z.N. (2013). Grinding-Aided Electrochemical Discharge Machining of Particulate Reinforced Metal Matrix Composites. *International Journal of Advanced Manufacturing Technology*, 68(9–12), 2349–2357.

[35] Liu, J.W., Yue, T.M. and Guo, Z.N. (2010). An analysis of the Discharge Mechanism in Electrochemical Discharge Machining of Particulate Reinforced Metal Matrix Composites. *Int. J. Mach. Tools and Manuf.*, 50(1), 86–96.

[36] Manna, A. and Kundal, A. (2015). An Experimental Investigation on Fabricated TW-ECSM Set-up during Micro Slicing of Nonconductive Ceramic. *International Journal of Advanced Manufacturing Technology;* 76, 29–37.

[37] McGeough, J.A., Khairy, A.U. and Munro, W.J. (1983). Theoretical and Experimental Investigation of the Relative Effects of Spark Erosion and Electrochemical Dissolution in Electrochemical Arc Machining, *CIRP Annals of Manufacturing Technology,* 32(1), 113–118.

[38] Mediliyegedara, T.K.K.R., De Silva, A.K.M., Harrison, D.K. and McGeough, J.A. (2005). New Developments in the Process Control of the Hybrid Electro Chemical Discharge Machining (ECDM) Process, *Journal of Materials Processing Technology,* 167(2–3), 338–343.

[39] Nesarikar, V.V., Jain, V.K. and Choudhury, S.K. (1995). Experimental Investigations into Electrochemical Spark Machining of Kevlar Epoxy Composites, *Proc. of Xth American Society of Precision Engineering, Annual Meeting*, Auxtin Texas, 292–295.

[40] Nguyen, K.H., Lee, P.A. and Kim, B.H. (2015). Experimental Investigation of ECDM for Fabricating Micro Structures of Quartz. *Int. J. Precis. Eng. Manuf.*, 16(1): 5–12.

[41] Yadav, V. (2014). Electrochemical Spark Micro-Machining, in Introduction to Micro-machining (edited by Jain V.K.), published by Narosa Publications, New Delhi.

[42] Peng, W. and Liao, Y. (2004). Study of Electrochemical Discharge Machining Technology for Slicing Non-Conductive Brittle Materials, *Journal of Materials Processing Technology,* 149(1–3), 363–369.

[43] Shankar, Prem, Jain, V.K. and Sundararajan, T. (1997). "Analysis of Spark Profiles during EDM Process", *Machining Science and Technology*, Vol. 1, No. 2, pp. 195–217.

[44] Singh, Y.P., Jain, V.K., Kumar, P. and Agrawal, D.C. (1996). Machining Piezoelectric (PZT) Ceramics using an Electrochemical Spark Machining (ECSM) Process, *Journal of Material Process Technology,* 58, 24–31.

[45] Skrabalak, G., Skrabalak, M.Z. and Ruszaj, A. (2004). Building of Rules Base for Fuzzy Logic Control of the ECDM Process, *Journal of Material Process Technology,* 149(1–3), 530–535.

[46] Tsuchiya, H., Inoue, T. and Miyazaki, M. (1985). Wire Electrochemical Discharge Machining of Glasses and Ceramics, *Bulletin of Japanese Society of Precision Engineering,* 19(1), 73–4.

[47] West, J. and Jadhav, A. (2007). ECDM Methods for Fluidic Interfacing through Thin Glass Substrates and the Formation of Spherical Micro Cavities, *Journal of Micromechanics and Microengineering,* 17(2), 403–409.

[48] Wüthrich, R. (2009). Micromachining using Electrochemical Discharge Phenomenon, UK: William Andrew Book Company.

[49] Wüthrich, R. and Hof, L.A. (2006). The Gas Film in Spark Assisted Chemical Engraving (SACE)—A Key Element for Micro-machining Applications, *International Journal of Machine Tool and Manufacture,* 46(7), 828–835.

[50] Yang, C.T., Song, S.L., Yan, B.H. and Huang, F.Y. (2006). Improving Machining Performance of Wire Electrochemical Discharge Machining by Adding SiC Abrasive to Electrolyte, *Int. J. Mach. Tools and Manuf.,* 46(15), 2044–2050.

[51] Yang, C.K., Cheng, C.P., Mai, C.C., Wang, A.C., Hung, J.C. and Yan, B.H. (2010). Effect of Surface Roughness of Tool Electrode Materials in ECDM Performance. *Int. J. Mach Tool Manuf.,* 50, 1088–96.

[52] Yang, C.K., Wu, K.L., Jung, J.C., Lee, S.M.D., Lin, J.C. and Yan, B.H. (2011). Enhancement of ECDM Efficiency and Accuracy by Spherical Tool Electrode. *Int. J. Mach. Tool Manuf.*, 51, 528–35.

SELF TEST QUESTIONS

Q1. Choose the most appropriate answer.

1. ECSM is applicable to,
   (a) electrically non-conducting materials only
   (b) electrically conducting materials only (c) both (a)&(b)
2. Mechanism of material removal of ECSM process is
   (a) electrochemical reaction (b) melting and vaporization
   (c) (a)&(b) both (d) none of these
3. During ECSM, sparking always takes place at,
   (a) cathode (b) anode
   (c) any one of these depending on the size of the electrode (d) none of these
4. Hydrogen gas bubbles are formed at
   (a) cathode (b) anode (c) bulk electrolyte (d) none of these
5. Oxygen gas bubbles are formed at
   (a) cathode (b) anode (c) bulk electrolyte (d) none of these
6. In ECSM, why don't you use NaOH concentration (by weight) beyond 23%?
   (a) No further change in the electrolyte electrical conductivity
   (b) conductivity starts decreasing
   (c) conductivity starts increasing but no enhancement in MRR
   (d) none of these

Q2. Answer in brief.

1. Explain the working principle of ECSM process.
2. Explain how the gas bubbles formation takes place during ECSM process?
3. What chemical reactions take place at anode, cathode and in the bulk electrolyte?
4. In ECSM process, tool is usually cathode and workpiece is electrically non-conductive. Is it necessary to have anode within the ECSM set-up?
5. Does ECSM require intermediate processing steps?
6. In ECSM, how the current density increases with an increase in the supplied voltage while the number of bubbles formed also increases?
7. What do you understand by the ECS trepanning operation?
8. What are the applications of ECS trepanning operation?
9. Why overcut increases in ECS drilling with an increase in the supplied voltage?
10. What will happen to the "kerf width" (increases or decreases) in ECS channelling operation along its depth if the thickness of the workpiece is increased?
11. Beyond a certain value of electrolyte temperature, why the performance of the ECSM process starts deteriorating?

12. Why do you have flowing electrolyte instead of still electrolyte during ECSM process?
13. Explain the mechanism of material removal in case of ECS drilling using the abrasive particles embedded drilling tool.
14. What do you understand by 'reverse polarity' in case of ECSM process?
15. Why do you use Perspex as a tool holding material in case of ECS drilling operation?
16. How can you prove that material removal in ECSM process takes place by mechanical means also, apart from melting and vaporization?
17. In ECSM process, under what conditions tool material gets deposited on electrically non-conductive workpiece material?

ACRONYMS

AMPs	Advanced Machining Processes
DC	Direct Current
EC	Electrochemical
ECM	Electrochemical Machining
ECDM	Electrochemical Discharge Machining
ECSM	Electrochemical Spark Machining
EDM	Electric Discharge Machining
MR	Material Removal
MRR	Material Removal Rate
PR	Penetration Rate
SACE	Spark assisted Chemical Erosion/Engraving/Etching

CHAPTER 18
AT-A-GLANCE
ELECTROCHEMICAL SPARK MACHINING (ECSM)

Introduction

- Hybrid process: ECM + EDM → ECSM/ECDM/SACE
- Capable to machine both: Electrically conducting and non-conducting materials.
- Applied voltage beyond a threshold value → sparking across a gas bubble.
- Sparking across gas bubbles: → melting and vaporization of workpiece material.
- In ECSM, Material removal from both tool and workpiece.
- No material removal from auxiliary electrode.
- Workpiece: Not an electrode.
- NaOH, NaCl, etc.: Electrolyte.

Electrochemical and other Reactions

At Cathode

- H_2 gas bubbles evolve. Sparking at cathode-electrolyte interface.
- NaOH is not consumed. Electrolyte concentration increases due to vaporization of water.
- Sparking voltage depends on → electrolyte concentration, H_2/O_2 gas properties, tool and work properties.
- Violent sparks → damage workpiece.

At Anode

- Made of non-dissolving material. O_2 gas bubbles are evolved.
- Large spacing between anode and cathode.

In the Bulk Electrolyte

- Reduction of electrolyte. At the onset of sparking: Transient momentary high current.

Mechanism of Sparking and Analysis of ECSM

- Valve theory used to explain sparking phenomenon.
- Spark energy and gas bubble diam calculated using FEM.

- Temperature distribution, material removal, overcut and penetration depth calculated using FEM.
- Simulated isotherms are close to the experimental ones.
- Crater shape obtained experimentally is approximately circular of 300 μm.

Electrochemical Spark Micromachining

- Crater shape and size depend on input parameters, and tool and work material properties.
- Electrolyte level in the chamber: About 1 mm above workpiece top surface level.
- Commonly used electrolyte: NaOH (15 to 20% concentration by weight).
- Used mainly for electrically non-conductive materials.
- Used in different configurations, with different electrolytes and work materials.

Analysis of ECSM of Different Types of Materials

Travelling Wire ECSM of Kevlar-Epoxy Composites

- Travelling wire ECSM operation for precision slicing.
- Flowing electrolyte helps in reducing temperature, and remove residues.
- Larger the specimen thickness, more non-uniform MR and lower MRR.
- Increase in voltage results in more H_2 gas bubbles generation leading to higher sparking rate which results in higher overcut.
- Beyond approximately 20% electrolyte concentration, MRR starts decreasing because k (conductivity) starts decreasing.

Drilling and Trepanning of Alumina and Quartz

- Electrically non-conducting, hard, brittle and temp. resistant materials.
- Machining performance deteriorates as machining depth increases.
- Spring fed cylindrical abrasive loaded electrode tool used.
- ECS spark trepanning operation. Vertical feed by a stepper motor.
- Slider attachment used for eccentric motion in ECS trepanning operation.
- Mixed electrolyte: KOH + NaOH.
- Process performance depends: electrolyte temperature and its contamination with the reaction products, and machining conditions.
- Higher machined depth and MRR with abrasive embedded tool as compared to bare tool.
- With the abrasive embedded tool, the mechanism of MR: melting + vaporization + mechanical shearing.

- In some special cases, MR can take place by chemical reaction also.
- MR takes place from the WP under the smaller electrode (anode or cathode).

Electrochemical Spark Micromachining

- Crater size in ECSM is around 0.2 mm. It can be used for micromachining.
- Different types of operations using ECSM.
- Cutting of microchannels in glass, etc. using ECSM for micro-mixing of fluids.
- Voltage range for general applications: 30 to 60 V.
- Continuous heating and cooling → micro-cracks leading mechanical MR.
- There are many avenues in ECSM for research activities.

Cutting of Micro channels

- Making of micro channels in quartz for fabricating a fluid micro mixer.
- Perspex as a tool holder material prevents lateral flow of current.
- Below 35 V there is no sparking and above 60 V the work piece cracks.
- Microchannels of different shapes are also generated.
- Material removal takes place not only by melting and evaporation, but also by mechanical means.
- ECSM process has been explored for micro channel fabrication on silicon.

Remarks

- Micromachining, copper deposition and surface treatment all are taking place simultaneously with AC supply source.
- Traces of copper are found in different percentages distributed across the micro channels during the use of AC power supply in ECSM.

INDEX

E